Photosynthetic pigments of algae

KINGSLEY S. ROWAN

Botany School, University of Melbourne

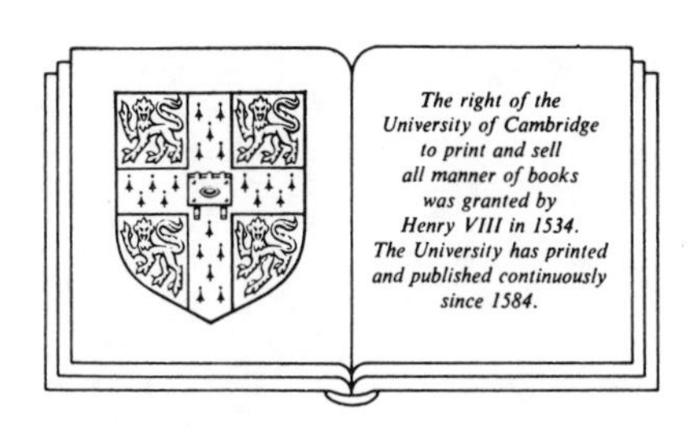

CAMBRIDGE UNIVERSITY PRESS

Cambridge

New York New Rochelle Melbourne Sydney

Published by the Press Syndicate of the University of Cambridge
The Pitt Building, Trumpington Street, Cambridge CB2 1RP
32 East 57th Street, New York, NY 10022, USA
10 Stamford Road, Oakleigh, Melbourne 3166, Australia

First published 1989

Printed in the United States of America

Library of Congress Cataloging-in-Publication Data
Rowan, K. S.

Photosynthetic pigments of algae/Kingsley S. Rowan.

p. cm.

Bibliography: p.

Includes index.

ISBN 0-521-30176-9

1. Algae – Composition. 2. Photosynthetic pigments.
3. Photosynthesis. 4. Primary productivity (Biology). I. Title.
QK565.R77 1989
589.3′19218 – dc19 88–21665
CIP

British Library Cataloguing in Publication Data

Rowan, Kingsley S.

Photosynthetic pigments of algae.

1. Algae. Photosynthetic pigments
I. Title
589.3′3342

ISBN 0-521-30176-9

To Helen

Contents

Preface

This book is intended for the phycologist, rather than the chemist or biochemist, and the mechanisms of photosynthesis have not been treated in detail. Pigments represent one of many properties of interest to the phycological taxonomist, and in earlier days seemed to provide simple rules for assigning algae to their respective classes, as exemplified by many textbooks of biology. Now, however, these rules have lost their precision because of the many exceptions that have arisen recently – in particular, the different forms of chlorophyll *c* currently known, the abnormal pigments produced in some members of a class through symbiosis, and the occasional occurrence of a pigment previously considered highly specific in its distribution (e.g., the rare reports of peridinin in red algae). Accordingly, I hope that this book will be helpful to phycological taxonomists interested in pigments as one of a number of properties used in classifying algae.

I would like to thank the following friends for advice and/or for supplying manuscripts before publication: Bob Andersen, Jan Anderson, Paul Broady, Tom Cavalier-Smith, Gustav Hallegraeff, Shirley Jeffrey, Arne Jensen, Tony Larkum, Sønnove Liaaen-Jensen, Geoff McFadden, Øjvind Moestrup, Charles O'Kelly, Hugo Scheer, Bill Woelkerling, and Simon Wright. I would also like to thank Andrew Staehlin for providing electronmicrographs, David Hill for drawing the absorption spectra of the cryptophyte biliproteins, Barbara Joyce for typing the text, Chris O'Brien for the photographs, Harry Swart for drawing the design for the jacket, and the Botany School at the University of Melbourne for providing me with facilities as a senior associate after my nominal retirement.

The Botany School
University of Melbourne

Kingsley S. Rowan

Abbreviations

ALA	δ-Aminolevulinic acid
APC	Allophycocyanin
ATP	Adenosine triphosphate
CD	Circular dichroism
CF	Coupling factor for photosynthetic phosphorylation
CP I	Reaction center of photosystem I (also known as CC I)
CPa	Reaction center of photosystem II (also known as CC II)
CV	Cryptoviolin
DEAE	Diethylaminoethyl ether
DMF	Dimethylformamide
DMSO	Dimethylsulfoxide
DNA	Deoxyribonucleic acid
E	Extinction
EM	Electronmicroscopy
F	Fluorescence
HPLC	High-performance liquid chromatography
IR	Infrared
K	Kilodalton
LDAO	Lauryldimethylamine oxide
LDS	Lithium dodecylsulfate
LHC I	LHCP of photosystem I
LHC II	LHCP of photosystem II
LHCP	Light-harvesting chlorophyll–protein
MgDVP	MgDivinylpheoporphyrin a_5 monomethyl ester
MS	Mass spectroscopy
NADPH	Reduced nicotinamide-adenine dinucleotide phosphate
NMR	Nuclear magnetic resonance

ODS	Octadecyl sulfate
ORD	Optical rotatory dispersion
P_{700}	Photochemically active chlorophyll of RC I
P_{680}	Photochemically active chlorophyll of RC II
PAGE	Polyacrylamide gel electrophoresis
PAR	Photosynthetically active radiation
PBS	Phycobilisome
PC	Phycocyanin
PCB	Phycocyanobilin
PE	Phycoerythrin
PEB	Phycoerythrobilin
PEC	Phycoerythrocyanin
PS I	Photosystem I
PS II	Photosystem II
PUB	Phycourobilin
RC I	Reaction center of PS I
RC II	Reaction center of PS II
RNA	Ribonucleic acid
S	Svedberg unit of sedimentation
S & S	Schleicher & Schull
SDS	Sodium dodecyl sulfate
TLC	Thin-layer chromatography

1

Role of the photosynthetic pigments

1.1. Introduction

The algae are unique in the Plant Kingdom in the variation of the photosynthetic pigments in each phylum and in many of the classes, for the remainder of the kingdom are monotonously similar. This diversity of the light-harvesting pigments implies that the common ancestor was primitive and that no close affinity exists between blue-green, red, brown, golden-brown, and green algae, to use their common names. This diversity of pigments has led to their recognition as taxonomic and phylogenetic markers, and this aspect will be discussed in Chapter 7. Measuring the relative and absolute concentration of light-harvesting pigments is important when one is examining the relationship of a species to its environment – specifically, the quality of the light in which it lives and the photon flux density. Estimating light-havesting pigments – in particular, chlorophyll *a* – is widely used to study the density of marine algae in relation to depth and area, and remote sensing by satellite is now being employed for this purpose. This book places all chlorophylls, phycobiliproteins, and carotenoids found in algae under the designation of photosynthetic pigments. Whereas all chlorophylls and biliproteins have a well-established function in harvesting light, the role of the carotenoids is less clear, for they may have two possible functions: (1) harvesting light energy for photosynthesis and (2) preventing destruction of activated chlorophyll by oxygen in light (Siefermann-Harms, 1987). The role of fucoxanthin, peridinin, and siphonaxanthin in light harvesting is clearly established, though this function for other carotenoids, such as β-carotene, is less certain. The carotenoids that have been found in light-harvesting pigment–protein complexes are shown in Table 6–2.

The pathway of evolution of living organisms has always interested biologists, and many schemes for algal phylogeny have been proposed. The distribution of accessory pigments accompanying the ubiquitous chlorophyll *a* provides important evidence for this pathway and led Christensen (1962, 1964) to divide the higher eukaryota into Chlorophyta (with chlorophyll *b*) and Chromophyta (without chlorophyll *b* but usually with chlorophyll *c*). However, neither phylum is monophyletic; in the Chlorophyta, the Euglenophyceae have few affinities with other classes, whereas the Chromophyta contain such unrelated classes as the Cryptophyceae, Dinophyceae, and Eustigmatophyceae. Nevertheless, many phycologists find the two phyla useful and the designations appear regularly in the literature. Some of the smaller or more recently defined classes such as the Raphidophyceae, and those in the Chlorophyta recently defined by Mattox and Stewart (1984) – the Ulvophyceae, Pleurastrophyceae, and Micromonadophyceae – will not be considered in detail, but the new class, Synurophyceae, unique among the Chromophyta in containing chlorophyll c_1 without chlorophyll c_2, is particularly interesting. In general, the classification used in this book is that of Liaaen-Jensen (1977, 1978), the major writer on carotenoids in algae.

This book is designed for the student at the senior undergraduate and postgraduate level interested in identifying and estimating photosynthetic pigments in algae; many of the methods discussed also apply to pigments in the remainder of the Plant Kingdom. The following are the major references covering one or more of the three types of pigment (shorter reviews are cited in the appropriate chapters): Zechmeister (1934), Karrer and Jucker (1948), Strain (1958), Goodwin (1965, 1976, 1980), Vernon and Seely (1966), Isler (1971), W. D. P. Stewart (1974), Hellebust and Craigie (1978), Jensen (1978), and MacColl and Guard-Friar (1987).

1.2. Early work on photosynthetic pigments in algae

Stiles (1925), Rabinowitch (1945), and Loomis (1960) have written valuable summaries of the sequence of observations that constitute our present knowledge of the chlorophylls, carotenoids, and biliproteins of the algae.

1.2.1. Lipid-soluble pigments

The brief description by Stokes (1864) of his experiments separating the green and yellow pigments shows that he had partially purified chlorophyll *c* and fucoxanthin from brown algae, had found green algae similar to higher plants, and knew that phycoerythrin was an "albuminous substance." In 1873, Sorby, while Secretary of the Royal Society, presented a detailed paper to them on his experiments, for which he coined the term "chromatology." He confirmed and extended Stokes's experiments, but also examined pigments using his elegant microspectroscope (Sorby, 1877). He compared the spectra of five of his fractions of carotenoids, naming them phycoxanthin (from *Oscillatoria:* oscillaxanthin or zeaxanthin?), pezizaxanthin, orange xanthophyll (from antheridia of *Fucus*: β-carotene; see Williams, 1967), xanthophyll (from *Porphyra vulgaris:* lutein), and yellow xanthrophyll (neoxanthin and violaxanthin), the latter two pigments showing a hypsochromic shift when treated with hydrochloric acid. He also compared the spectra of the chlorophyll fractions, naming them "blue chlorophyll" (chlorophyll *a*), "yellow chlorophyll" (chlorophyll *b*), and chlorofucine (chlorophyll *c*), the last showing the two peaks of chlorophyll *c* at the red end of the spectrum. He extracted Stokes's "third yellow substance" from brown algae, naming it "fucoxanthine." He carried out simple experiments on the effect of light intensity on the concentration of pigments in *Fucus serratus,* comparing the amount per unit fresh weight of blue chlorophyll, orange xanthophyll, fucoxanthine, and chlorofucine in the alga growing in exposed and shaded sites. His report of traces of chlorofucine in red algae probably reflected epiphytic diatoms, as avoiding this contamination with natural material requires considerable care.

No further important advance in identifying xanthophylls was made until Tswett (1906a,b) reported separating pigments by differential elution on columns of solid adsorbent, naming the process "chromatic adsorption analysis." He recognized three or four yellow compounds (xanthophylls) in addition to carotene, and prepared relatively pure samples by eluting the pigments from the stationary phase. At about the same time, Willstätter and his co-workers began experiments leading to the characterization of chlorophylls *a* and *b* and carotene. They also isolated "xanthophyll" with empirical formula $C_{40}H_{56}O_2$; although they were prepared to believe that isomers might exist, they believed that the components of the mixture of pigments separated by Tswett were derived from their "xanthophyll" by alteration during extraction. Schertz (1925) subsequently found that "xanthophyll" prepared by the method of Willstätter contained only one pigment when analyzed by use of Tswett's chromatographic column, thus apparently confirming that some of Tswett's pigments were artifacts of extraction. But as Willstätter prepared his "xanthophyll" by transferring from 85% acetone to petroleum spirit and then removing contaminants by washing the hypophase with aqueous acetone, this step may well have removed most of the violaxanthin and neoxanthin (Strain, 1938). In addition, the heating step used in crystallizing "xanthophyll" may have destroyed these epoxy pigments, as they are chemically less stable than the components of "xanthophyll," which must have been essentially lutein and some zeaxanthin. Such was Willstätter's prestige that his views were presented in textbooks (Stiles, 1925, 1936; Miller, 1931; Thomas, 1935) until the late 1930s, when the detailed experiments by Strain (1936, 1938) vindicated and improved Tswett's chromatographic techniques. We now accept the view that there are a number of different xanthophylls in extracts from photosynthetic tissue.

The prestige of Willstätter also delayed recognition of chlorophyll *c* as a natural pigment. Stokes (1864) described it as a "third green substance," whereas Sorby (1873), as already mentioned, recorded the spectrum and named it "chlorofucine." Tswett (1906a,b) separated it by column chromatography, but with the exception of Wilschke (Strain and Manning, 1942), other workers for many years believed, along with Willstätter and Page (1914), that it was an artifact of extraction. Stiles took this view still in 1925. Eventually Strain and colleagues (Strain and Manning, 1942; Strain, Manning, and Hardin, 1943) prepared pigment extracts from diatoms, dinoflagellates, and brown algae, using a number of different techniques all producing chlorofucine which they renamed chlorophyll *c*. After their two papers were published, the natural occurrence of chlorophyll *c* was not queried again. However, the properties of chlorophyll *c* isolated in different laboratories varied, and Jeffrey (1969) summarized these discrepancies in spectra and extinction coefficients in five preparations. Dougherty et al. (1966, 1970) had described crystalline preparation of chlorophyll *c* from *Nitzschia closterium,* prepared according to the method of Strain and Svec (1966), as a mixture of monomethyl esters of magnesium tetra- and hexa-dehydropheoporphyrin a_5, and Jeffrey (1968b, 1969) had resolved chlorophyll *c* prepared by her method (Jeffrey, 1963) from *Sargassum* into two spectrally distinct compounds, using a special polyethylene powder (Dow Chemical Co.) for thin-layer chromatography. She named these chlorophyll c_1 and c_2. Aggregation of chlorophyll c_2, previously studied by Jeffrey and Shibata (1969), occurred when water was added to methanol solutions, or when solutions in ether were chilled (5° to −20°C) for some weeks. This aggregated form of chlorophyll c_2 was named chlorophyll c_3 and could be readily separated from c_2 on TLC plates. She believed that differences in band ratios of chlorophylls c_1 and c_2, variable distribution in the algae, and aggregation explained differences in absorption spectra and extinction coefficients in previous publications. When another suitable batch of polyethylene became available, Dougherty et al. (1970) showed that chlorophyll c_1 was the tetradehydro- and c_2 the hexadehydro-pheoporphyrin a_5. Several other laboratories have confirmed these structures (Jackson, 1976). Jeffrey (1972) then crystallized the two forms and recalculated the extinction coefficients that are now used for calculating concentrations of chlorophyll c_1 and c_2 in solution (Jeffrey and Humphrey, 1975). The chlorophyll c_3 described above must not be confused with the new chlorophyll c_3 recently found by Vesk and Jeffrey (1987).

Whether or not chlorophyll *d* is an artifact of extraction is also debatable. Manning and Strain (1943) extracted chlorophyll with methanol from 18 species of red algae and examined the ratio of absorbance in the extract at E_{max} (665 nm) and at 700 nm, where they saw a distinct shoulder in the absorption spectra prepared from some species. They believed that a value for $E665/E700$ below 50 indicated that a second pigment was present with an E_{max} at about 700 nm (the ratio for pure chlorophyll *a* was 90). By subtracting the absorption spectrum for pure chlorophyll *a* from that of the extract, they calculated a value of E_{max} of 695 nm for the hypothetical compound. This pigment was then purified in extracts from *Erythrophyllum delesserioides* by phase separation and column chromatography; extracts prepared with and without heating were identical. Strain (1958) subsequently detected chlorophyll *d* in a number of red algae collected in Australia. Zhou, Zheng, and Tseng (1974) detected chlorophyll *d* in a difference spectrum between a green and a red alga. Although chlorophyll *d* certainly exists in extracts, it might be formed during extraction, for Holt (1961) cast doubts on its natural occurrence after finding that the 2-desvinyl-2-formyl derivative of chlorophyll *a* was identical to chlorophyll *d*. O'hEocha (1971) concluded that chlorophyll *d* occurred naturally, primarily on the basis of evidence from the work by Strain. Of more recent reviewers, Meeks (1974) has accepted O'hEocha's opinion, but Govindjee and Braun (1974) and Jackson (1976) prefer to be uncertain.

1.2.2. Biliproteins

Phycocyanin was first extracted as a "sky-blue" solution from a slimy deposit of blue-green algae by von Esenbeck (1836), who suggested that it was related to the proteins. Kutzing (1843) prepared extracts of phycoerythrin from red algae, subsequently described as "albuminous" by Stokes (1864). Sorby (1877) prepared water-soluble fractions from both *Oscillatoria* and *Porphyra,* recorded their absorption maxima using his microspectroscope, and observed the red fluorescence of the "phycocyan" fraction from both species and the orange fluorescence of "pink phycoerythrin" from *Porphyra*. He gave a diagram of a five-band spectrum for the extract from *Prophyra*, similar to that seen in a recent paper by Kikuchi, Ashida, and Hirao (1979). Independently, Schutt (1888a) examined extracts from *Ceramium*, plotting what appears to be the first recorded absorption spectra of solutions of phycoerythrin and the lipid-soluble pigments. In a second paper (1888b), he examined the fluorescence of phycoerythrin. Molisch (1894) confirmed that the pigments were proteinaceous, previously doubted because they did not give a biuret reaction.

Lemberg (see Lemberg and Legge, 1949) isolated the major prosthetic groups from the apoproteins (phycoerythrobilin and phycocyanobilin), naming them "phycobilins." Haxo, O'hEocha, and Norris (1955) isolated a third type of pigment, allophycocyanin, as a minor fraction in red algae, and we shall see that it is now considered an essential part of the light-harvesting mechanism. O'hEocha (1958) suggested the term "biliprotein" to cover this class of photosynthetic pigment. O'Carra, O'hEocha, and Carroll (1964) found another prosthetic group in phycoerythrin firmly bound to the protein which they named phycourobilin. Bryant, Glazer, and Eiserling (1976) discovered a fourth type of biliprotein, phycoerythrocyanin, in place of phycoerythrin in a few blue-green algae; this pigment contained a new type of phycobilin (CV) as a prosthetic group. The various types of phycobilins will be discussed in Chapter 5.

1.3. Harvesting of light energy by photosynthetic pigments

1.3.1. Action spectra

As all algae contain chlorophyll *a*, this alone could have been the only pigment concerned in light harvesting, with the other pigments – green, red, orange or yellow – not being directly involved. The first step in demonstrating which pigment absorbs the light driving a biological process is to measure an "action spectrum," that is, to plot the amount of the activity of the process concerned against the wavelength of inducing light. Light below saturation must be used at all wavelengths, or plateaus would obscure peaks in the action spectrum. The first action spectra for photosynthesis measured over a hundred years ago by Engelmann (see Blinks, 1964) were made with crude methods, but they provided the first evidence that the pigments conferring their color on red, blue-green, green, and brown algae did in fact absorb the light involved in photosynthesis.

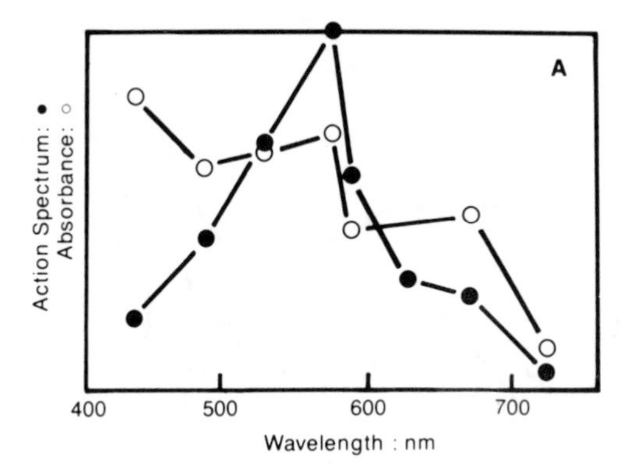

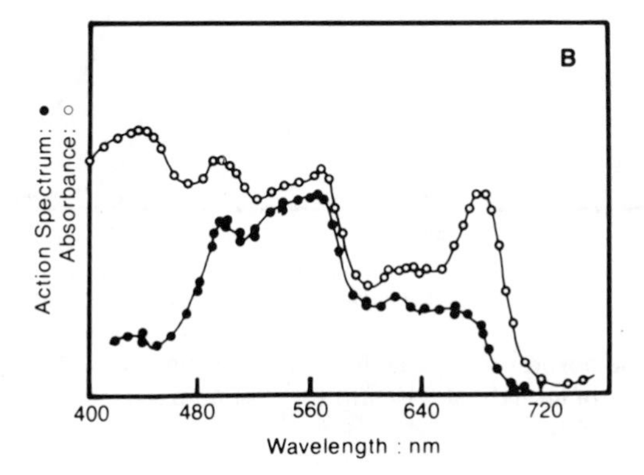

Fig. 1–1. The thallus absorption (○) and photosynthetic action spectra (●) of the red algae *Porphyra,* as determined by Engelmann in 1883–4 (A) and by Haxo and Blinks (1950) (B). (From Rowan, 1981, by permission.)

These experiments were not improved until Haxo and Blinks (1950) developed their platinum electrode for measuring oxygen evolution by membranous and unicellular algae. Although with this design of electrode chamber the rate of diffusion of carbon dioxide tended to limit the rate of evolution of oxygen, the apparatus provided clear evidence for the role of the various pigments in light harvesting in the different groups of algae. Figure 1–1 shows the action spectrum for photosynthesis in the red alga, *Porphyra,* measured by the methods of Engelmann and of Haxo and Blinks (Rowan, 1981). These action spectra show little photosynthesis above about 680 nm, whereas the peak of absorption of light by the thallus is at this wavelength. The paper by Haxo and Blinks (1950)

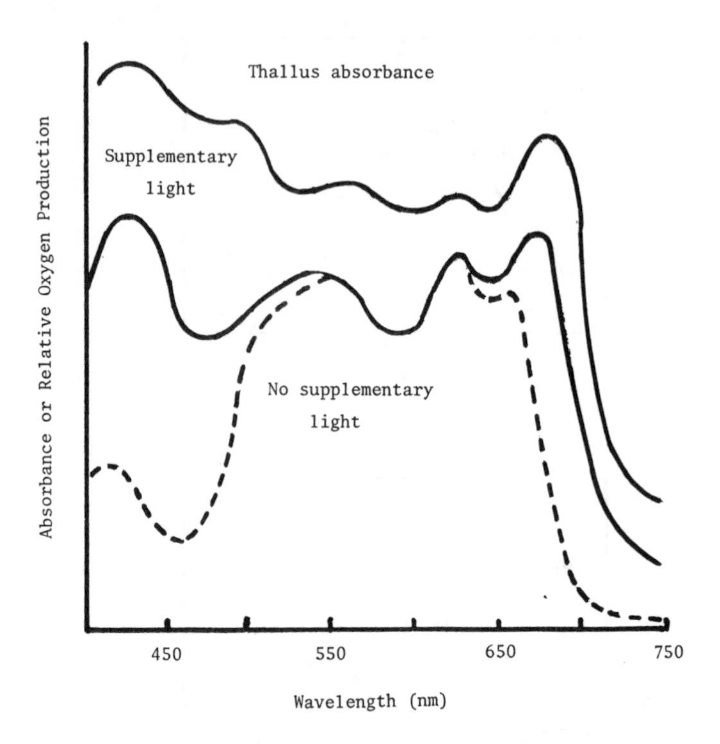

Fig. 1–2. The absorption and action spectra (with and without supplementary background light at 546 nm) for *Porphyra perforata*. (Redrawn from Fork, 1963.)

resulted in several years of discussion about the role of chlorophyll *a* in the red algae, finally resolved by Emerson's (1957) description of the enhancement effect of supplementary light on the rate of photosynthesis (Fig. 1–2), and by the so-called "Z" scheme of Hill and Bendall (1960; see Fig. 1–3).

The nature of this energy transfer has been discussed recently (Knox, 1975, 1977; Canaani and Gantt, 1980; Pearlstein, 1982), but whether it is best described as Förster resonance transfer or exciton migration, the requirements for transfer by either mechanism are that distance between molecules must be small (Förster resonance transfer is inversely proportional to the sixth power of the distance), and, in heterogeneous transfer, the fluorescence of the donor molecule should overlap the absorption spectrum of the acceptor molecule.

The evidence for transfer of energy from the so-called accessory pigments (chlorophylls *b* and *c*, biliproteins, and carotenoids) to chlorophyll

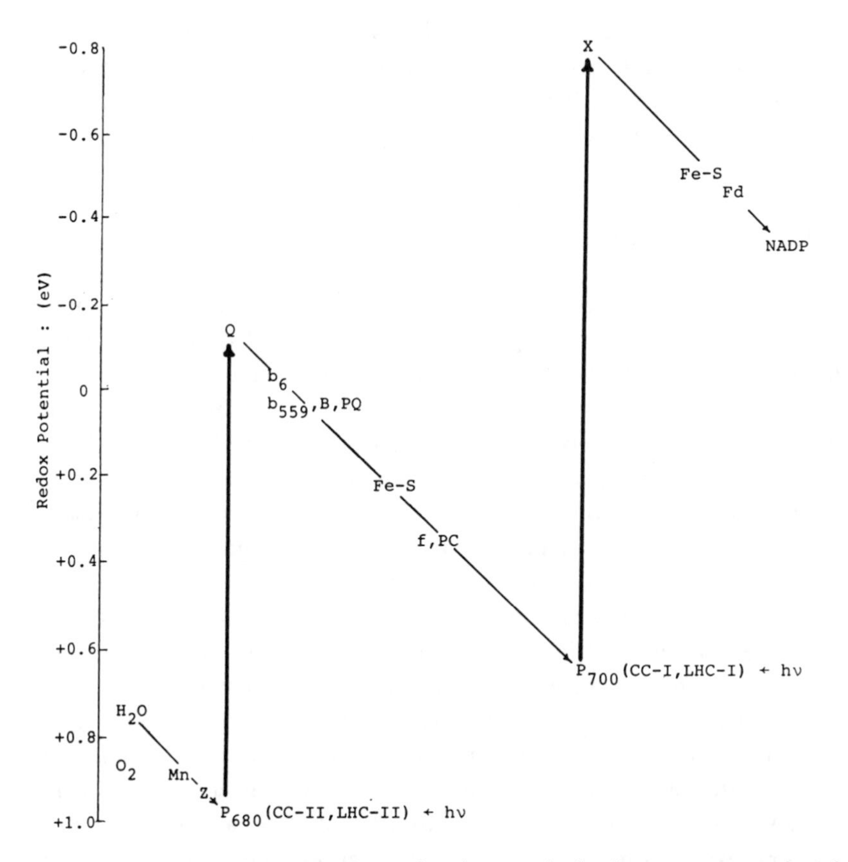

Fig. 1–3. The "Z" scheme for photosynthetic electron transport based on that proposed by Hill and Bendall (1960), with compounds arranged vertically according to their redox potentials. The mechanism of coupling this electron transfer with synthesis of ATP, according to the chemiosmotic hypothesis, is shown diagrammatically in Fig. 1–6. Z, primary electron donor of PS II; b_6, b_{559}, and f, cytochromes; Q, a fluorescence quenching quinone, primary electron acceptor of PS II; B, a quinone, secondary electron acceptor of PS II; PQ, plastoquinone; Fe-S, Rieske Fe-S; PC, plastocyanin; X, electron acceptor of PS I; Fd, ferridoxin.

a is based on the observation that light energy absorbed by the accessory pigments (detected from the photosynthetic action spectrum of the organism) was reradiated almost entirely as the fluorescence known to be derived from chlorophyll *a*, not at the wavelengths known for the accessory pigments. Duysens (1951) demonstrated this by showing that the fluorescence emitted from the red alga, *Porphyra lacineata,* irradiated at a wavelength specific for chlorophyll *a* (420 nm) was lower than that at a wavelength of light absorbed by all pigments (546 nm). This led French and Duysens (Duysens, 1951) to propose that the pathway of energy transfer in red algae was phycoerythrin → phycocyanin → chlorophyll *a*.

1.4. Light-energy absorption and transfer

The allowed energy states of the pigment molecule determine the shape of the absorption spectrum, because only those light quanta containing the energy required to drive electrons from the ground state to the permitted excited singlet states will be absorbed by the molecule; the basic principles involved have been described by Govindjee and Govindjee (1974, 1975) and Whittingham (1974). Whereas in atoms the absorption is in the ultraviolet region of the spectrum and the absorption lines are sharp, rotation and vibration of nuclei within a molecule provide a range of permitted energy states spaced on either side of the maximum, thus accounting for the relatively broad peaks of absorption spectra of colored molecules. When light falls on a molecule of chlorophyll *a*, the quanta coinciding with the energy of the higher excited singlet states will be those causing absorption in the blue region of the spectrum; however, the amount of photochemistry any quantum can carry out is independent of wavelength, because the electrons driven to the higher excited singlet states in the blue region of the spectrum rapidly fall back to the level of the first excited singlet state, and it is from this energy level that photochemistry, fluorescence, or transfer to another molecule occurs.

As mentioned previously, a range of pigments (chlorophylls, carotenoids, and biliproteins) absorb photosynthetically active radiation in the cell, but, in general, most do not reradiate this energy as fluorescence or donate it to the primary reactions of photochemistry. Instead, the energy is transferred to the specialized chlorophylls (P_{700} and P_{680}) of the two reaction centers (Fig. 1–3) by a series of homogeneous (between similar pigments) and heterogeneous (between different pigments) transfers. That a large array of pigment molecules is involved in funneling light energy to the photochemical reaction was first appreciated by Emerson and Arnold (1932), who calculated that a unit of about 2400 molecules of chlorophyll was required for the production of one molecule of oxygen. As we now know that eight electron transfers are required for this process, a unit containing 300–400 chlorophyll molecules seems the probable size (Whittingham, 1974).

1.5 Physical distribution of the pigments in the thylakoid membrane

All light-harvesting pigments are now known to be bound to protein (Larkum and Barrett, 1983) and are physically distributed around the reaction centers of PS I and PS II (Fig. 1–3). These pigment–protein associations are believed to contribute to the particles seen in electronmicrographs of freeze-fractures and freeze-etching of thylakoid membranes. Appression between thylakoid membranes occurs in all orders of the algae except those containing phycobilisomes – that is, the Cyanophyceae and the Rhodophyceae. Appression of thylakoids gives rise to granal stacks in the classes of the Chlorophyta, and appression between two to three thylakoids occurs in the other classes. The possible fracture-faces seen in electronmicrographs of cells containing both appressed and nonappressed thylakoids are shown diagrammatically in Fig. 1–4 (Staehelin, Armond and Miller, 1977), using the nomenclature of Staehelin (1976). This scheme is based on the fact that all biological membranes consist of a protoplasmic (P) and exoplasmic (E) leaflet and that each leaflet has a fracture face (F) and a true surface (S). Locations in appressed (stacked) regions (s) and unappressed (unstacked) regions (u) are designated by lowercase suffixes. The technique of freeze-fracture is based on the splitting of a membrane along the plane of the hydrophilic interior. Kaplan and Arntzen (1982) and Staehelin and DeWit (1984) have summarized evidence for the nature of the particles seen on the two fracture faces, EF and PF. Although PS II activity (Fig. 1–3) can occur in the absence of the large particles (14 nm) on the EF face (Larkum and Barrett, 1983), they concluded that these large particles (Fig. 1–5) were a PS II core complex with some but perhaps not all of the light-harvesting pigment–protein LHC II. The nature of the pigment–protein complexes will be considered in Chapter 6. The components of the small PF particles (9 nm) is less certain; they were originally identified as PS I centers, but because the ratio of PF to EF particles may be 3:1 and the ratio of activity of PS I to PS II is only about unity, they appear to contain proteins other than those associated with PS I, such as the cytochrome b_6–f complex, coupling factor CF_0, or part of LHC II. Kaplan and Arntzen (1982) have presented a stylized model of these components embedded in the thylakoid membrane (Fig. 1–6).

In the algae without appressed thylakoids (Cyanophyceae and Rhodophyceae), it is well established that phycobilisomes (PBS) are arranged on the thylakoid lamellae in rows about 50 nm apart (reviewed by Gantt,

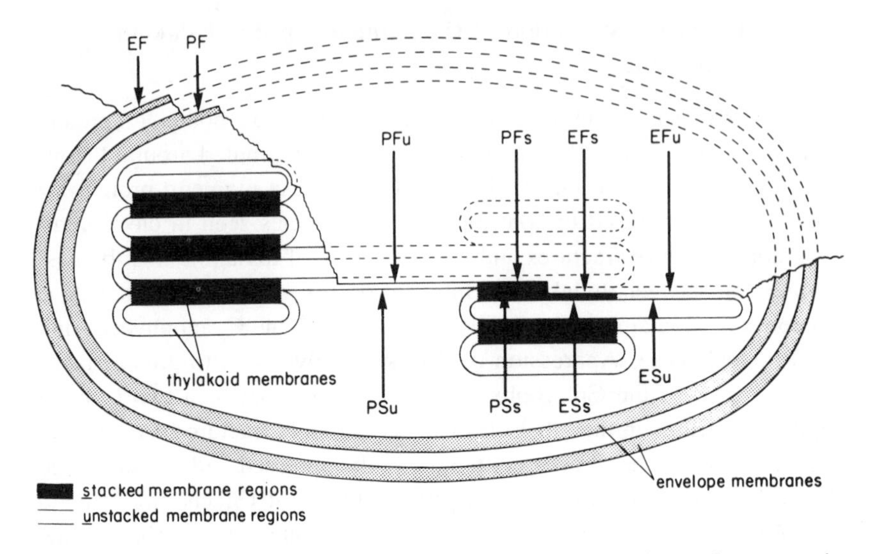

Fig. 1–4. Diagram illustrating the nomenclature for surfaces seen in electronmicrographs made from thylakoid membranes following freeze-fracture or freeze-etching. P, Protoplastmic leaflet (cytoplasmic or stromal); E, exoplasmic leaflet (external or lumenal); F, internal fracture faces of membranes; S, surfaces of membranes seen after freeze-fracture or freeze-etching; u, regions of unstacked membranes; s, regions of stacked membranes. (From Staehelin et al., 1977, by permission.)

1980; Golecki and Drews, 1982; and Larkum and Barrett, 1983); Lefort-Tran, Cohen-Bazire, and Pouphile (1973) found that the large particles on the EF fracture face were also arranged in rows opposite the rows of the PBS. (In nonappressed thylakoids, the large particles on the EF fracture face are only about 10 nm in diameter, not 14 nm, and the small particles on the PF fracture face are 7, not 9 nm.) As the large (10-nm) particles are thought to be the PS II units, this arrangement is consistent with efficient energy transfer from PBS to PS II. Giddings, Wasmann, and Staehelin (1983) have prepared fracture faces in the membranes of the cyanelles of *Cyanophora,* clearly showing the rows of large particles (9.4 nm) on the EF face and complementary grooves on the PF face between the small particles (7.3 nm) (Fig. 1–7). The distance between rows on the EF face was 46 nm, typical of the spacing in true Cyanophyceae and Rhodophyceae. They also show rows of fan-shaped PBS on the surface of the thylakoids (PS), and, where the PBS have fallen off, arrays of low-profile particles, apparently EF particles seen from the outside of the membrane; these profiles showed that the particles contained two sub-

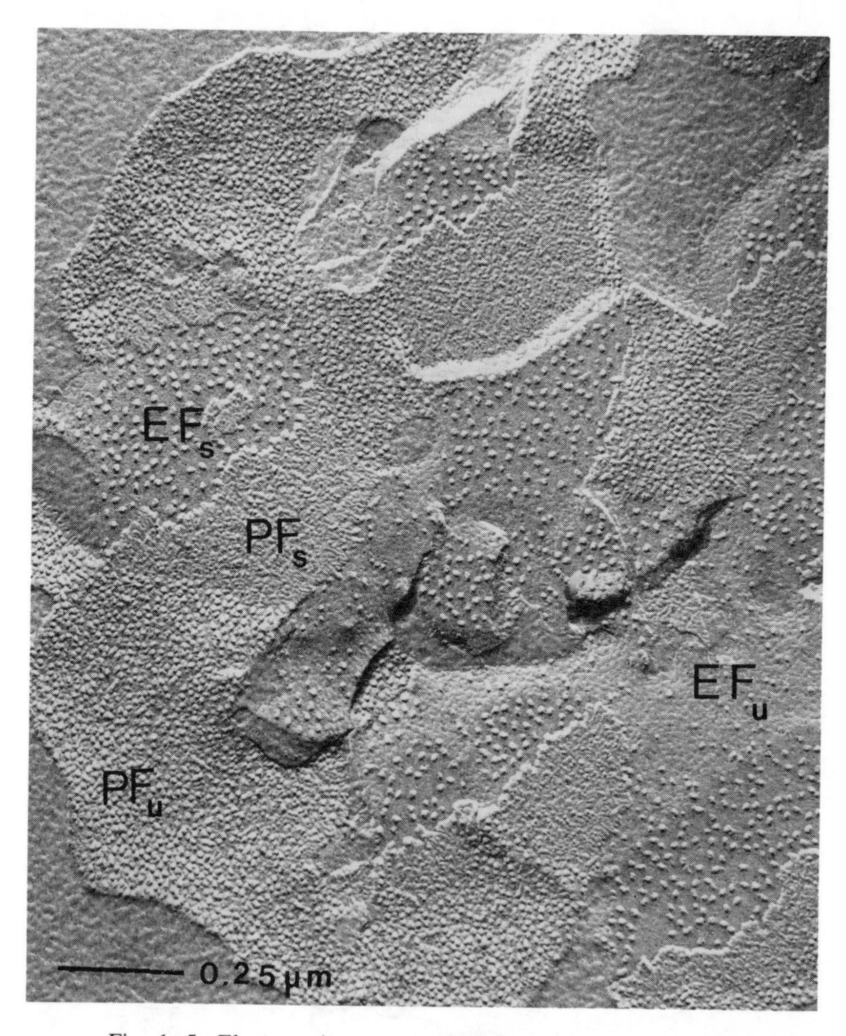

Fig. 1–5. Electronmicrograph of a thylakoid membrane freeze-etched in a divalent cation solution to preserve granal stacking. See Fig. 1–4 for nomenclature of the fracture faces. (From Kaplan and Arntzen, 1982, by permission; electronmicrograph supplied by Dr. L. A. Staehelin.)

units. This work confirms the spatial relationship previously postulated between PBS and the EF particles by previous authors (Lefort-Tran et al., 1973, Lichtlé and Thomas, 1976; Golecki, 1979; Wollman, 1979; Gantt, 1980). They also found what appeared to be CF_1 coupling subunits on the PS face when PBS were removed from it. A diagram of the struc-

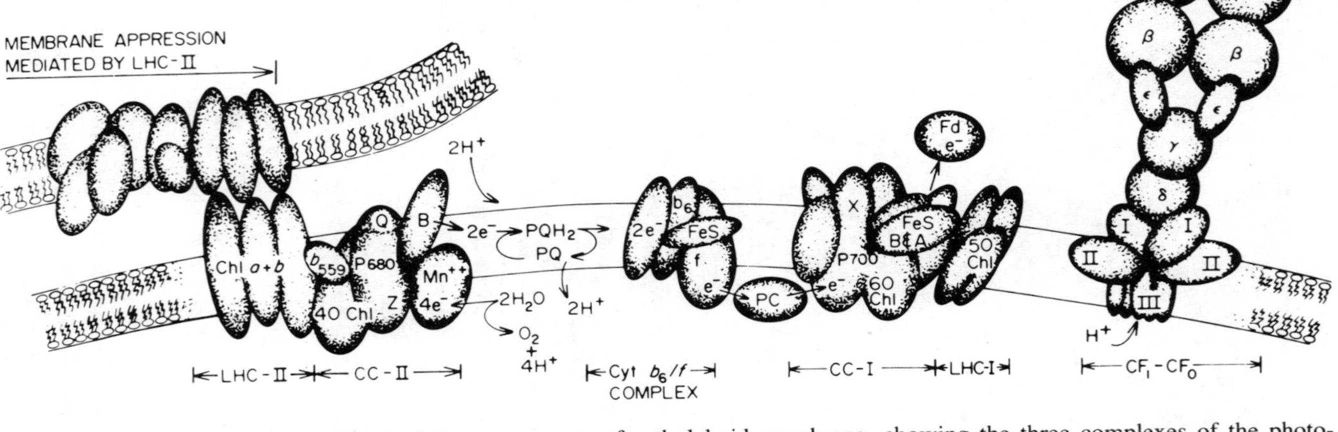

Fig. 1–6. A stylized model of the components of a thylakoid membrane, showing the three complexes of the photosynthetic electron-transport chain depicted in the "Z" scheme in Fig. 1–3 (CC-I and -II are CP1 and CPa, respectively, using the nomenclature of Chapter 6), and the ATPase (right) involved in ATP synthesis. (From Kaplan and Arntzen, 1982, by permission.) Magnification 50,000×.

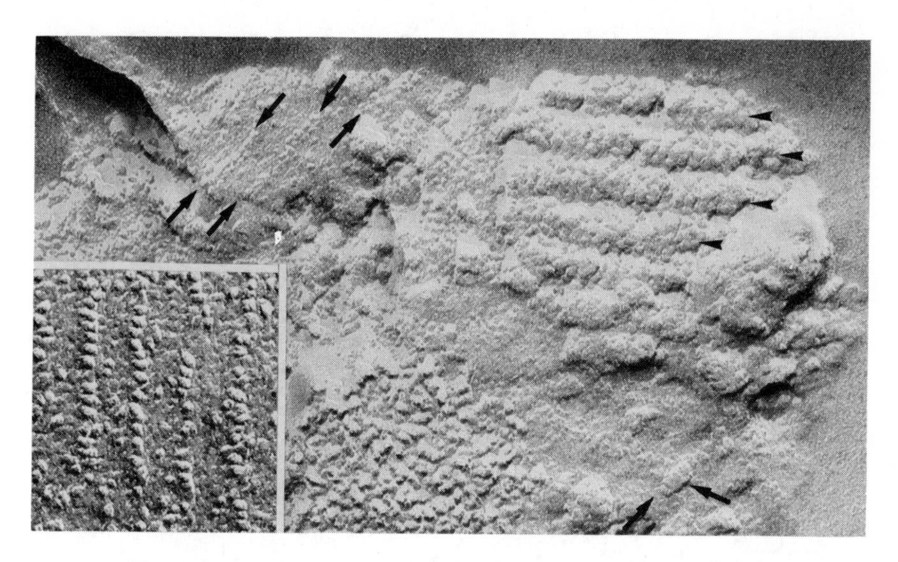

Fig. 1–7. Electronmicrograph of a thylakoid membrane from *Cyanophora paradoxa* after freeze-etching, showing rows of phycobilisomes (arrowheads). Arrows show underlying lines of low-profile particles, each consisting of two subunits (see Fig. 1–8). Inset shows rows of particles on the E face of the thylakoid membrane following freeze-fracture. Particles are spaced at 10 nm and sometimes show the two subunits seen in the freeze-etching. (From Giddings et al., 1983, by permission; electronmicrographs supplied by Drs. T. H. Giddings and L. A. Staehelin.) Magnification 150,000×.

ture of the thylakoid they propose is shown in Fig. 1–8. However, the corresponding arrays of PBS and EF particles have not been found in all experiments (Staehelin et al., 1977) in some species of the Rhodophyceae, though in one of them, *Spermathamnion,* Gantt and Conti (1966a) had previously found regular arrays of PBS and EF particles. Lack of corresponding arrays does not necessarily mean that PBS and EF particles do not coincide on each side of the membrane.

Gantt (1980) has reviewed evidence that the PBS form a functional energy-transmitting unit with the PS II unit, and in her laboratory, Gantt and colleagues have prepared particles from the red alga, *Porphyridium cruentum,* containing both PBS and PS II activity (Clement-Metral, Gantt, and Redlinger, 1985; Chereskin, Clement-Metral, and Gantt, 1985). These particles contain only 2% of the total chlorophyll (Chereskin et al., 1985), consistent with the amount of antenna chlorophyll expected (Clement-Metral et al., 1985), the remainder presumably being PS I chlorophyll. These particles did not show any PS I activity, as judged by low-

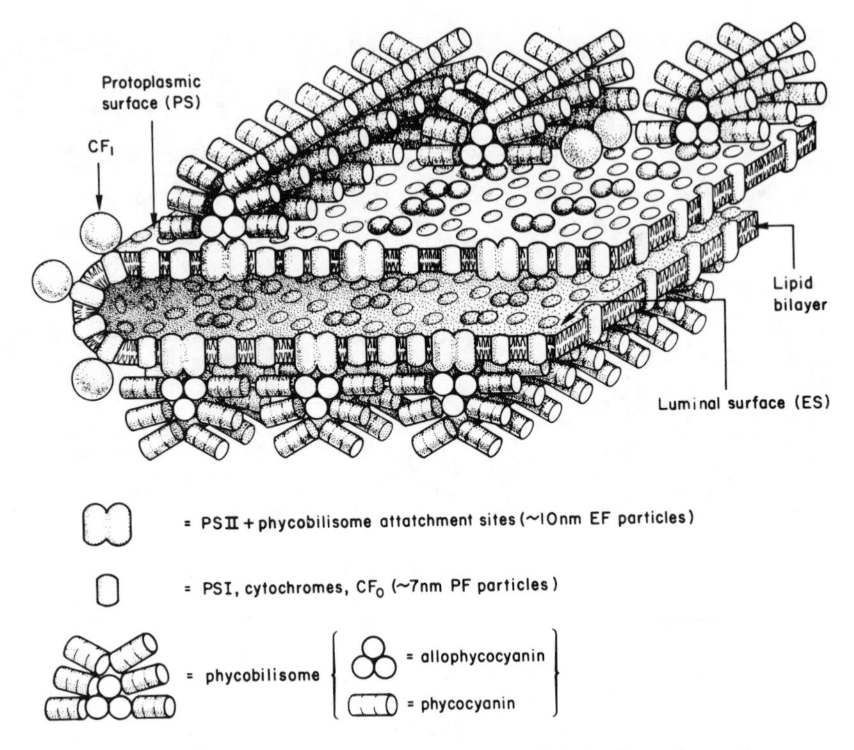

Fig. 1–8. A model of the thylakoid membrane from *Cyanophora paradoxa* showing the phycobilisomes attached to 10-nm EF particles. CF_0 and CF_1, subunits of coupling factor. For other nomenclature, see Fig. 1–4. (From Giddings et al., 1983, by permission; diagram provided by Dr. L. A. Staehelin.)

temperature fluorescence (Clement-Metral et al., 1985) and absence of P_{700}. Lundell, Glazer, Melis, and Malkin (1985) have isolated a PS I particle from *Synechococcus* 6301 with molar ratios for chlorophyll *a*/P_{700} of 130 and for carotenoid/P_{700} of 16. Manodori, Alhadeff, Glazer and Melis (1984) found that in cells of wild-type *Synechococcus* 6301, excitation of APC at 620 nm activated PS II fifteen times as much as PS I; they argued that this means that the small amount of PS I activated at 620 nm is due to direct absorption by PS I chlorophylls and not to energy transfer from PBS.

If all PBS and EF particles were combined to form a PS II activity particle, as suggested above, they would occur in equal numbers per unit area of thylakoid. Although ratios of unity are sometimes reported, earlier estimates of the PBS/EF particle ratio have varied from 2 to 0.2 (Gantt,

1980; Kursar and Alberte, 1983). Diner (1979) found evidence for a ratio of 0.5, and Kursar, Van der Meer, and Alberte (1983b) found ratios of less than unity, suggesting that up to six RC II units could be associated with one PBS in some Rhodophyceae and Cyanophyceae, depending on the size of the PBS. Manodori et al. (1984) found the ratio to be about 0.5 for wild-type *Synechococcus* 6301, but about unity for the mutant AN 112. This suggests that in wild-type *Synechococcus* two RC II were bound to one PBS, but, as the integrated absorption of light by PBS was 2.5 times higher than in the mutant, and the absorption of light by RC II was also correspondingly higher, they believed that half the RC II units were not coupled with PBS, though what their function was remained uncertain.

In the cyanophyte, *Gloeobacter violaceus,* the cytoplasmic membrane is the only one in the cell and carries the photosynthetic apparatus (Rippka, Waterbury, and Cohen-Bazire, 1974). Freeze-fracture of the cytoplasmic membrane has identified 11-nm particles in other algae without appressed thylakoids (Guglielmi, Cohen-Bazire, and Bryant, 1981). Immediately inside the cytoplasmic membrane an electron-dense layer (80 nm) appeared to contain the PBS, shown as rod-shaped elements in electron-micrographs and containing biliproteins typical of other cyanophytes.

The relationship of the biliproteins of the Cryptophyceae to the thylakoid membranes is uncertain, but Lichtlé, Duval, and Lemoine (1987) have isolated a photosynthetically active PE–xanthophyll–PS II complex enriched in CC II from *Cryptomonas rufescens* and interpret small geometrical units attached to thylakoid vesicles as PE.

Not all carotenoids are involved in light harvesting and not all are contained in the thylakoid membranes. Jeffrey, Dounce, and Benson (1974) have identified xanthophylls in the chloroplast envelope of higher plants, and they may well occur there in the envelopes of chloroplasts of eukaryotic algae. Some Cyanophyceae contain a wide range of xanthophylls (Chapter 4), but they do not necessarily harvest light energy. Murata, Sato, Omata, and Kuwabara (1981) and Omata and Murata (1983b) have separated the cell envelope (containing the cell membrane), cytoplasmic membranes, and thylakoid membranes from *Anacystis nidulans:* β-Carotene and zeaxanthin comprised a large proportion of the carotenoids in the thylakoid membranes, with lesser amounts of nostoxanthin and caloxanthin, and this fraction contained virtually all the chlorophyll *a*. Negligible β-carotene occurred in the cytoplasmic membranes, though they contained a high proportion of the total carotenoid as zeaxanthin and caloxanthin. Krogmann and associates (Holt and Krogmann, 1981; Diversé-Pierluissi and Krogmann, 1988) have extracted water-soluble carotenoid–protein complexes from four species of Cyanophyceae, though the amount of the zeaxanthin–protein that they found in *A. nidulans* could account for only a few percent of the zeaxanthin found by Murata et al. (1981).

In *Synechocystis* PCC 6714, Omata and Murata (1984) again found that nearly half of the total carotenoid in the thylakoid membranes was β-carotene, with zeaxanthin and a myxoxanthophyll-like carotenoid each contributing about 20%. The cytoplasmic fraction contained considerably more zeaxanthin and myxoxanthophyll-like carotenoid per unit protein than the thylakoid fraction. Unlike *Anacystis nidulans,* 90% of the carotenoid of the cell-wall fraction of *Synechocystis* was an unidentified, acetone-insoluble acidic carotenoid. Resch and Gibson (1983) found carotenoids in the cell-wall fraction from three species of Cyanophyceae but did not identify them.

2

Extraction and separation of lipid-soluble pigments

As the methods used for extracting chlorophylls and carotenoids are similar, these will be considered together in this chapter, though often the workers concerned are interested in only one of the two.

2.1. Techniques for extracting lipid-soluble pigments from algae

The more recent references to methods for extracting lipid-soluble pigments from algae include papers by Strain (1954, 1958), Goodwin (1955, 1980), Holden (1965, 1976), Strain and Svec (1966, 1969), Davies (1965, 1976), Jensen (1966, 1978), Aasen and Liaaen-Jensen (1966), Hager and Stransky (1970a), Britton and Goodwin (1971), Liaaen-Jensen (1971), Liaaen-Jensen and Jensen (1971), Kjøsen and Liaaen-Jensen (1972), Johansen, Svec, Liaaen-Jensen, and Haxo (1974), Berger, Liaaen-Jensen, McAlister, and Guillard (1977), Svec (1978), Krinsky and Welankiwar (1984), and Pechar (1987).

Many of the references mentioned above stress that hydrolysis (including the action of chlorophyllase), isomerization, allomerization of the chlorophylls, and oxidation can occur in the process of extracting lipid-soluble pigments; light, oxygen, acidity of the cell sap, and high temperature variously contribute to these changes and, with the occasional exception of heat when extracting from certain algae, should be avoided as far as possible.

The method used for extracting pigments from algae will depend on the reason for extracting the pigments, the type of alga (microphyte or macrophyte), and the amount of tissue available. If the only information required is the type of chlorophyll present (as when one is distinguishing green from yellow-green algae), chlorophyllase, oxidation, or allomerization will not distort the distribution, unless one is trying to detect chlorophyll *d*, which can possibly be produced by oxidation of chlorophyll *a* during extraction (Holt, 1961).

When quantitative measurements of chlorophylls are made, obviously precautions must be taken to prevent degradation of most kinds. If a dichromatic equation is being used to measure the ratio of chlorophyll *a* to *b*, degradation to pheophytins must be prevented, though chlorophyllase (see Sec. 2.1.1, below) will not interfere because the absorption spectra of the chlorophyllides are similar to those of the parent molecules. However, when one is using quantitative chromatography (Jeffrey, 1981), chlorophyllase must be inhibited with acetone if possible, though the chlorophyllides might well be pooled with the equivalent chlorophyll if they appear. However, some workers have reported chlorophyllide, even when extracting with 90% acetone. Pheophytin found in extracts may not represent faulty extraction, as any dead or senescent tissue will contain it, and traces act as light-harvesting pigments (Chapter 6). It could possibly be detected in severely senescent tissue by direct spectroscopy, as it can be in leaves of higher plants (Moore and Rowan, unpublished). Degradation of chlorophyll and carotenoids is difficult to inhibit during extraction from cells containing strongly acid sap, such as *Desmarestia* (Seely, Duncan, and Vidaver, 1972) or *Globospira* (Rowan, unpublished), since adding alkali to the extracting medium is not effective. The 5,6-epoxy carotenoids are vulnerable to acidic adsorbents such as silica gel, and they must be protected by neutralizing the acidity of the stationary phase (see Sec. 2.2.3C).

Algae vary considerably in the ease with which pigments can be extracted from them and Daley, Gray, and Brown (1973a) have proposed a series in order of increasing ease of extraction: detrital and fecal material → laucustrine algae → freshwater sediments → marine algae. Thus methods appropriate for one species may not be suitable for another.

2.1.1. Extracting solvents

Most workers have stressed that the initial solvent should be water-miscible (usually acetone or methanol) and used at about 20 ml per gram fresh weight (Strain and Svec, 1969) so that the water in the cell sap is diluted rapidly, thus preventing degradation of the pigments. The workers in the Argonne laboratories (Strain and Svec, 1966, 1969) added petroleum spirit (20–40°C) and/or diethyl ether to the methanol to speed removal of the pigments from the aqueous phase during extraction The mixtures they used were methanol–petroleum spirit (2:1 or 3:1), methanol–diethyl ether–petroleum spirit (5:2:1), or methanol–diethyl ether (9:1 or

7:3). Acetone has been used for many years as the solvent for extracting pigments from higher plants, and Barrett and Jeffrey (1964, 1971) have shown that at concentrations above 90% it inhibits the enzyme chlorophyllase, responsible for the hydrolysis of chlorophylls *a* and *b* to the respective chlorophyllides. Unfortunately, increasing numbers of workers have found it inefficient in extracting the lipid-soluble pigments, though we have found it satisfactory with some macrophytes, such as *Padina* and *Zonaria* (Phaeophyceae), *Griffithsia* (Rhodophyceae), some species of *Caulerpa* (Chlorophyceae), and *Rivularia* (Cyanophyceae).

Jeffrey and colleagues (Jeffrey, 1968a, c, 1974, 1981; Jeffrey, Sielicki, and Haxo, 1975; Jeffrey and Hallegraeff, 1980a; Jeffrey and Wright, 1987) have found 100% acetone satisfactory for extracting from some microalgae, and others also have used it (Allen, Goodwin, and Phagpolngarm, 1960; Riley and Wilson, 1967; Williams, 1967; Kleinig, 1969; Walton et al., 1970; Hager and Stransky, 1970a; Ricketts, 1971a; Nitsche, 1973; Smallidge and Quackenbush, 1973; Lewin, Norris, Jeffrey, and Pearson, 1977; Benson and Cobb, 1981). Bjørnland (1982, 1983, 1984), Bjørnland and Tangen (1979), and Bjørnland, Borch, and Liaaen-Jensen (1984) made the first extraction with pure acetone to inhibit chlorophyllase, followed by acetone–methanol (3:7) whereas Withers et al. (1978a) and Fiksdahl, Withers, Guillard, and Liaaen-Jensen (1984b) used 100%, then 90% and 80% acetone with sonication. When inhibition of chlorophyllase was not important (i.e., in extraction of chlorophyll *c* or carotenoids), Ricketts (1966a, 1967c), Jeffrey (1968b, 1969, 1972), Mandelli (1968), Garside and Riley (1969), Cheng, Don-Paul, and Antia (1974), Caron, Jupin, and Berkaloff (1983), and Haxo (1985) used 90% acetone, whereas Weber and Czygan (1972) used 85% acetone. Wojciechowski et al. (1988) homogenized phytoplankton isolated on glass-fiber filters in 90% acetone, and then held for 12 h at 4°C in darkness.

Inhibiting possible effects of chlorophyllase is often less important than efficient extraction of the full range of pigments in the tissue, and many workers have shown that acetone is less efficient than other solvents (Sand-Jensen, 1976; Garside and Riley, 1969; Seely et al., 1972; Marker, 1972; Shoaf and Lium, 1976; Holm-Hansen and Riemann, 1978; Marker, Crowther, and Gunn, 1980; Nusch, 1980; Duncan and Harrison, 1982; Pechar, 1987). Although methanol can allomerize chlorophylls during extraction (see Fig. 3–1), many workers have used it, alone (Thomas and Goodwin, 1965; Kleinig and Egger, 1967; Antia and Cheng, 1977, 1982, 1983; Yokahama, Kageyama, Ikawa, and Shimura, 1977; Yokohama, 1981a,b, 1983; Establier and Lubian, 1982; Araki, Oohusa, Omata, and

Murata, 1984; Haxo, 1985; Farnham, Blunden, and Gordon, 1985) or in combination with other solvents (Hirayama, 1967; Loeblich and Smith, 1968; Whittle and Casselton, 1969; Francis, Strand, Lien, and Knutsen, 1975; Pechar, 1987). When concerned only with carotenoids, the Trondheim school have used mixtures of methanol–acetone in ratios of 3:2 (Johansen et al., 1974), 1:1 (Buchecker, Liaaen-Jensen, Borch, and Siegelman, 1976; Berger et al., 1977; Foss, Guillard, and Liaaen-Jensen, 1984), 3:7 (Hertzberg and Liaaen-Jensen, 1966a,b, 1969a; Bjørnland and Tangen, 1979), 1:2 (Francis et al., 1975), and 1:3 (Pennington, Haxo, Borch, and Liaaen-Jensen, 1985). Jeffrey (1963, 1969) soaked *Sargassum* (Phaeophyceae) in methanol, followed by maceration with an equal volume of acetone added; this mixture has proved more effective than either solvent alone for extracting pigments from a number of macrophytes in this laboratory.

Pechar (1987) compared 90% acetone, 90% acetone–methanol (5:1), and methanol, using three methods (i.e., cold or hot extraction after homogenization or hot extraction) and found considerable variation. Extracting pigments from corals and clams is difficult; Jeffrey (1968b) extracted pigments from the zooxanthellae layer chiseled from coral, using methanol for 2 h with $MgCO_3$ present, whereas Jeffrey and Haxo (1968) froze crushed coral tissue containing zooxanthellae before extraction with methanol. Bowles, Paerl, and Tucker (1985) measured how efficient 100% acetone, 90% acetone, 90% methanol, and acetone–methanol (1:1) were for extracting pigments from a range of phytoplankton. No one solvent was best for all planktonic communities. Chapman (1966a,b; Chapman and Haxo, 1966) extracted carotenoids with ethanol. Renstrom, Borch, Skulberg, and Liaaen-Jensen (1981) removed chlorophylls from *Haematococcus* by soaking in 20% aqueous acetone overnight, next day extracting carotenoids with acetone–methanol (7:3).

DMSO and, to a lesser extent, DMF have been used for extracting lipid-soluble pigments from algae. Volk and Bishop (1968) found DMF efficient for extracting pigments from some species of algae, as it is with higher plants (Moran and Porath, 1980), but it was less efficient than DMSO. Reger and Krauss (1970) used DMSO–methanol for extracting chlorophylls from *Chlorella*. Seely et al. (1972) used DMSO for extracting pigments from kelps (Phaeophyceae), followed by acetone, without macerating the tissue. They found that the major part of the chlorophyll *c* and fucoxanthin were removed with the DMSO, and that the remaining chlorophylls and xanthophylls were removed with acetone. DMSO–acetone removed pigments from all Chlorophyta tested but did not do so for many Rhodophyceae. Wheeler (1980) modified the above method for extracting pigments from *Macrocystis* (Phaeophyceae). Shoaf and Lium (1976) found DMSO to be superior to 90% acetone for extracting chlorophylls from some of the Chlorophyta, though acetone was equally effective with some Bacillariophyceae and Cyanophyceae. Stauffer, Lee, and Armstrong (1979), comparing the efficiency of acetone, methanol, acetone–DMSO (1:1), and methanol–DMF (1:1), found acetone–DMSO to be most efficient for Cyanophyceae, Bacillariophyceae, and Chlorophyta, with methanol being equally efficient only with the Cyanophyceae. DMF also extracted the pigments efficiently, but raised the absorbance of chlorophyll when the DMF was added to a methanolic solution.

Burnison (1980) found that heating to 65°C for 10 min in a heating block improved extraction of chlorophylls from glass-fiber filters when using DMSO without the grinding employed by other workers when utilizing other solvents. Duncan and Harrison (1982) found methanol to be more effective as a second solvent following DMSO than was acetone; in some species such as *Fucus* (Phaeophyceae), chlorophyll *c* but little chlorophyll *a* was removed by the DMSO so that the chlorophyll *c* spectrum could be clearly seen at 580 and 630 nm. Unfortunately this differential extraction does not occur with all Phaeophyta. For quantitative analysis, sequential extraction with DMSO and methanol requires the use of separate dichromatic equations for DMSO–water (Seely et al., 1972) and methanol (Jeffrey and Haxo, 1968). Speziale, Schreiner, Giammatte, and Schindler (1984), comparing efficiencies of DMF, DMSO–acetone (1:1), and 90% acetone for extracting chlorophylls from phytoplankton, found that (1) all solvents were effective for Chrysophyceae and flagellate Chlorophyta, (2) DMF and DMSO–acetone were more efficient than acetone for Chlorophyta and Cyanophyceae, and (3) DMF was better than DMSO–acetone for chlorococcalean species. Stauber and Jeffrey (1988) found pigments difficult to extract from a few species of diatoms, for which they used acetone:DMSO (1:1).

In this laboratory we have used DMSO for screening pigments in Cyanophyceae and unicellular Chlorophyceae and Tribophyceae. Small pellets following centrifugatigon at 3000 *g* were resuspended in 2–3 ml DMSO and vigorously shaken, using a vortex mixer; the extract was then recovered by further centrifugation. Some species of the Tribophyceae required further extraction with methanol or acetone, though usually two extractions with DMSO were effective. Occasionally, none of these procedures were efficient, even with maceration using an Ultra-Turrax blender; freezing and thawing did not improve extraction, and sonication was probably required. The amount of pigment remaining in the cells can be tested by suspending the pellet after extraction in water and examining the absorption spectrum using a diffusion plate (Shibata, 1958).

Porra and Grimme (1974) have described a novel method for extracting and estimating chlorophylls in *Chlorolla* with an alkaline pyridine reagent (2.1 M pyridine in 0.35 M NaOH) when acetone, methanol, and methanol–DMSO were ineffective. No doubt this method would be equally effective with other recalcitrant species. Waaland, Waaland, and Bates (1974) extracted biliproteins and chlorophyll *a* from the same batch of tissue of the red alga, *Griffithsia,* by first blending the tissue in water with a glass homogenizer to extract the biliproteins; the pellet precipitated by centrifugation (25,000 *g* for 20 min) was then treated with 80% acetone to extract the chlorophyll with 100% efficiency.

2.1.2. Physical methods

Algae range from large plants to nanoplankton, and thus techniques used for extracting pigments from them will vary. Many of the macroalgae are tough, and we find that they are usually cut into small pieces and blended in various power-driven macerators, possibly preceded by freezing and thawing or boiling for a few minutes. Strong solvents such as DMSO can be used with some tissues without maceration. Pigments are often surprisingly difficult to extract from microalgae and require the same techniques as macroalgae described above. These methods are described in subsequent sections. Microalgae must be collected by filtration or centrifugation, and the particular method used will depend on whether the cells are suspended in water or cultured on solid media. Algae cultured on slopes of solid media can be scraped carefully from the surface; removing pieces of the medium can be avoided by using a wide, straight-ended spatula or by suspending the cells via washing the slope in extracting medium such as acetone, methanol, or DMSO. Ideally, macroalgae used for pigment extraction should be grown in unialgal culture; otherwise contaminating epiphytes (macroalgae or phytoplankton such as dinoflagellates or diatoms) will contribute to the distribution of pigments found. Extraction without degradation of the pigments depends on the speed used in removing pigments from sources of degradation. Strain (Sec. 2.1.2, above) has extracted pigments into a mixture of methanol and petroleum spirit so that the pigments can move rapidly into the nonaqueous phase, where they are separated from any water-soluble agent causing degradation, such as enzymes, acids, or oxidizing agents.

A. Macerators

Often extracting pigments requires rapid meçhanical disruption of cells, in both macroalgae and phytoplankton. The best modern macerators have sharp blades rotating inside fixed plates, drawing the tissue across the shearing blades by centrifugal force, and sucking fresh suspension from below. These macerators, such as the Ultra-turrax or Polytron, are much more efficient than domestic-type blenders such as the Waring used by Goodwin (1955), Jensen (1978), and Bjørnland (1984), and heat from the electric motor is not conducted to the shearing blades. Various sizes of probe are available, and volumes between 3 and over 300 ml can be used. The risk that sparks from the commutator will ignite the volatile solvents used is also much lower with the macerator than with domestic blenders, because with the latter, the solvent can leak or splash down onto the electric motor. Older, top-drive laboratory blenders (Jeffrey, 1962, 1963) are free from this fault. Even the modern macerators cannot cope with large intact tissue, and the macroalgae must be cut into small pieces before blending. For any of the blenders, sand and stones must be removed to prevent damage to the shearing blades. A further advantage of the modern blenders is that the probes can be readily dismantled for cleaning. Whittle and Casselton (1969), Hager and Stransky (1970a), and Braumann and Grimme (1979) have used glass beads in homogenizers of various brands for extracting pigments from microalgae, whereas Walton et al. (1970) and Caron et al. (1983) used a French press to disrupt cells of *Phaeodactylum* (Bacillariophyceae) before extracting chlorophylls with 90% acetone.

Many workers have used hand- or power-operated homogenizers, usually made in glass, such as the Ten Broeck or Potter-Elvehjem, for extracting pigments from phytoplankton (Jeffrey 1968a,d, 1974; Jeffrey et al., 1975; Ramus, Beale, Mauzerall, and Howard, 1976a; Carneto and Catoggio, 1976; Jeffrey and Vesk, 1977; Ramus, Lemons, and Zimmerman, 1977; Holm-Hansen and Riemann, 1978; Withers, Vidaver, and Lewin, 1978; Jeffrey and Hallegraeff, 1980a; Beer and Levy, 1983; Stauber and Jeffrey, 1988) or from plastics such as Teflon (Lorenzen, 1967; Shoaf and Lium, 1976; Sartory and Grobbelaar, 1984). Many macroalgae are too tough for extracting pigments with homogenizers of the above types, but Waaland et al. (1974), Ramus and colleagues (Ramus et al., 1976, 1977; Rosenberg and Ramus, 1982; Ramus, 1983; Ramus and van der Meer, 1983), and Yokohama et al. (1977) have used them for extracting pigments from macroalgae such as *Griffithsia* with softer tissue. A pestle and mortar, used with or without an abrasive, such as acid-washed sand, is effective for extracting pigments from small samples (Allen, French, and Brown, 1960; Jeffrey, 1976b; Keast and Grant, 1976; Eskins, Schoefield, and Dutton, 1977; Jensen, 1978; Falkowski and Owens, 1980; Benson and Cobb, 1981; Farnham, Blunden, and Gordon, 1985).

B. Sonication

Efficient extraction using sonication requires good contact between the cells to be disrupted and the probe of the instrument; therefore, the container used should be barely larger than the probe, thus spreading the suspension of cells in a thin layer around it. Nelson (1960) was the first worker to use sonication for extracting pigments from phytoplankton, reporting 11–30% more chlorophyll *a* in extracts using a 9-kV sonic oscillator for up to 9 min than with overnight extraction in 90% acetone at 4°C. Subsequently, Mann and Meyers (1968), Garside and Riley (1969), Daley and Brown (1973), Daley, Gray, and Brown (1973a), Withers et al. (1978a,b), Abaychi and Riley (1979), Fiksdahl, Withers, and Liaaen-Jensen (1984), Manodori and Melis (1984), Paerl, Lewin, and Cheng (1984), Wright and Shearer (1984), Bowles et al. (1985), Farnham et al. (1985), and Stauber and Jeffrey (1988) have used sonication, although these authors did not compare this approach with other methods of extraction. Chang and Rossmann (1982) reported that grinding was sometimes more effective than sonication with mixed populations of phytoplankton, though effectiveness varied with the composition of the harvest. In some experiments, they found sonication no more effective than soaking unruptured cells in acetone. Daley and Brown (1973) sometimes lyophilized suspensions before extracting the pigments by sonication. Jeffrey et al. (1975) found that sonication improved extraction with some but not all of 22 dinoflagellate species. Daley et al. (1973a) designed an apparatus for extracting pigments from filters by disintegration with an ultrasonic probe (Sonifyer, Model 125), followed by filtration; a cooling jacket provided control of temperature during disintegration. The method proved more efficient than shaking with glass beads, mechanical grinding, or disintegration in a sonic bath, and was the only method not producing artifacts during extraction (chlorophyll *a′* or allomerized chlorophyll).

C. Soaking and osmotic shock

Efficient extraction of pigments does not always require mechanical disruption, and soaking with or without shaking in an appropriate solvent has proved effective with some species of algae. Seely et al. (1972) found DMSO effective without shaking, though Wheeler (1980) used occasional agitation, and we have found a vortex mixer effective for microalgae. When collecting cells by centrifugation, Jeffrey (1968a) extracted the pigments from the pellets by multiple extractions with 90% acetone or methanol; in earlier experiments when the extraction was continued overnight (Jeffrey, 1961, 1965), phytinization occurred. She preferred to suspend marine phytoplankton in pure water for 30 min before treating with acetone, thus extracting about 98% of the pigments rapidly. Riley and Wilson (1965) also used lysis in pure water before extracting from phytoplankton before shaking with acetone at greater than 80%. They allowed ground cells to stand in 90% acetone in darkness overnight, but only found pheophytin *a* in extracts from aged culture.

Jeffrey (1968c) extracted pigments from endosymbiotic chlorophytes in brain coral by soaking small pieces chiseled from the coral in methanol in darkness for 1–2 h with $MgCO_3$ present, using several changes of methanol until no further pigment was extracted. Sardana and Mehkotra (1979) shook lyophilized cells in three different solvents in darkness for up to 3 h, finding 1 h sufficient.

D. Heating

Oxidation, hydrolysis, and isomerization are the main reactions degrading pigments during extraction, and some workers have used heat to inactivate the enzymes responsible. Mann and Myers (1968) heated diatoms suspended in 90% acetone to 55°C for 30 s before sonication, whereas Strain and associates boiled microalgae in water (Aitzetmuller et al., 1969) or Na_2HPO_3 (Strain et al., 1970) for 2 min before chilling and extracting pigments with methanol and diethyl ether. Although heating can inactivate some enzymes, it can increase isomerization and pyrolysis (Strain, 1954; Strain and Svec, 1966). Brief heating seems most valuable when one is making large-scale extractions where degraded products formed during extraction can be removed during subsequent purification (Jeffrey, 1972). Jeffrey and Wright (1987) dipped *Sargassum* fronds in boiling seawater for 20 s, chilled in ice-cold seawater for 10 s, and immersed in methanol until the first traces of carotenoid were released. Pigments were then extracted by transferring the fronds to 100% acetone for 5–10 min. Jensen (1978) has recommended heating for 20 s in steam or boiling water to inhibit lipoxygenase in some red algae, and Speziale et al. (1984) has boiled fronds of *Fucus* in seawater for 10 s before extracting pigments with pure acetone. Holden (1965) suggested that fatty-acid oxidation could be involved in bleaching of chlorophyll, though this was not due simply to action of lipoxygenase.

E. Freezing and lyophilization

Workers such as Yokohama et al. (1981a, 1983) and Withers et al. (1978a), who froze tissues before extraction of pigments, did not suggest that this was a deliberate step to improve extraction; rather, it appears to have been used for avoiding degradation when tissues were being held after harvest or were being transported over long distances (Allen et al., 1960). Daley et al. (1973a) found no loss of chlorophyll after freezing for two months, and Gieskes and Kraay (1983a) held plankton on filters for three months at −20°C without loss of pigments, but Whittle and Casselton (1969) warned that long freezing could cause phytinization. Other workers have used freezing as an aid to efficient extraction. Jeffrey and Haxo (1968), who found pigments difficult to extract from zooxanthellae symbiotic in corals and clams, used freezing for 4–8 h before extracting with methanol, and Jeffrey (1969) froze fronds of *Sargassum* at −20°C for 30 min before extracting with methanol and acetone. Tett, Kelly, and Hornberger (1975) used freezing followed by extraction with boiling methanol for extracting pigments from algae on rocks and in soft sediments. Lewin et al. (1977) froze phytoplankton to −80°C using dry ice–ether mixtures and then extracted pigments at −20°C with 100% acetone. Marker (1980) found that freezing increased the efficiency of extraction from *Cladophora* with methanol; Benson and Cobb (1981) froze 50-g samples of *Codium fragile* for 30 min at −20°C and dehydrated them in methanol before extraction, whereas Renstrom et al. (1981) froze *Haematococcus* prior to extracting pigments with acetone and methanol.

Daley et al. (1973a) considered that lyophilization did not degrade chlorophyll unless the cells were exposed to high intensities of light, but Bjørnland et al. (1984) blamed furanoid rearrangement of antheraxanthin to mutatoxanthin on using lyophilized material, as this reaction did not occur in fresh material. However, lyophilization has been used regularly in the Trondheim laboratories (Liaaen-Jensen, 1971; Liaaen-Jensen and Jensen, 1971) on microalgae without any obvious isomerization. Garside and Riley (1969) found that lyophilization did not improve extraction of pigments from phytoplankton.

F. Filtration

Filtration is used either for removing cell debris after maceration of macroalgae or for collecting microalgae suspended in water.

Macroalgae. If the extract is required for analysis by spectroscopy using chromatic equations, most cuvettes used need no more than 3 ml, but the extract should be optically clear without any trace of opalescence, usually tested by measuring the adsorption at 730–50 nm. Filtering through Whatman No. 50 paper usually gives a clear filtrate, but if not, high-speed centrifugation may remove the opalescence, though some workers have found this method to be ineffective. As light-scattering causes broadening and displacement of absorption maxima, it must be avoided if chromatic equations are to be used accurately. The small volume required does not take long to collect, but the filtration should be made at room temperature in darkness; a chilled filtrate will frost the glass of the cuvette, causing light scattering. If the extract is to be concentrated for chromatography by phasing into diethyl ether (Sec. 2.2.3), coarse filter paper such as Whatman No. 1 or 541 (41H) can be used since any colloid passing the paper will remain in the hypophase after partitioning. This filtration may take longer than a few minutes and should be carried out in the cold as well as in darkness.

Microalgae. Collecting microalgae efficiently by filtration requires rapid rates, high retention of cells on the filter, and efficient elution of pigments from the filter. Daley et al. (1973a) tested six filters for filtration rate and retention: Gelman Alpha-6-Metricel (1) and Epoxy Versapor (2), Millipore Solvinert UH (3), Reeve Angel 984 H (4), and Whatman GFC (5) and GFA (6), all of which are inert in the solvents used for extracting pigments. All but the coarse Whatman filter (GFA) gave efficient retention, but the Gelman filters gave rates of only 20% of the Whatman GFC filter, and pigments were not easily extracted from the Versapor filter. Glass-fiber filters found efficient by other workers were not only Whatman GFC (Long and Cooke, 1971; Jeffrey, 1974; Hallegraeff, 1976; Holm-Hansen and Riemann, 1978; Gieskes and Kraay, 1983a) but also Gelman AN-450 (Hallegraeff, 1977) and A-E (Falkowski and Owens, 1980), Reeve Angel 984H (Long and Cooke, 1971; Holm-Hansen and Riemann, 1978; Stauffer et al., 1979), S & S (Faust and Norris, 1982), S & S No. 16 and GF 92 (Schanz, 1982), and Sartorius SM 13400 (Sartory and Grobbelaar, 1984). Sheldon (1972) has plotted percent retention against particle diameter for Millipore (cellulose ester), Whatman GFC and GFA, Reeve Angel 984H, the Selas Flotronics metal membranes, and the GE Nuclepore polycarbonate membrane of a range of pore sizes. The relationship between median retention size and nominal pore size differed considerably; Flotronics and Nuclepore retention agreed well with nominal pore size, Millipore filters all tended to have similar retention size, no matter what the nominal pore size was, and the glass-fiber filters all had median sizes less than 1 μm. Recently evidence has been accumulating that nanoplankton form a significant fraction of the phytoplankton in oceanic waters, and therefore, uncritically accepting a filter may lead to serious error in estimating biomass (Smith et al., 1985).

MF Millipore HA filters, though retaining cells efficiently (Humphrey, 1960; Holm-Hansen and Riemann, 1978; Garside and Riley, 1969; Jensen and Sakshaug, 1973), are soluble in the extracting solvents, and the cellulose triacetate from which they are made must be precipitated with diethyl ether before chromatography (Jensen and Sakshaug, 1973). Garside and Riley (1969) found that filters of a reconstituted cellulose (Sartorius Cella) were equally efficient as glass-fiber filters but more difficult to handle, and that all membrane filters tested were slower in filtering than glass-fiber filters. Their claim that up to 5% of cells pass GFC filters is consistent with the work by Sheldon (1972) mentioned previously. Although glass-fiber filters appear more efficient than membrane filters for preparing extracts of pigments, Fournier (1978) considered that they did not possess the advantages of the latter for preparing cells for other purposes. Marker, Nusch, Rai, and Riemann (1980) have summarized the findings of a workshop held in 1978, recommending Whatman GFC filters rather than membrane filters, though the latter may retain slightly smaller cells.

Many workers have used $MgCO_3$ with the filters (Jensen and Sakshaug, 1973; Jensen, 1978), either as a filter-aid (Humphrey, 1960; Garside and Riley, 1969; Jeffrey, 1974; Carneto and Catoggio, 1976; Sand-Jensen, 1976; Hallegraeff, 1976, 1977; Abaychi and Riley, 1979; Lubian and Establier, 1982; Gieskes and Kraay, 1983a) or to neutralize acids liable to degrade pigments (Chang and Rossmann, 1982; Wood, 1985) although Strain and Svec (1966) and Svec (1978) warned that $MgCO_3$ does not neutralize acids sufficiently rapidly. Sand-Jensen (1976) found that $MgCO_3$ was more efficient as a filter-aid when it was added to the filter before the cell suspension rather than simultaneously with it, but that it lowered the efficiency of extraction after freezing. Riemann (1978a), Holm-Hansen and Riemann (1978), and Nusch (1980) found no advantage in using $MgCO_3$ as a filter-aid, whereas Daley et al. (1973a) found that, in addition, it absorbed about 80% of the chlorophillides in the extract. Marker and Jinks (1982) found that when $MgCO_3$ was used to neutralize methanolic extracts after acidification, substantial errors resulted through adsorption of pheophorbides.

When the microalgae have been collected on the filter pads, pigments have been extracted from them by sonication of the broken pads (Garside and Riley, 1969; Daley et al., 1973a), by mechanical or hand grinding (Long and Cooke, 1971; Jensen and Sakshaug, 1973; Jeffrey 1974, 1976b; Carneto and Catoggio, 1976; Jensen, 1978; Holm-Hansen and Riemann, 1978; Jeffrey and Hallagraeff, 1980a; Falkowski and Owens, 1980; Hallagraeff, 1981; Faust and Norris, 1982; Sartory and Grobbelaar, 1984), or by simple soaking (Holm-Hensen and Riemann, 1978; Gowen, Tett, and Wood, 1982). Chang and Rossmann (1982) found grinding the glass-fiber filters more efficient than soaking or sonication. Sartory and Grobbelaar (1984) concluded that ethanol and methanol were superior to acetone and DMSO in extracting pigments from Sartorius filters and that boiling the filter in ethanol (90–95%) or pure methanol gave complete extraction. They claimed that homogenization or sonication did not improve extraction. Wood (1985) found no loss of chlorophyll *a* when filters coated with $MgCO_3$ were held in $CHCl_3$–methanol (2:1) in darkness for 10 days at room temperature or 4°C.

G. Centrifugation

Centrifugation can be used for the same purposes as filtration, that is, for removing cell debris and for collecting microalgae. Throndsen (1978) has discussed methods for collecting phytoplankton by centrifugation, and the methods apply equally well to suspensions of culture algae suspended in liquids. Two types of centrifuge are used: (1) cup type, either swing-out or angle-head, and (2) continuous flow.

Cup-type centrifuges. The centrifugal force used should be sufficient to form a firm pellet at the base of the tube without rupturing the cells of delicate microalgae, since this might lead to chemical changes in the pigments. The minimum sedimentation required should be determined by trial and error. When one is recovering an extract in plastic tubes, the solvent should not react with the material of the tubes; makers' catalogs specify tolerance of tubes to solvents and should be consulted. If the extract is required for quantitative spectroscopy, the sedimentation should be sufficient to give a solution free of all opalescence, and the pellet should be resuspended in solvent until all pigment is removed from it. The pellet is readily resuspended by first dispersing on the walls of the tube with the end of a slightly smaller test-tube or a glass rod (conical tubes) before more solvent is added. With small tubes (up to 15 ml), a vortex mixer is convenient for extracting pigments from pellets of microalgae or from debris of macroalgae. If an efficient solvent is used, microalgae should not need maceration.

Continuous-flow centrifuges. Centrifuges of this type, available for collecting phytoplankton from large volumes of water, include the simple cream-separator type and the continuous counterflow type (Throndson, 1978). Commercial types of the former, with flow-rates of about 30 $l \cdot h^{-1}$, have been described by Throndson (1978); individual instruments have been made by Davis (1957) and by the Workshop of the Botany School, University of Melbourne.

The dense suspension of cells formed after continuous centrifugation on these instruments can be conveniently collected as a soft pellet by use of a small hand-driven bench centrifuge. Recently, Donze and de Groot (1982) have described a modification to a commercial spin-dryer collecting cells in a honeycomb mat made of silicone rubber, with a flow-rate of 200 $l \cdot h^{-1}$, much higher than that of any commercial instrument. The rates of the commercial flow-through systems using the Beckman centrifuges JE-6B and CF-32Ti are only 6 and 9 $l \cdot h^{-1}$, respectively, but the Sorvall RC2-B, employing a different design, gives flow-rates of 30 $l \cdot h^{-1}$ at 48,000 *g*, using the regular type SS-34, 8 × 50 ml head. Gross, Stroz, and Britton (1975) used a Sorval KSB continuous-flow apparatus at 17,000 *g* at the same rate of flow. Hallegraeff (1977) used the MSE 18 centrifuge at 7000 *g* with a flow-rate of 60 $l \cdot h^{-1}$. Without specifying details of performance of the centrifuges, Nitsche (1973) used an MSE, whereas Bjørnland and colleagues (Bjørnland and Tangen, 1979; Tangen and Bjørnland, 1981; Fiksdahl, Bjørnland, and Liaaen-Jensen, 1984a; Fiksdahl, Withers, Guillard, and Liaaen-Jensen, 1984b) used a Kahlisco model 903-15, and Withers, Fiksdahl, Tuttle, and Liaaen-Jensen (1981), Foss et al. (1984), and Pennington et al. (1985) used a Sharples.

2.2. Separation and purification

Chromatography has replaced the older methods based on phase separation used by earlier workers such as Willstäter and Stohll (see Davies, 1976) and these will not be described in detail. In investigations of carotenoids only, removal of chlorophylls and other organic acids by saponification is widely practiced.

2.2.1. Saponification

Saponification involves adding KOH dissolved in methanol to the crude extract and incubating in darkness at room temperature, thus forming potassium salts of the organic acids, including chlorophyll *a* and *b*, which are now water-soluble. The details of the methods used vary among workers: Liaaen-Jensen and Jensen (1971) and Jensen (1978) recommended evaporating the crude extract to dryness, dissolving the pigments in a small volume of diethyl ether, and adding an equal volume of 10% methanolic KOH. After 1–2 h in darkness, the unsaponified pigments are recovered by adding 2–5% NaCl solution in a separating funnel, leaving the saponified pigments and other acids in the hypophase and the unsaponified pigments in the epiphase of diethyl ether, which can be concentrated by evaporation under a stream of N_2 gas and applied to a chromatogram. Davies (1965, 1976) recommended dissolving the pigment in ethanol or methanol, then adding 1/10 of a volume of 60% aqueous KOH, and finally either heating the alkaline mixture for 5–10 min in the dark in a bath of boiling water, or holding the mixture in the dark at room temperature. Both treatments should be carried out under N_2 gas. Goodwin (1955) added 1/10 of a volume of the KOH but held the mixture in darkness overnight.

Numerous slight modifications to the above methods have been used (Chapman and Haxo, 1963; Thomas and Goodwin, 1965; Aihara and Yamamoto, 1968; Aitzetmuller et al., 1969; Britton and Goodwin, 1971; Francis et al., 1975; Antia and Cheng, 1977; Brown and McLachlan, 1982; Lubian and Establier, 1982). Saponification should not be used when the few carotenoids unstable in alkali (i.e., astaxanthin, fucoxanthin, and peridinin) are present in the extracts. Cholnoky et al. (1969) considered that isomerization to cis isomers tended to occur during saponification, though Pennington et al. (1985) found none. Some workers have added KOH directly to the extracting solvent (Hiyama, Nishimura, and Chance, 1969).

2.2.2. Separation by partition

The variable polarity of the lipid-soluble pigments allows partition of mixtures of them into an epiphasic (upper) and hypophasic (lower) fraction; the two-phase system usually contains petroleum spirit in the epiphase and 90% methanol in the hypophase (Goodwin, 1955; Davies, 1965,

1976; Britton and Goodwin, 1971). Carotenes, chlorophylls *a* and *b*, and xanthophylls containing one epoxide, keto, or hydroxy group, and esters and methylated pigments generally move to the epiphase, whereas the remaining xanthophylls move to the hypophase. Alkali-labile xanthophylls such as fucoxanthin and peridinin will be in the hypophase, allowing esterification of the epiphasic chlorophylls (Liaaen-Jensen, 1971). Monoketo and monohydroxy xanthophylls can then be removed from the epiphase by phasing into 95% methanol. As an alternative to saponification, alkali-labile xanthophylls can be transferred from petroleum spirit into an equal volume of 84% methanol, leaving chlorophylls and some carotenoids in the epiphase from which chlorophylls can be removed by esterification (Liaaen-Jensen, 1971). Xanthophylls from any of these methanolic hypophases can be recovered by addition of diethyl ether and water. The workers in the Trondheim laboratories fractionated carotenoids into epiphasic and hypophasic fractions in their studies of the pigments in the blue-green algae (Hertzberg and Liaaen-Jensen, 1966a,b, 1969a,b, 1971; Francis, Hertzberg, Andersen, and Liaaen-Jensen, 1970). Parsons (1963) used a two-phase system of a hexane-rich epiphase and an aqueous acetone-rich hypophase to separate chlorophyll *c* (hypophase) from carotenoids and other chlorophylls in the epiphase. This was a preliminary step for the estimation of chlorophyll *c* (Chapter 3). Other phytyl-free chlorophyll derivatives would move with the chlorophyll *c* (chlorophyllides and pheophorbides), and this method was used by Ricketts (1966a) to isolate MgDVP. White, Jones, and Gibbs (1963) separated pheophorbides and chlorophyllides from chlorophylls and pheophytins by phasing solutions of all pigments dissolved in diethyl ether with 0.01 M KOH, thus collecting the phytyl-free derivatives in the aqueous-rich hypophase. The phase test (Seely, 1966), in which 30% methanolic KOH is poured under a solution of chlorophyll in diethyl ether, to form a transitory brown ring when chlorophyll is present, is used for testing allomerized chlorophyll since this oxidized form of chlorophyll remains in the epiphase with the carotenoids while the esterified chlorophylls pass to the aqueous hypophase.

2.2.3. Chromatography

Chromatography in its various forms is now the major method of separating and purifying the lipid-soluble pigments, often preceded by saponification when only carotenoids are to be examined or measured. Chromatographic separations are based either on differential adsorption of a mixture of compounds between a stationary phase and a moving phase (in columns, paper, or thin layers), or on the differential partition of the mixture between a stationary liquid phase (supported on an inert material in a column, fine tube [HPLC], filter paper, or thin layer) and a moving phase. These are known as adsorption and partition chromatography, respectively. Partition chromatography on paper sometimes requires equilibration between the vapor of the liquid forming the stationary phase and the paper; Hanes and Isherwood (1949) shortened the time for equilibration by attaching a soft-iron wire in a sealed glass tube to the end of the descending paper chromatogram, oscillating the paper from its support by revolving a magnet outside the tank. Chromatography is known as "reverse phase" when the support (paper, thin-layer, or powder) is impregnated with a liquid giving the reverse type of retention to the untreated support. Kost (1988) has recently edited a handbook on the chromatography of plant pigments.

Pigments are not sufficiently concentrated in the initial extracts to apply directly to chromatograms. Most workers concentrate and partially purify the initial extracts by adding a small volume of a solvent that is immiscible with water (e.g., diethyl ether, ethyl acetate, or petroleum spirit) to the initial extract in alcohol or acetone, and then adding water to form an epiphase in which the pigment is concentrated and from which some of the more polar contaminants are removed in the water-rich hypophase; usually a 10% NaCl solution is used to prevent excess emulsification at the interface and in the hypophase. Diethyl ether forms a two-phase system with water, but it is 15% miscible with water at laboratory temperatures, and thus the volume of the extract can be reduced by washing with excess volumes of NaCl solution. A separating funnel is usually used for larger amounts, but on a small scale (Jeffrey, unpublished) the ether-rich phase can be collected by adding the NaCl solution to the mixture of extracting solvent and diethyl ether in a small volumetric flask (25 or 50 ml) so that the epiphase containing the pigments rises into the neck of the flask; from here it can be removed with a Pasteur pipette or applied directly to a chromatogram. If necessary, the ether phase can be concentrated with a stream of N_2 gas. If a dry sample of extract is required, the water in the ether is removed by adding anhydrous Na_2SO_4 or NaCl (note that Jeffrey [1968a] has warned that Na_2SO_4 can adsorb pigments, undesirable for quantitative work), and the water is thus removed as water of crystallization. Alternatively, excess ethanol can be added and the water evaporated under reduced pressure as the 95% ethanol–5% water azeotrope, or the water can be removed by freezing.

A. *Column chromatograph*

This method is still essentially that developed by Tswett (1906a,b, 1911) in which a glass column is filled with adsorbent, a concentrated solution of the pigments is applied to the head of the column, and pigments are eluted in successive bands by washing one or more organic solvents through the column, with or without gentle suction (up to 0.5 atm). Strain and associates refined Tswett's original method (Strain, 1934, 1938, 1958; Strain and Svec, 1966, 1969; Strain, Sherma, and Grandolfo, 1968; Strain and Sherma, 1972; Svec, 1978). They claimed that column chromatography is as fast as many TLC systems, protects pigments from air, and can be used on a large scale, taking up to 1 g of chlorophyll when large columns are used (Strain and Svec, 1966; Svec, 1978). The column can be prepared by packing dry adsorbent into a glass column with a glass sinter, or by pouring a slurry of powder into the tube and allowing the adsorbent to settle into a compact column, though by the latter method columns may shrink and channel when sucked dry (Svec, 1978). Columns used range from 5 mm in diameter for analytical work to 10 cm in diameter and about 35 cm long for preparative work. Columns provide considerable advantages over TLC for preparative work, as the calculations by Strain et al. (1968) show. Many adsorbents and developing solvents have been used (Strain and Svec, 1969; Jensen and Liaaen-Jensen, 1971; Holden, 1976; Svec, 1978) though sucrose columns developed with 0.5 to 2% propan-1-ol in petroleum spirit (60–80°C) seem the most satisfactory in the Argonne laboratory and separate pigments in extracts from all algal classes (Strain and Svec, 1969; Svec, 1978). Strain and Svec (1966) stressed that diethyl ether must be dry if used rather than acetone for eluting a column, as shown by Perkins and Roberts (1964). The adsorbent used in the column and the developing solvent will depend on the pigment separation required. Strain and Svec (1969), Britton and Goodwin (1971), and Davies (1976) have discussed the major factors affecting column chromatography – in particular, variations in the efficiency of partition and adsorption systems in relation to the type of pigments examined. Thus partition sorbents, such as polypropylene, separate xanthophylls with different polar groups (hydroxy, keto, and epoxy), but not pigments differing only in the position of double bonds, such as α- and β-carotene. The latter are better separated by using surface-active, polar adsorbents, such as activated alumina or MgO. Though these adsorbents are valuable for such purposes as these, Strain (1938, 1958), Strain and Svec (1966), Strain, Sherma, and Grandolfo (1967), Davies (1976), and Holden (1976) have pointed out that inorganic adsorbents, silica gel in particular, are liable to decompose chlorophylls and some carotenoids, though they may well be satisfactory for other pigments. Strain and Svec (1969) and Svec (1978) listed adsorbents and eluting solvents commonly used for column chromatography of lipid-soluble pigments and provided a table of solvents in order of eluting power or polarity. Liaaen-Jensen and Jensen (1971) and Davies (1976) graded adsorbents in order of increasing activity and published a table of solvents in order of eluting power. Mixed beds of adsorbents are sometimes used, usually including diatomaceous earth as a filter acid. Davies (1976) divided methods of column chromatography into three basic types.

Zone chromatography. The single solvent used must separate all the pigments while they are still in the column; they then must be extruded and the separated pigments dissolved in suitable solvents (Strain, 1958; Liaaen-Jensen, 1971). An adsorbent retaining its shape on extrusion, such as sucrose or $CaCO_3$, is thus essential. Alternatively, the bands may be collected as they are eluted from the column.

Stepwise elution chromatography. The column is washed with solvents of increasing polarity; Britton and Goodwin (1971) have tabulated a series of solvents eluting carotenoids from alumina (Brockmann Grade III) in a preliminary separation into five classes, increasing in polarity at each step. Each class could then be resolved by a series of separations on different TLC plates. The components used in the two systems described above are shown in Table 2–1.

Wun, Rho, Walker, and Litsky (1979) have adsorbed pigments from phytoplankton at the head of a column of Amberlite XAD-1 resin; the cells were collected on a Whatman GF/F filter paper at the top of the column, from where pigments were eluted with 7% ammonia in methanol. The chlorophylls were bound at the head of the column while the carotenoids were washed through with the ammoniacal methanol (the ammonia improves the retention of chlorophyll at the head of the column). The chlorophylls were then eluted with benzene, evaporated to dryness by use of a rotary evaporator, and redissolved in 90% acetone. Omata and Murata (1983a) separated chlorophyll *a* and *b* from carotenoids by using a second column of DEAE-Sepharose CL-6B, and separated chlorophyll *a* from *b* by using a second column of Sepharose CL-6B. They found that only small columns were required, and high flow-rates could be used. Subsequently, Araki et al. (1984) changed the eluting solvents

Table 2–1. *Systems used for column chromatography of lipid-soluble pigments extracted from algae*

Adsorbent	Solvent	Separation	Reference
Zone chromatography			
Sucrose	0.5–2% Propan-1-ol in petroleum spirit	General	Strain & Svec (1969); Svec (1978)
	2% Propan-1-ol in petroleum spirit (40–60°C)	MgDVP	Ricketts (1966a)
MgO–Hyflo Supercel (1:1)	2% Acetone in *n*-hexane	Carotenes	Hertzberg & Liaaen-Jensen (1966b)
Seasorb–Celite (1:1)	10% Propan-1-ol in petroleum spirit	Weakly bound xanthophylls	Aitzetmuller et al. (1969)
Cellulose powder	20–30% Acetone in petroleum spirit	Hypophasic carotenoids	Hertzberg & Liaaen-Jensen (1966b)
	Pyridine	Oscillaxanthin	
	0.5% Propanol-1-ol in petroleum spirit (60–80°C)	Chlorophyll *c*	Jeffrey (1972)
Deactivated alumina	2% Diethyl ether in petroleum spirit	β-Carotene	Hertzberg & Liaaen-Jensen (1966a)
	7% Acetone in petroleum spirit	Myxoxanthophyll	
	25–30% Acetone in petroleum spirit	Epiphasic carotenoids	
Deactivated alumina (Brockmann Grade III)	5% Diethyl ether in petroleum spirit (30–60°C)	Carotenes	Gross et al. (1975)
Activated MgO–Celite (1:1)	10% Propan-1-ol in petroleum spirit	Loroxanthin Neoxanthin	Aitzetmuller et al. (1969)
Polyethylene PRX-1025	Acetone	Chlorophyll c_1 & c_2	Jeffery (1972)
Stepwise elution chromatography			
Sucrose	0 → 3% Acetone in petroleum spirit	General	Allen et al. (1960)
MgO–Celite–Hyflo Supercel	Acetone → petroleum spirit (40–60°C)	Xanthophylls	Ricketts (1966b, 1967a)
	Propan-1-ol → petroleum spirit (40–60°C)	Siphonaxanthin	Ricketts (1971a)

MgO–Celite (2:1)	Petroleum spirit (40–60°C) → 4% methanol in acetone	General	Whittle & Casselton (1975a)
MgO–Kieselguhr (1:2)	Petroleum spirit (40–60°C) → acetone	General	Whittle & Casselton (1975b)
MgO–Celatom (1:1)	5 → 20% Acetone in *n*-hexane	Preliminary separation	Loeblich & Smith (1968)
Cellulose	10% Acetone in petroleum spirit → >30% acetone in petroleum spirit	Hypophasic carotenoids	Hertzberg & Liaaen-Jensen (1969a); Francis et al. (1970)
S & S No. 123 Cellulose–silica gel G (4:1)	5% Propan-1-ol in acetone → petroleum spirit	Preliminary separation	Bjørnland & Aguilar-Martinez (1976)
Deactivated alumina (Brockmann Grade III)	Diethyl ether → petroleum spirit	Siphonaxanthin	Walton et al. (1970)
	Petroleum spirit → 1% diethyl ether in petroleum spirit → 20% benzene in petroleum spirit	Carotenes	Britton & Goodwin (1971)
	30 → 50% Diethyl ether in petroleum spirit	Monohydroxy xanthophylls	
	Diethyl ether → 5% ethanol in diethyl ether	Dihydroxy & epoxy xanthophylls	
	Diethyl ether → petroleum spirit	General	Withers et al. (1978a)
Deactivated alumina	Petroleum spirit → 30 → 70% diethyl ether in petroleum spirit → 10% methanol in diethyl ether → methanol–acetic acid	Xanthophylls	Lubian & Establier (1982)
Activated alumina	10% → 15% → 70% Diethyl ether in petroleum spirit	Carotenes	Lubian & Establier (1982)
Silica gel II–Celatom (1:1)	5 → 20% Acetone in *n*-hexane	Purification of carotenoids	Loeblich & Smith (1968)
Microcel C	10 → 80% Acetone in petroleum spirit	Preliminary separation	Chapman (1966a)
$ZnCO_3$–diatomaceous earth (3:1)	0 → 25% Acetone in petroleum spirit	Carotenoids	Chapman (1966a)
$ZnCO_3$–Celite	0 → 100% Acetone in petroleum spirit	Carotenoids	Whittle & Casselton (1975a)
$CaCO_3$–Celatom (1:1)	5 → 20% Acetone in *n*-hexane	Purification of carotenoids	Loeblich & Smith (1968)
$CaCO_3$–$Ca(OH)_2$–diatomaceous earth (2:2:1)	0 → 20% Acetone in benzene	Preliminary separation	Chapman (1966a)
Polyethylene	85 → 95% Methanol	Chlorophyll *c*	Chapman (1966b)

so that carotenoids, chlorophyll *a*, pheophytin *a*, and chlorophyllide *a* could be separated by using a single column of DEAE-Sepharose.

High-performance liquid chromatography. This system will be considered in Section 2.2.3D.

B. Paper chromatography

Sestak (1958, 1965, 1980) has reviewed methods of separating chloroplast pigments by paper chromatography. Although usually replaced now by TLC and HPLC, paper chromatography, in addition to providing a method for teaching students the principles of chromatography, is still preferred by some workers for quantitative analysis (Jensen, 1978). Sestak (1958) reviewed the early attempts to separate the chloroplast pigments by so-called capillary analysis, culminating in the radial chromatography by W. G. Brown (1939), who separated several zones of pigments using hard blotting paper held between two 6-inch glass plates, applying the extract in CS_2 through a small hole in the center of the upper plate and developing by adding drops of CS_2. Strain and Sherma (1972) have published a photograph of one of Brown's chromatograms. Following the development of paper chromatography as we know it today by Martin and Synge (1941), several groups published experiments in 1952 showing separations of chloroplast pigments (Sestak, 1958).

Factors regulating R_f values. The important factors regulating the R_f values observed with a given solvent system in paper chromatography are given in the following three sections.

Components and proportions of the developing solvent. Early chromatograms were developed with single organic solvents (Brown, 1939; Asami, 1952), and monochlorobenzene, a solvent of intermediate polarity, gave almost complete separation of chloroplast pigments on S & S 2043b paper (Bauer, 1952). Markus (see Sestak, 1958) and Bauer (1952) were first to develop chromatograms with mixtures of organic solvents; mixtures of relatively polar and nonpolar solvents were usually used from then on (Sestak, 1958). As Sestak (1965, 1980) subsequently pointed out, the original solvents listed in his first review have not changed, only variations in the proportions in which they were used. Sestak (1965) stated that the mixtures with a small proportion of polar to nonpolar solvents separate chlorophylls badly but carotenoids well; however, if separation of chlorophylls was improved by adding more polar solvents, some chlorophylls and xanthophylls did not separate. He claimed that the best one-dimensional solvent was the complex mixture used by Hager (1955): ligroin–benzene–chloroform–acetone–propan-2-ol (50:35:10:0.5:0.17). The developing solvents in all forms of paper chromatography are shown in the tables in Sestak's reviews (1958, 1965, 1980), whereas Strain and Svec (1969) and Svec (1978) also list some of the same systems, mostly those used in the Agronne laboratories. Strain and Svec (1969) have listed some of the more common organic solvents in order of decreasing polarity. Sestak (1958) has cited numerous references on the effect of quantity of pigment applied to the paper and suggested 5 and 15 μg chlorophyll as the load for No. 1 and No. 3 Whatman paper, respectively (Sestak, 1965). Strain and Sherma (1969) have demonstrated the disadvantage of loading too much extract onto Whatman No. 1 paper, showing that the more slowly running spots tend to move faster, thus reducing separation between them.

Degree of equilibration between paper and solvent. The dimensions of the tank or test-tube used can alter the R_f values, and identical R_f values cannot be expected in differently shaped tanks (Sestak, 1958). Using a descending chromatography tank, Holden (1962) found improved separation of chlorophylls *a* and *b* and degradation products, using petroleum spirit–benzene–acetone (4:1:0.5) when the paper was equilibrated overnight with the solvent. Hanes and Isherwood (1949) reduced the time necessary for equilibration by attaching a soft-iron wire in a sealed glass tube to the end of descending chromatograms and oscillating the paper by rotating a magnet outside the tank.

Temperature. In general, variation in temperature alters the running speed of a chromatogram but usually does not alter the R_f value (Sestak, 1965), though Sestak (1980) quoted an exception to this reported in a thesis by Popova, where the R_f values of chlorophyll increased with increasing temperature. Holden (1962) ran her chromatograms at a fixed temperature (20 ± 2°C) and suggested that the solvent composition might need to be altered at higher or lower temperature. Temperature gradients across a tank will cause crooked fronts but can be detected by placing marker spots on either side of the paper.

One-dimensional paper chromatography. As mentioned above, Sestak (1958) reviewed early reports of one-dimensional chromatography of the photosynthetic pigments. Resolution of all chlorophylls from the major xanthophylls in extracts from higher plants or algae is usually incomplete, though Hager (1955) did achieve this by using a complex solvent and a vacuum tank as the chromatography chamber; his system does not seem to have been used by anyone else, however. The use of circular paper gives some improvement, though Strain, Sherma, Benton, and Katz (1965c) were unable to separate chlorophyll *b* and violaxanthin with it. Strain, Sherma, Benton, and Katz (1965a) used one-dimensional chromatography to examine the effects of high loading and colorless contaminants; trailing of spots could not be avoided but could be improved by running lines of the extract the whole width of the paper. Chlorophylls c_1 and c_2 have not been resolved on paper. The main use of one-dimensional paper chromatography on unimpregnated paper seems to be in screening for chlorophyll *b* or *c*, since these pigments sometimes provide useful taxonomic evidence, as in separating Tribophyceae from Eustigmatophyceae, and in identifying the Prochlorophyceae. The system also resolves chlorophylls from their degradation products (Holden, 1962; Chiba et al., 1967).

Impregnated paper. Papers may be impregnated either with a liquid immiscible in the running solvent (paper–partition chromatography) or with a solid such as kieselguhr, silica gel, or sugar (paper–adsorption chromatography). Paper–partition chromatography includes so-called reversed-phase chromatography (Strain, 1953; Strain and Sherma, 1969) in which the strongly nonpolar pigments are strongly adsorbed in the nonpolar liquid bound to the paper, thus reversing the order in which a mixture of pigments move as compared with a simple paper–adsorption chromatogram. Paper soaked in liquid is hard to handle, and the stationary liquid phase is sometimes eluted with the spots of pigments, thus interfering with subsequent spectroscopy or rechromatography (Sestak, 1965). This type of chromatography has not been used recently. Angapindu, Silberman, Tantnatana, and Kaplan (1958) used circles of Whatman No. 1 paper impregnated with paraffin oil to separate pigments extracted from *Dunaliella* and *Ecklonia* with methanol as solvent. This method appears to have overcome the tailing seen in conventional one-dimensional chromatograms. Strain and Sherma (1969) dipped paper in the hypophase of a dimethylformamide–petroleum spirit mixture (i.e., the polar phase) and, after equilibration, developed the chromatogram with the epiphase rich in petroleum spirit. However, this method did not give complete separation of the xanthophylls of higher plants, and thus it is no longer used.

Paper–adsorption chromatography has been used by Jensen and coworkers, employing several types of impregnated filter paper for separating lipid-soluble pigments, utilizing the method of Rutter (1948) for circular paper. The most widely used has been S & S No. 287 kieselguhr paper, usually with acetone–petroleum spirit developing solvents (Table 2–2) (Jensen, 1959, 1963, 1966, 1978; Jensen and Liaaen-Jensen, 1959). Jensen (1959) could separate cis–trans isomers, and also lutein from zeaxanthin by using this paper. Paper impregnated with Al_2O_3 (S & S No. 667 or 288, made specially for Jensen, 1960) was effective in separating the carotenes. This paper no longer appears in the maker's catalogs. Jensen and Aasmundrud (1963) used commercial paper impregnated with $CaCO_3$ (S & S No. 996) or Whatman No. 1 paper soaked in 18% aqueous sucrose solution for chromatography of the chlorophylls. These two systems separated chlorophylls *a*, *b*, and *d* (synthetic: 2-desvinyl-2-formyl chlorophyll *a*) (Holt and Morley, 1959) and pheophytin *a* and *b*; chlorophyll *c* and pheoporphyrin *c* were not resolved, but the combined spot close to the start line was well separated from the other compounds. This seems the only reference to the probable R_f value of chlorophyll *d* on paper chromatograms.

Other workers have also used impregnated papers; Jeffrey (1968d) found that kieselguhr-impregnated paper gave smaller spots than plain paper for the one-dimensional chromatography used in her previous methods (Jeffrey, 1961). Sherma (1971) used commercial papers impregnated with silica gel or $Al(OH)_3$ with petroleum spirit–acetone as developing solvent, whereas Sherma and Strain (1968) used ion-exchange papers with petroleum spirit–propan-1-ol. Both systems separated carotenoids but were not satisfactory for separating chlorophylls.

Two-dimensional paper chromatography. Although other systems had been used previously (see Sestak, 1958), Jeffrey (1961) was the first to develop a method separating pigments extracted from the major eukaryotic algae by two-dimensional chromatography on Whatman 3MM paper. Several organic solvents were added to petroleum spirit (60–80°C) in concentrations up to 4%; 4% propan-1-ol was chosen as the most satisfactory for developing the first dimension. The solvent for the second dimension was 30% chloroform in petroleum spirit (Lind, Lane, and Gleason, 1953). Chlorophyll *c*, pheophytin *a*, and derivatives of chlorophyll *a* were resolved, but degradation products of chlorophyll *b* did not appear on the chromatograms, consistent with the greater stability of this pigment. For good resolution, solvent mixtures were prepared fresh

Table 2–2. *R_f values of lipid-soluble pigments on chromatography paper impregnated with kiesleguhr (S & S No. 287) (× 100)*

Pigment	% Acetone in petroleum spirit								% Acetone in hexane				Reference
	0	2	5	10	15	20	30	50	2	5	10	20	
Aphanizophyll	–	–	–	–	–	–	–	42	–	–	–	–	Hertzberg & Liaaen-Jensen (1966b)
Astaxanthin	–	–	–	44	–	–	–	–	–	–	–	–	Johansen et al. (1974)
	–	–	–	57	–	–	85	–	–	–	–	–	Jensen & Liaaen-Jensen (1959)
Canthaxanthin	1	22	38	–	–	–	–	–	–	–	–	–	Halfen & Francis (1972)
	–	24	–	–	–	–	–	–	–	–	–	–	Norgard et al. (1974a)
	–	–	–	–	–	–	–	–	87	–	–	–	Berger et al. (1977)
α-Carotene	78	–	–	–	–	–	–	–	–	–	–	–	Norgard et al. (1974b)
β-Carotene	95	98	98	98	–	–	–	–	–	–	–	–	Jensen & Liaaen-Jensen (1959)
	78	98	99	–	–	–	–	–	–	–	–	–	Halfen & Francis (1972)
	98	–	–	–	–	–	–	–	–	–	–	–	Norgard et al. (1974a)
	–	–	–	–	–	–	–	–	–	–	90	–	Berger et al. (1977)
γ-Carotene	68	–	–	–	–	–	–	–	–	–	–	–	Jensen & Liaaen-Jensen (1959)
Chlorophyll *a*	–	–	41	–	–	–	–	–	–	–	–	–	Jensen & Liaaen-Jensen (1959)
	–	–	35	–	–	–	–	–	–	–	–	–	Liaaen-Jensen & Jensen (1971)
Chlorophyll *b*	–	–	21	–	–	–	–	–	–	–	–	–	Liaaen-Jensen & Jensen (1971)
Chlorophyll *c*	–	–	0	–	–	–	–	–	–	–	–	–	Jensen (1966)
α-Cryptoxanthin	–	–	68	–	–	–	–	–	–	–	–	–	Hertzberg & Liaaen-Jensen (1966a)
	–	–	–	–	–	–	–	–	–	75	–	–	Berger et al. (1977)
Cryptoxanthin diepoxide	–	–	–	–	–	–	–	–	–	–	87	–	Berger et al. (1977)
Diadinoxanthin	–	–	–	68	–	–	–	–	–	–	–	–	Jensen (1966)
	–	–	–	68	–	–	–	–	–	–	–	–	Liaaen-Jensen & Jensen (1971)
	–	–	–	49	–	–	–	–	–	–	–	–	Johansen et al. (1974)
Diatoxanthin	–	–	–	62	–	–	–	–	–	–	–	–	Jensen (1966)
	–	–	–	62	–	–	–	–	–	–	–	–	Liaaen-Jensen & Jensen (1971)
	–	–	–	–	–	–	–	–	–	–	36	–	Berger et al. (1977)

Echinenone	–	–	79	–	–	–	–	–	–	–	–	–	Hertzberg & Liaaen-Jensen (1966a)
	–	–	81	–	–	–	–	–	–	–	–	–	Hertzberg & Liaaen-Jensen (1966b)
	18	65	78	–	–	–	–	–	–	–	–	–	Halfen & Francis (1972)
	–	–	–	–	–	–	–	–	80	–	–	–	Berger et al. (1977)
Fucoxanthin	–	–	–	53	–	–	–	–	–	–	–	–	Jensen (1966)
	–	–	–	40	–	81	–	–	–	–	–	–	Jensen & Liaaen-Jensen (1959)
	–	–	–	48	–	–	–	–	–	–	–	–	Liaaen-Jensen & Jensen (1971)
	–	–	–	–	–	–	–	–	–	–	–	77	Berger et al. (1977)
Isozeaxanthin	–	–	–	76	–	–	–	–	–	–	–	–	Jensen (1966)
Lutein	–	–	–	75	–	–	–	–	–	–	–	–	Jensen (1966)
	–	–	39	72	–	91	–	–	–	–	–	–	Jensen & Liaaen-Jensen (1959)
Lycopene	53	86	–	–	–	–	–	–	–	–	–	–	Jensen & Liaaen-Jensen (1959)
Mutatochrome	–	55	–	–	–	–	–	–	–	–	–	–	Hertzberg & Liaaen-Jensen (1967)
Myxoxanthophyll	–	–	–	–	–	–	–	58	–	–	–	–	Hertzberg & Liaaen-Jensen (1966b)
	–	–	–	–	–	–	52	–	–	–	–	–	Francis & Halfen (1972)
Neoxanthin	–	–	–	–	26	–	–	–	–	–	–	–	Jensen (1966)
	–	–	–	–	–	66	–	–	–	–	–	–	Liaaen-Jensen & Jensen (1971)
Oscillaxanthin	–	59	–	–	–	–	–	–	–	–	–	–	Francis & Halfen (1972)
Peridinin	–	–	–	57	–	–	–	–	–	–	–	–	Johansen et al. (1974)
Peridinol	–	–	–	43	–	–	–	–	–	–	–	–	Johansen et al. (1974)
Pyrrhoxanthin	–	–	41	–	–	–	–	–	–	–	–	–	Johansen et al. (1974)
Pyrrhoxanthinol	–	–	45	–	–	–	–	–	–	–	–	–	Johansen et al. (1974)
Siphonaxanthin	–	–	–	17	–	56	–	–	–	–	–	–	Ricketts (1971a)
Taraxanthin	–	–	–	40	–	–	–	–	–	–	–	–	Jensen (1966)
Vaucheriaxanthin	–	–	–	–	–	31	–	–	–	–	–	–	Norgard et al. (1974a)
Vaucheriaxanthin ester	–	–	–	50	–	–	–	–	–	–	–	–	Norgard et al. (1974a)
Violaxanthin	–	–	18	44	–	83	–	–	–	–	–	–	Jensen & Liaaen-Jensen (1959)
Zeaxanthin	–	9	30	59	–	87	–	–	–	–	–	–	Jensen & Liaaen-Jensen (1959)
	–	–	–	73	–	–	–	–	–	–	–	–	Jensen (1966)
	–	–	–	72	–	–	–	–	–	–	–	–	Liaaen-Jensen & Jensen (1971)

each day, and if separation was complete within 2 h, no degradation of chlorophyll occurred, as judged by the absence of pheophytins in chromatograms of the extracts from fresh cultures, though pheophytin *a* always occurred in extracts from aged cultures. For quantitative measurements, spots containing chlorophylls were cut out, eluted in acetone, and estimated by spectroscopy. This system did not separate chlorophylls c_1 and c_2. This method subsequently proved useful for screening phytoplankton (Jeffrey, 1965), with chlorophyll *b* from Chlorophyta and chlorophyll *c* from Bacillariophyceae and Chrysophyceae used as markers. Jeffrey (1968c) has also utilized the method for screening pigments from siphonous green algae endophytic in coral. The R_f values found by Jeffrey in the two dimensions are shown in Table 2–3. An essentially similar system was used by Strain, Sherma, Benson, and Katz (1965b) and Strain and Svec (1966) on Whatman No. 1 paper. Ikamori and Arasaki (1977) have used a similar two-dimensional system with hexane rather than petroleum spirit as the major developing solvent. Jeffrey and Allen (1967) have modified a chromatobox for running two-dimensional chromatograms in a ship at sea.

C. *Thin-layer chromatography*

Although known in the late 1930s, TLC owes its present popularity to Stahl (1965), and Geiss (1987) has recently published a monograph on the use of this technique. Thin-layer chromatograms can run in one or two dimensions, and soaking the material of the layer in organic liquids allows reverse-phase chromatography by partition of pigments between a stationary phase and a moving phase. The particular advantage of TLC over paper chromatography of the chloroplast pigments lies in the linearity of the adsorption isotherms compared to the convex isotherms with paper as adsorbent; thus with suitable adsorbents and solvents, tailing of spots often found on paper does not occur and the spreading of spots is minimal. Although the adsorption isotherms obtained with TLC are often linear over a larger concentration range than with paper chromatography, the amount that can be applied varies considerably with different layers. The compactness of the resulting spot on the TLC plate will be a function of the rate of establishing adsorption equilibrium of the pigment between layer and solvent, and the degree of tailing of the spot will depend on the linearity of the adsorption isotherm (Randerath, 1968). Sestak (1967, table 1) has shown that the advantages of TLC outweigh those of paper chromatography for separating chloroplast pigments; possible degradation

Table 2–3. *R_f values of pigments from eukaryotic algae separated by two-dimensional chromatography on Whatman 3MM paper in the first (1) and second (2) dimension*

Pigment	Jeffrey (1961)		Ricketts (1970)	
	4% Propan-1-ol (1)	30% $CHCl_3$ (2)	2% Propan-1-ol (1)	20% $CHCl_3$ (2)
α-Carotene	95–96[a]	93–96	97	97
β-Carotene	95–96	94–96	–	–
Chlorophyll *a*	76–84	21–36	–	–
Chlorophyll *b*	56–63	10–16	–	–
Chlorophyll *c*	10–20	0	–	–
Diadinoxanthin	54–71	44–54	–	–
Diatoxanthin	57	60	–	
Dinoxanthin	54–73	44–48	–	–
Fucoxanthin	49	28	–	–
Lutein	74–80	73–86	64	73
Neoxanthin	32–36	5–8	–	–
Peridinin	49–51	19–23	–	–
Prasinoxanthin	–	–	32	45
Siphonaxanthin	30	5	3	10
Siphonein	58	38	19	43
Taraxanthin	–	–	41	60
Violaxanthin	64–65	48–50	40	62
Zeaxanthin	–	–	52	64

Note: The solvent shown is dissolved at the stated percentage in petroleum spirit (60–80°C).
[a]R_f values × 100.

of pigments by the adsorbent of the layer is the only serious disadvantage, and, as we will see below, choosing an inert adsorbent or neutralizing its acidity can overcome this problem. The separation of chlorophylls and degradation products by TLC has been reviewed by Sestak (1967, 1982), Svec (1978), and Dolphin (1983).

One advantage of TLC over paper chromatography lies in the wide choice of adsorbents. Silica gel G (Stahl, 1965) is the most easily handled, being relatively impervious to moisture, and it can be stored for a relatively long time. Unless the climate is humid, it can be used without activation at high temperature, by drying overnight at room temperature. Unfortunately, silica gel is slightly acidic, causing degradation of chlorophylls and some carotenoids; the problem must be overcome for quantitative work and for accurate spectroscopic work on eluted pigments. Bacon (1966) has shown artifacts formed from chlorophyll *a* and *b* when chromatographed on silica gel G, probably due to oxidation, since gassing the chromatogram with N_2 reduced the amount formed, whereas Schaltegger (1965) reported chlorophyll's changing color while on the plate. However, Garside and Riley (1969) could not detect any decomposition of any pigments when using the alkaline solvents of Riley and Wilson (1965), who overcame the acidity by including the alkali diethylamine or formdimethylamide in the developing solvent of petroleum spirit (60–80°C)–ethyl acetate–diethylamine, 58:30:12 (v/v) (Riley and Wilson, 1965) or 55:32:13 (v/v) (Riley and Wilson, 1969). Garside and Riley (1969) found that Merck silica gel PF 254 gave better resolution of the xanthophylls than silica gel G. These additives appear to have been sufficiently alkaline to neutralize the acidity of the layer without altering chlorophylls or fucoxanthin. Garside and Riley (1969) also added diethylamine (1%) to the diethyl ether in which pigments were stored at low temperature for any length of time. Keast and Grant (1976) buffered their silica gel G in 0.2 M borax buffer (pH 8), finding this superior to the 1% KOH used by Whitfield and Rowan (1974). Whittle and Casselton (1975a,b), Gibbs (1979), and Gillan and Johns (1980) have also used diethylamine as a component of their developing solvents with silica gel G, and in this laboratory we have used silica gel G suspended in the borax buffer of Keast and Grant (1976) with the alkaline solvents of Riley and Wilson (1965). For chromatography of the carotenoids, Bjørnland and Aguilar-Martinez (1976) used layers of silica gel G mixed with $CaCO_3$ or $Ca(OH)_2$, whereas Hager and Stransky (1970a) used a strongly alkaline layer containing $CaCO_3$–MgO–$Ca(OH)_2$, which has proved efficient in separating chemically similar carotenoids such as lutein and zeaxanthin. In spite of the risks of decomposition, many workers still use silica gel G without buffering, though an equal number are now using layers of less reactive organic materials such as sucrose and cellulose (Sestak, 1982). Precoated commercial TLC sheets are now widely used for chromatography of lipid-soluble pigments (Sestak, 1982, table 1). These commercial layers require different solvents versus layers prepared on glass sheets in the laboratory, and Sherma and Lipstone (1969) list solvents that can be used with a variety of these plates.

Sestak (1967; 1982, table 4) has listed papers describing separation of chlorophylls using reverse-phase chromatography, where layers of kieselguhr, silica gel, sugars, or cellulose have been impregnated with various oils. Daley et al. (1973a) developed reverse-phase chromatography involving low-, medium-, and high-range systems for resolving pheophytins, chlorophylls and pheophorbides, and chlorophyllides, respectively. They found this method better than previous methods based on cellulose, silica gel, or kieselguhr (Daley, Gray, and Brown, 1973b). They claimed to have resolved 29 compounds, chlorophylls or derivatives, and to have overcome difficulties due to elution of the oil with the pigments when attempting quantitative analysis, by using a sensitive direct-scanning fluorometer. Sherma and Latta (1978) overcame this difficulty by using commercial chemically bonded C_{18} plates, though the solvents used did not separate clearly all the pigments of higher plants. Isaksen and Francis (1986) tested 12 mixtures of petroleum spirit (40–160°C), acetonitrile, and methanol, finding solvents containing 50–60% methanol most effective; however, these tests did not include many xanthophylls common in algae and still did not separate lutein and isozeaxanthin (and, presumably, zeaxanthin).

Two-dimensional systems have been widely used with adsorption TLC, though not with reverse-phase chromatography (Sestak, 1982), and are useful in resolving the large number of pigments found in extracts from algae. However, the two runs must be completed rapidly if decomposition of chlorophylls is to be prevented. The factors affecting the R_f values of pigments run on TLC are essentially similar to those for paper chromatography (see Sec. 2.2.3B).

Thin-layer chromatography has proved valuable for quantitative estimation of chlorophylls in harvested phytoplankton, because these natural populations contain senescent and dead cells and their remnants, and thus the extracts from them are liable to contain decomposition products interfering with accurate estimation of chlorophylls *a*, *b*, and *c* using trichromatic equations (see Chapter 3). Jeffrey (1968a) used thin layers of sucrose in the first TLC method that she developed. All components of this system had to be kept anhydrous and plates could not be spread with a commercial spreader of the Desaga type, though we have used the Quickfit spreader in this laboratory without trouble. The solvents generally used were, in the first dimension, 0.8% propan-1-ol in petroleum spirit (60–80°C); and in the second dimension, 20% chloroform in petroleum spirit (60–80°C). If the air was either too dry or too damp, the R_f value of chlorophyll *a* rose or fell from 0.75 and was adjusted by lowering or raising the amount of the polar components. The meticulous control of water vapor required seems to have deterred other workers, since no reference to its use by other than Jeffrey and co-workers (Jeffrey, 1968b; Lewin et al., 1977) appears in the literature (Sestak, 1982).

Jeffrey also used plates prepared from Mackerey-Nagel cellulose (MN 300), requiring less protection from water vapor but giving more diffuse spots and lower recovery on elution. The sucrose plate did not resolve the slow-running chlorophyllides and, if they were present, the pheophorbides; these had to be eluted and rechromatographed on cellulose, using petroleum spirit (60–80°C)–acetone (Bacon and Holden, 1967) or 2% propan-1-ol in petroleum spirit (60–80°C) (Barrett and Jeffrey, 1974); chlorophyll c_1 and c_2 and their pheophorbides were separated by running again on polyethylene plates. The careful control of water vapor obviously posed problems for working in extreme ranges of humidity.

Jeffrey (1968a) had used cellulose layers as an alternative to sucrose and subsequently developed a method using only cellulose (Jeffrey, 1974) for TLC plates, with 25% chloroform in petroleum spirit (60–80°C) in the first dimension and 2% propan-1-ol in petroleum spirit in the second dimension. Running solvents in this order (cf. Jeffrey, 1968a) ensured that the polar lipids otherwise interfering with the original first solvent were removed by the chloroform before the second dimension was run. Mackerey-Nagel cellulose as supplied gave variable results in later work, but could be used if soaked in 50% chloroform in petroleum spirit (60–80°C) for 1–2 h, followed by washing in petroleum spirit (Jeffrey, 1981). The powder was thoroughly dried overnight in a fume hood and stored in airtight containers. When layered on small glass plates (10 × 8.5 cm) in aqueous suspension, air-dried for 30 min and heated at 95°C for 1 h, the layers could be stored in a desiccator for some weeks before requiring reactivation at 90°C for 30 min. After the washing with chloroform–petroleum spirit, the cellulose was more adsorbent, and the polarity of the solvent was increased to propan-1-ol–petroleum spirit (2.5:97.5) for the first dimension and chloroform–petroleum spirit–acetone (30:70:0.5) for the second dimension.

When only small amounts of extract were availble, Jeffrey concentrated the extract to a small volume by evaporating with N_2 gas and applying to the plates with a micropipette, drying with a gentle steam of N_2. As mentioned earlier, two-dimensional chromatography may take so long that chlorophylls decompose before the separation is finished. Using small spots on small plates with fast-running solvents, Jeffrey (1981) found that the two runs took only 4 and 3 min, respectively. The spots containing the chlorophylls and carotenoids were then scraped from the plates and eluted with acetone (chlorophylls *a* and *b*), methanol (chlorophyll *c*), or ethanol (carotenoids). With the use of a sensitive spectrophotometer, very small amounts of the pigments could be analyzed (see Chapter 3).

The TLC systems used in the major laboratories examining lipid-soluble pigments in Norway (Trondheim and Oslo), Australia (Jeffrey), and Munich (Hager) are shown in Table 2–4 with the R_f values in 14 solvents.

Table 2–4a. *Components of thin layers and chromatographic solvents used to obtain R_f values given in Table 2–4b*

Components of thin layers	Components of chromatographic solvents (by volume)	
a. SG	1. A–H(30:70)	8. PS–A–B–Pr-2-ol (69.5:25:4:1.5)
b. SG–$CaCO_3$(1:1)	2. A–H(40:60)	9. PS(100–140°C)–A–$CHCl_3$–M (50:50:40:1)
c. SG-$Ca(OH)_2$–MgO–$CaSO_4$ (10:4:3:1)	3. A–H(45:55)	10. PS(60–80°C)–Pr-1-ol (99.2:0.8)
d. $CaCO_3$–MgO–$Ca(OH)_2$ (30:6:4)	4. PS–A–Pr-2-ol (73:25:2)	11. PS(60–80°C)–$CHCl_3$ (80:20)
e. Sucrose	5. PS–A–Pr-2-ol (84:15:1)	12. PS(60–80°C)–Pr-1-ol (97.5:2.5)
f. Cellulose powder	6. PS–A(70:30)	13. PS(60–80°C)–$CHCl_3$–A (70:30:0.5)
	7. PS–A–B–Pr-2-ol (75:20:4:1)	14. PS(60–80°C)–D–E (58:30:12)

[a]*Abbreviations:* A, acetone; B, benzene; D, diethylamine; E, ethyl acetate; H, *n*-hexane; M, methanol; Pr-1-ol, propan-1-ol; Pr-2-ol, propan-2-ol; PS, petroleum spirit; SG, silica gel G.

Sources: (b,c) Bjørnland & Aguilar-Martinez (1976); (d) Hager & Stransky (1970a); (e) Jeffrey (1968a); (f) Jeffrey (1981).

Table 2–4b. R_f *values of lipid-soluble pigments from algae with various TLC layers and solvents (× 100)*

Pigment	a[a]				b			c		d	e		f	
	1[b]	2	3	14	6	7	8	4	5	9	10	11	12	13
Alloxanthin	44	–	60	–	–	–	–	–	–	20	–	–	–	–
Anhydrodiatoxanthin	–	50	–	–	–	–	–	–	–	–	–	–	–	–
Antheraxanthin	39	–	–	31–35	37	–	–	48	–	52	–	–	–	–
Aphanizophyll	–	–	–	–	–	–	–	–	–	5	–	–	–	–
Astaxanthin	40	–	–	–	–	–	–	–	–	–	73	76	–	–
Caloxanthin	–	–	–	–	–	–	–	–	–	24	–	–	–	–
Canthaxanthin	36	–	–	–	–	–	–	–	–	83	–	–	–	–
α-Carotene	98	–	–	–	100	100	100	–	–	97	97	97	–	–
β-Carotene	93–95	82–84	–	95	–	100	100	–	–	92	92–97	92–97	–	–
γ-Carotene	–	–	–	–	–	–	–	–	–	71	–	–	–	–
ϵ-Carotene	–	–	–	–	–	–	–	–	–	96	–	–	–	–
Crocoxanthin	–	58	70	–	–	–	–	–	–	75	–	–	–	–
α-Cryptoxanthin	–	–	–	–	60	–	–	–	52	–	–	–	–	–
β-Cryptoxanthin	–	–	–	–	60	–	–	–	33	–	–	–	–	–
Cryptoxanthin diepoxide	–	–	–	–	–	–	–	–	–	82	–	–	–	–
Cryptoxanthin epoxide	–	–	–	–	–	–	–	–	–	79	–	–	–	–
Diadinoxanthin	–	35–36	–	35–36	–	34–37	59	–	–	41	48–56	60–75	63	89
Diatoxanthin	–	36	–	38	–	44–45	67	–	–	28	56–69	75–83	–	–
Dihydroprasinoxanthin epoxide	–	40	–	–	–	–	–	–	–	–	–	–	–	–
Dinoxanthin	–	35–38	–	36	–	–	–	–	–	–	46–56	47–65	–	–
Echinenone	52	–	–	–	–	–	–	–	–	88	–	–	–	–
Eutreptiellanone	–	80	–	–	–	–	–	–	–	–	–	–	–	–

Fucoxanthin	30	39	–	48	–	26–32	–	–	–	72	34–35	50–52	43–49	40–50
Fucoxanthinol	10	16	–	–	–	–	–	–	–	–	–	–	–	–
Heteroxanthin	–	10	–	10	–	–	–	–	–	12	–	–	–	–
Hexadehydro-β-caroten-3-ol	–	70	–	–	–	–	–	–	–	–	–	–	–	–
19′-Hexanoyloxyfucoxanthin	–	–	–	–	–	26	–	–	–	–	–	–	–	–
3′-Hydroxyechinenone	–	–	–	–	–	–	–	–	–	80	–	–	–	–
Loroxanthin	–	25	–	–	–	–	–	–	–	–	–	–	–	–
Lutein	42	42	–	40–42	45	–	–	45	–	55	68	84	71	70
Monadoxanthin	–	–	65	–	–	–	–	–	–	–	–	–	–	–
Myxoxanthophyll	–	–	–	–	–	–	–	–	–	3	–	–	–	–
Neoxanthin	15–22	15	–	23	–	–	25	–	–	49	–	–	–	–
Oscillaxanthin	–	–	–	–	–	–	–	–	–	6	–	–	–	–
Peridinin	18–41	24	–	39	–	20–27	–	–	–	–	21–27	20–28	36	34
Prasinoxanthin	–	46	–	–	–	–	–	–	–	–	–	–	–	–
Pyrrhoxanthin	–	52	–	–	–	–	–	–	–	–	–	–	–	–
Siphonaxanthin	–	20	–	–	–	–	–	–	–	–	–	–	–	–
Siphonein	–	40	–	–	–	–	–	–	–	–	–	–	–	–
Taraxanthin	–	–	–	–	–	–	–	–	–	75	–	–	–	–
Uriolide	–	42	–	–	–	–	–	–	–	–	–	–	–	–
Vaucheriaxanthin	54	–	–	7	–	–	–	–	–	19	–	–	–	–
Vaucheriaxanthin ester	–	–	–	56	–	–	–	–	–	69	–	–	–	–
Violaxanthin	54	–	–	27–37	–	–	–	–	–	74	28	53	46	61
Zeaxanthin	39–42	40–46	–	–	45	–	–	25	–	39	–	–	–	–

[a]Components of thin layers (a–f); see Table 2–4a.
[b]Components of chromatographic solvents (1–14): see Table 2–4a.

Table 2–5 shows special TLC systems separating pigments often not separated by those shown in Table 2–4. Although R_f values are not reproducible, their value lies in showing which pigments can be separated by the different systems. Using the new high-performance TLC plates from Merck (RP-8), Jeffrey and colleagues (Jeffrey and Wright, 1987; Vesk and Jeffrey, 1987; Wright and Jeffrey, 1987) separated fucoxanthin from derivatives, and chlorophyll c_3 from chlorophyll c_1 and c_2, and Suzuki, Saitoh, and Adachi (1987) separated chlorophyll *a* and *b* from derivatives (Table 2–5), using RP-18 plates.

D. *High-performance liquid chromatography*

This method of chromatography using small-diameter columns, fine particle size, and rapid flow-rate is now widely used; its theory is discussed in articles by Simpson (1976) and Snyder and Kirkland (1979). It has the advantage of speed and sensitivity, in addition to protecting pigments from degradation by oxygen, and it can be used for preparative chromatography. Schwartz and von Elbe (1982) have reviewed the systems used for separating plant pigments; few separate all chlorophylls and carotenoids in mixtures containing more than ten pigments and none completely separates all pigments (Wright and Shearer, 1984). Roy (1987) has published tables summarizing methods of separating chloropigments by the different methods of HPLC. Although many workers are interested in both carotenoids and chlorophylls, the analysis of chloropigments alone can be simplified by detecting them by absorbance or fluorescence at the red end of the spectrum where the carotenoids are transparent (Straub, 1980; Brown, Hargrave, and MacKinnon 1981; Shioi, Fuka, and Sasa, 1983; Bidigare, Kennicutt, and Brooks, 1985; Farnham et al., 1985; Gieskes and Kraay, 1986) though carotenoids have been removed by precipitating them with dioxane (Scholz and Ballschmiter, 1981; Watanabe et al., 1984) or removing them with Sep-Pak C_{18} Bondapak cartridges (Sartory, 1985). Adding ion-pairing reagents such as tetrabutylammonium acetate (Mantoura and Llewellyn, 1983) improves separation of the dephytylated chloropigments and does not affect their absorption spectra. Although HPLC can resolve carotenoids and chlorophyllous pigments into sharp peaks, certain separations, as with TLC, are not always possible with some of the methods used – in particular, lutein and zeaxanthin; diadinoxanthin, dinoxanthin, and fucoxanthin; and chlorophyll c_1 and c_2. An outstanding resolution has been achieved by Wright and Shearer (1984), who separated a mixture of 44 chlorophyllous and carotenoid pigments. Wright

(1987) has used this method for measuring the abundance of 27 pigments in the upper 200 m of the water column at 62 sites at Prydz Bay near Mawson in the Antarctic. Elution from the columns may be isocratic (using one solvent mixture for the whole separation) or gradient (step or continuous), and stopped-flow spectrophotometry can be used to test purity and identify peaks resolved on the columns (Krinsky and Welankiwar, 1984). Both adsorption and reverse-phase partition chromatography have been used. Table 2–6 shows systems of both types used for separating algal pigments.

Adsorption chromatography. The advantage of this system is that isocratic elution of pigments has sometimes been used, thus simplifying the apparatus required, but, in contrast to reverse-phase chromatography, the extracts must be free of water when applied to the columns. In addition, the inorganic adsorbents used in this type of column, such as silica gel, have been criticized for causing degradation of chlorophylls (Braumann and Grimme, 1981), as they have in other types of chromatography (2.2.3A and C), though Abaychi and Riley (1979) appear to have prevented degradation in their silica gel columns by using the alkali diethylamine in their first eluting solvent, as in Riley's TLC system (Riley and Wilson, 1965), and Watanabe et al. (1984) have claimed that no degradation occurred when a silica gel column was used for separating chloropigments. Gieskes and Kraay (1983a) also used 0.25% diethylamine in their eluting solvent. As Table 2–6 shows, few recent workers have used adsorption chromatography for lipid-soluble pigments.

Reverse-phase partition chromatography. Octadecyl–silica columns are generally employed and usually require gradient elution, though Goeyens et al. (1982), using a fluorometric detector, separated at least 16 chloropigments and derivatives, utilizing a carefully calculated mixture of methanol, acetone, and water for isocratic elution. The systems of Mantoura and Llewellyn (1983) and Wright and Shearer (1984) have been the most successful in separating mixtures of chloropigments and carotenoids. The latter give details of several precautions which, if followed, improve efficiency of separation: For example, preliminary purification with C_{18} Sep-Pak cartridges prevented clogging of the chromatography columns, and using two RCH-100 columns of ODS in series rather than a single column usually gave better resolution and taller peak-heights. They found that replacing the mixture of methanol and ethyl acetate used by Eskins and Dutton (1979) with acetonitrile, which has lower viscosity,

Table 2–5. *Special solvent systems for resolving lipid-soluble pigments by TLC*

Pigment	Layer	Solvent	Reference
α-Carotene, β-Carotene, Lutein, Zeaxanthin	Al_2O_3–MgO (3:1)	4% Ethyl acetate in *n*-hexane	Chapman (1966b)
	Silica gel G–Al_2O_3 (1:1)	35% Acetone in *n*-hexane	Chapman (1966a,b) Chapman & Haxo (1966)
	Polyamide	Methanol–methyl ethyl ketone–H_2O (5:5:1)	Ricketts (1967c)
	$CaCO_3$–MgO–Al_2O_3 (30:6:4)	Petroleum spirit (100–140°C)–acetone–benzene–propan-2-ol (75:20:4:1)	Hager & Stransky (1970a,b)
Diadinoxanthin, Diatoxanthin, Fucoxanthin	Silica gel G	15% Propan-1-ol in *n*-hexane	Chapman (1965)
Siphonaxanthin, Neoxanthin	Cellulose impregnated with 10% plant oil in petroleum spirit	Methanol–acetone–H_2O	Kleinig (1969)
	Cellulose	Methyl ethyl ketone–*n*-hexane (1:4)	Yokohama (1981a,b)
	Kieselgel	Petroleum spirit–acetone (7:3)	Yokohama (1983)

Chlorophyll a, b, & c and degradation products	Cellulose	20% Acetone in petroleum spirit (60–80°C)	Bacon & Holden (1967) Jeffrey (1968a)
Chlorophyll c_1,(c_2+c_3) MgDVP	Polyethylene Dow Chemical Co. (3796-4-23) Koppers PRX-1025	90% Acetone 100% Acetone 90% Acetone	Jeffrey (1969) Jeffrey & Wright (1987) Jeffrey (1972)
Chlorophyll c Chlorophyllides Pheophorbides	Silica gel PF_{254}	Petroleum spirit–ethyl acetate–dimethylformamide (1:1:2)	Garside & Riley (1969)
10-C Epimeric chlorophylls and oxidation products	Sucrose	1% Propan-2-ol in petroleum spirit	Sahlberg & Hynninen (1984)
Fucoxanthin and derivatives Chlorophyll c_3, (c_1 + c_2)	Merck RP-8	90% Methanol	Jeffrey & Wright (1987) Vesk & Jeffrey (1987) Wright & Jeffrey (1987)
Chlorophyll a, b, and derivatives	Merck RP-18 F_{254_s}	Ten solvents	Suzuki et al. (1987)

Table 2–6. *Systems used for separating lipid-soluble pigments by HPLC*

Reference	Column	Solvents	Wavelength (nm)		Algal material
			Abs.	*F*	
Isocratic					
Jacobsen (1978)	Silica gel H	Acetone–ligroine (65–75°C) (2:8)	Calibration with standards		Phytoplankton
Shoaf (1978)	Partisil PX5 1025 ODS-2	95% Methanol	654	—	Phytoplankton
Abaychi & Riley (1979)	Partisil 10	Petroleum spirit (60–80°C)–acetone–DMSO – diethylamine (75:23.25:1.5:0.25)	440	—	*Dunaliella*
Liebezeit (1980)	5-μm Spherisorb ODS	98% Methanol	—	580	Phytoplankton
Gieskes & Kraay (1983a)	μ-Porasil	Petroleum spirit (60–80°C)–acetone–DMSO–diethylamine (75:23.25:1.5:0.25)	436	660	*Rhodomonas* *Isochrysis* *Fucus*
Nells & DeLeenheer (1983)	7-μm Zorbox	Acetronitrile–dichloromethane–methanol (70:20:10)	430	—	*Spirulina*
Paerl et al. (1983)	Ultrasphere ODS-18RP	Methanol–acetonitrile (9:1)	360,440,475	663	*Microcystis*
Korthals & Steenbergen (1985), Burger-Wiersma et al. (1986)	5-μm Spherisorb	Hexane–acetone (8:2)	440	—	Phytoplankton

Pennington et al. (1985)	5-µm Spherisorb	15–18% Acetone in *n*-hexane (with 1% methanol)	445–450	—	Cryptophyceae
Murray et al. (1986)	5-µm C-18RP	Ethyl acetate–methanol–H_2O (45:39.5:5.5)	436	667	Phytoplankton
Nakayama et al. (1986)	Finepak C_{18} ODS	97% Methanol	440	—	*Bryopsis*
Otsuki et al. (1987)	Unisil Q C18 + TSK-Gel G3000SW	100% Methanol	430	670	Chlorophyll *a* derivatives
		100% Methanol	470	660	Chlorophyll *b* derivatives
		92% Methanol	440	630	Chlorophyll *c* derivatives
Suzuki et al. (1987)	5-µm TSK–Gel ODS 80 TM	95% Ethanol (5 mM NaCl)	425	—	Chlorophyll *a* and *b* and derivatives
Fiksdahl & Liaaen-Jensen (1988)	5-µm Spherisorb	Hexane	465	—	*Euglena viridis*
Reverse-phase: step gradient					
Eskins et al. (1977)	37–75-µm Bondapak C_{18} Porasil B	80% Methanol, 90% methanol, 95% methanol, 97.5% methanol, 100% methanol, 10% diethyl ether in methanol, 50% diethyl ether in methanol, 75% diethyl ether in methanol	440	—	*Nitzschia*
Braumann & Grimme (1979)	Sorb Sil 60-D10	Methanol–ethanol (1:1), methanol	440	—	*Chlorella*
	C_{18} MN				

Table 2–6. (*Cont.*)

Reference	Column	Solvents	Wavelength (nm)		Algal material
			Abs.	*F*	
Abaychi & Riley (1979)	Partisil 10	Petroleum spirit (60–80°C)–acetone–DMSO–diethylamine (75:23.25:1.5:0.25), Petroleum spirit (60–80°C)–acetone–DMSO–diethylamine (30:40:27:3)	440	—	Pheophytin Chlorophyll *c* *Phaeodactylum*
Falkowski & Sucher (1981)	RP 8	90% Methanol, 98% methanol	—	Corning 2030	*Skeletonema*
Bidigare et al. (1985)	10-μm Radial PAK C_{18}	Methanol–propan-2-ol–H_2O (8:1:1), methanol	—	670	Phytoplankton
Sartory (1985)	5-μ Resolve C_{18} RP	(A) 97% Methanol; (B) 97% acetone A, 77% A + 23% B, then isocratic	—	Filters	Sediments
Trees et al. (1986)	10-μm Radial-PAK C_{18} RP	Methanol–ion-pairing soln.–H_2O (80:10:10), methanol	—	670	Phytoplankton
Vernet & Lorenzen (1987)	Zorbax C-18 ODS	90% Methanol, methanol, ethanol	410,440	—	Chlorophyll and degradation products in sediments and water column

Reverse phase: continuous gradient					
Fiksdahl et al. (1978a)	5-µm Spherisorb	0 → 40% Acetone in hexane containing 1% methanol	445–450	—	Carotenoids
Davies & Holdsworth (1980)	Partisil 10 ODS C18 µPorasil µBondapak	Methanol–H_2O (70:30 → 95:5) Methanol–H_2O (60:40)	440,650	—	*Pavlova* *Phaeodactylum* *Amphidinium* Mixed populations *Chlorella*
Braumann & Grimme (1981)	10-µm Lichrosorb 8RP 10-µm Sil 60 RP 18	Methanol–acetonitrile (1:3) in H_2O 75% → 100%, then isocratic	445	—	
Brown et al. (1981)	10-µm Brownlee 8RP	Methanol–H_2O (95:5) → methanol–acetone–H_2O (75:20:5)	—	550	Sediments
Gieskes & Kraay (1983b)	10-µm Lichrosorb RP18	(A) 70% Methanol (B) 20% Ethyl acetate in methanol A → 25% B → 95% B	426	660	Phytoplankton
Mantoura & Llewellyn (1983)	5-µm Hypersil ODS	(A) Methanol–ion-pairing soln.–H_2O (8:1:1) (B) Acetone–methanol (2:8) A → B, then isocratic	440	600	*Phaeodactylum* *Gyrodinium* *Oscillatoria* *Tetraselmis*
Friedman & Alberte (1984)	Whatman C_{18} RP	70% Methanol → 100% ethyl acetate	450	—	Diatoms
Paerl (1984)	Ultrasphere ODS-18RP	(A) Methanol–acetonitrile (9:1) A → acetone–A (6:4) → A	440,475	663	Cyanophyceae

Table 2–6. (*Cont.*)

Reference	Column	Solvents	Wavelength (nm) Abs.	Wavelength (nm) *F*	Algal material
Paerl et al. (1984)	Ultrasphere ODS-18RP	(A) Methanol–acetonitrile (9:1) A → acetone–A (6:4) → A	440,475	663	*Prochloron*
Wright & Shearer (1984)	5-μm Rad-Pak A (2 cols.)	90% Acetonitrile → ethyl acetate	405 + 436	—	Phytoplankton
Bowles et al. (1985)	Ultrasphere ODS-18RP	(A) Methanol–acetonitrile (9:1) A → acetone–A (6:4) → A	440,475	663	Cyanophyceae, Chlorophyceae, Diatoms
Farnham et al. (1985)	Bondapak C_{18}	Methanol–H_2O (92:8) → 100% methanol	—	580	Tribophyceae, Chlorophyceae
Knight & Mantoura (1985)	3-μm Hypersil ODS		440	600	Symbiotic algae in Foramenifera
Alberte & Andersen (1986)	Porasil C_{18} RP	70% Methanol → 100% ethyl acetate	450	—	Chromophyta
Murray et al. (1986)	5-μm C_{18} RP	90% Methanol (10 min), then ethyl acetate	436	667	Phytoplankton
Wilhelm et al. (1986)	5-μm Hypersil ODS	(B) Methanol–acetonitrile (1:1) → 95% B→ 99% B	440	—	Prasinophyceae

Gieskes & Kraay (1987)	Novapak C_{18}	(A) 70% Methanol in P buffer, pH 7 → (B) Ethyl acetate–methanol (2:8) 100% B at 20 min	436,658	670	Phytoplankton
Lichtlé et al. (1987)	Du Pont Zorbax ODS	0.1% → 40% Dichloromethane in acetonitrile–methanol (7:3)	445	—	Carotenoids in pigment–protein complexes
Wilhelm (1987)	3-μm Hypersil ODS	(B) Methanol–acetonitrile–ethyl acetate 70% B → 85% B → 100% B	440	—	Prasinophyceae
Fawley (1988)	5-μm Alltech C_{18} Econosphere	(B) Methanol–acetonitrile (1:1) 0 → 80% B: 10 min; 80% B → 98% B: 10 min; 98% B: 1 h	440	—	*Pavlova gyrans*

kept back-pressure in the columns to acceptable levels. Using a spectrophotometer with a 15-μl flow-cell allowed them to measure pigment spectra without stopping the flow through the column. Absorbances were measured at both 405 and 436 nm and were integrated to give greater sensitivity. Systems used by other workers are shown in Table 2–6. Fawley (1988), using Alltech Econosphere C_{18} as the stationary phase, has resolved chlorophyll c_1 and c_2 and believes that the success of the system lies in the chemistry of the packing because similar solvents have not resolved the two pigments when other columns are employed. Carter, Flett, and Gibbs (1988) have described a method for correcting baseline irregularities in gradient elution, using methanol–water–ethyl acetate as solvent.

3

Chlorophylls and derivatives

3.1. Distribution and chemistry

Although other photosynthetic pigments mask or modify the green color of chlorophyll in the cells of most classes of algae, all algae contain chlorophyll *a* as a large fraction of their photosynthetic pigments; hence it is not surprising that chlorophyll *a* is the essential molecule for converting the energy of electromagnetic radiation to the thermochemical energy driving the biochemical processes of photosynthesis. This photochemical activation of the photosynthetic pigments is the only true endergonic reaction in photosynthesis, the other reactions of the electron-transport chain and the so-called dark reactions being exergonic. Thus the subsequent reactions in which numerous molecules in the cell are formed from the phosphorylating power (ATP) and reducing power (NADPH) generated by the operation of the electron-transport chain within the thylakoid membranes involve loss of some of this free energy initially captured from light.

3.1.1. Biosynthesis

The synthesis of chlorophyll *a*, reviewed by Castelfranco and Beale (1983), Beale (1984), and Rüdiger and Benz (1984), has been studied intensively, with δ-aminolevulinic acid (ALA) regarded as the first committed precursor in most organisms. In higher plants and algae, ALA incorporated into chlorophyll is formed by the so-called C_5 pathway, not by ALA synthase, though the latter pathway is also found in *Euglena* for ALA incorporated into nonplastid heme (Weinstein and Beale, 1983). Gutman (1984) has doubted that intact ALA is incorporated into chlorophyll *a* in *Euglena*. Protoporphyrin IX is formed from ALA by a series of reactions; insertion of Mg into the center of the ring system by the action of a chelatase, and ring closure to form the isocyclic ring found in all chlorophylls lead to the synthesis of the important intermediate, MgDVP. Synthesis of chlorophyll *a* continues with formation of chlorophyllide *a* followed by phytylation through esterification with geranylgeraniol and hydration. Chlorophyllase, originally believed important in catalyzing incorporation of the phytyl chain into chlorophyll *a*, is now believed important in chlorophyll catabolism (Castelfranco and Beale, 1983; Beale, 1984).

Rebeiz and his associates (Rebeiz et al., 1983) have proposed a six-branched biosynthetic pathway for a number of derivatives of chlorophyll *a* in green plants, including a mixture of monovinyl and divinyl derivatives. They believe that this multiplicity of biosynthetic routes may be responsible for position and orientation of certain species of chlorophyll *a* in the thylakoid membranes. They also have proposed that heterogeneity of chlorophyll *a* in the reaction centers is essential for efficient charge separation.

In chlorophyll *b*, a formyl group replaces methyl at C-3 on the macrocyclic ring of chlorophyll *a* (Fig. 3–1), possibly by formation of a hydroxymethyl intermediate between chlorophyll *a* and *b* (Beale, 1984). Possibly this conversion occurs on chlorophyllide *a* rather than chlorophyll *a*, thus forming chlorophyll *b* by phytylation of chlorophyllide *b*. We know even less of the pathways of synthesis of chlorophyll c_1 and c_2. Beale (1984) has suggested synthesis of chlorophyll c_2, the more common form, by dehydrogenation of the propionate at C-7 of MgDVP, and alternate pathways of synthesis of chlorophyll c_1, either by reduction of the vinyl group at C-4 of MgDVP to give protochlorophyllide followed by dehydration of the propionate at C-7, or by reduction of the methyl group at C-4 of chlorophyll c_2. In view of the recent discovery that the Synurophyceae contain chlorophyll c_1 without c_2 (Andersen and Mulkey, 1983), the former pathway seems more probable.

3.1.2. Distribution within the divisions and classes of the algae

Table 3–1 shows the usual distribution of the chlorophylls in 15 algal classes, adopting the type of classification used by Christensen (1962, 1964) and Liaaen-Jensen (1978). Several exceptions to this scheme will be discussed in later sections. The original Class Xanthophyceae Allorge ex Fritsch (Fritsch, 1935) was divided into Xynthophyceae and a new class, Eustigmatophyceae Hibberd & Leedale (Hibberd and Leedale, 1970, 1971, 1972), based on the structure of the zoospores. Subsequently, Hibberd has replaced the name Xanthophyceae by Tribophyceae Hibberd, conforming with the International Code of Botanical Nomenclature.

In addition to chlorophylls listed in Table 3–1, extracts from algae sometimes contain derivatives of the parent chlorophylls. Although small

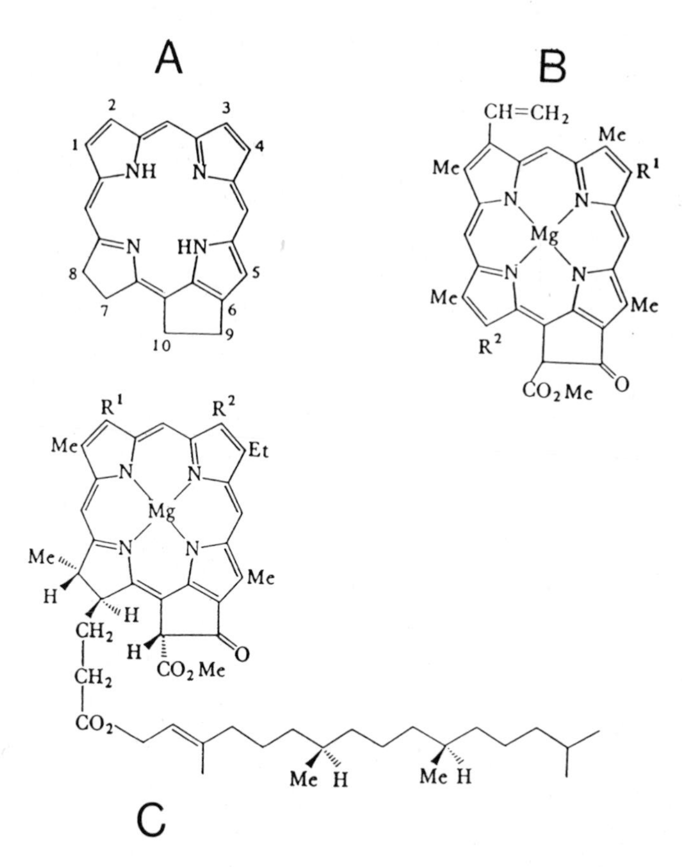

Fig. 3–1. Chemical structure of chlorophyllous pigments. (A) Phorbin. The basic nucleus of the chlorophyllous pigments, showing the numbering system used and the fifth exocyclic ring. (B) Chlorophyll c_1 (R^1 = Et; R^2 = $CH{=}CH{-}CO_2Me$), chlorophyll c_2 (R^1 = $CH{=}CH_2$; R^2 = $CH{=}CH{-}CO_2Me$), and MgDVP (R^1 = $CH{=}CH_2$; R^2 = $CH{=}CH{-}CO_2H$). (C) Chlorophyll *a* (R^1 = $CH{=}CH_2$; R^2 = Me), chlorophyll *b* (R^1 = $CH{=}CH_2$; R^2 = CHO), and chlorophyll *d* (R^1 = CHO; R^2 = Me). (After Jackson, 1976.)

amounts of some of these derivatives are now believed to harvest light, their presence is usually due to dead cells within the sample or to degradation occurring during extraction (Sec. 3.1.4).

A. *Chlorophyll a*

All algae contain chlorophyll *a*, though photosynthetic bacteria (other than the Cyanophyceae) contain one or more chemically related bacteriochlorophylls, and in some Archebacteria, rhodopsin is the light-harvesting pigment (Woese, 1981).

Table 3–1. *Distribution of chlorophyllous pigments in the divisions (Christensen, 1962, 1964) and classes of the algae*

Division	Class	Chlorophyllous pigment						
		a	*b*	c_1	c_2	c_3	*d*	MgDVP
Prokaryota								
Cyanophyta	Cyanophyceae	+	–	–	–	–	–	–
Prochlorophyta	Prochlorophyceae	+	+	–	–	–	–	–
Eukaryota								
Rhodophyta	Rhodophyceae	+	–	–	–	–	±	–
Chromophyta	Cryptophyceae	+	–	–	+	–	–	–
	Dinophyceae	+	±[a]	±[b]	+	–	–	–
	Chrysophyceae	+	–	±[c]	±[d]	±	–	–
	Synurophyceae	+	–	+	–	–	–	–
	Prymnesiophyceae	+	–	+	+	±	–	–
	Bacillariophyceae	+	–	+[e]	+	±	–	–

	Tribophyceae	+	−	+	+	−	−	−
	Eustigmatophyceae	+	−	−	−	−	−	−
	Phaeophyceae	+	−	+	+	−	−	−
Chlorophyta	Euglenophyceae	+	+	−	−	−	−	−
	Chlorophyceae	+	+	−	−	−	−	−
	Prasinophyceae	+	+	±[f]?	−	−	−	±[f]?
	Charophyceae	+	+	−	−	−	−	−

[a]Strain Y-100 contains chlorophyll *b* through symbiosis with a chlorophyte (Watanabe et al., 1987).

[b]Most Dinophyceae containing chlorophyll c_1 also contain fucoxanthin or a chemically related pigment (Table 4–2), though *Gambierdiscus toxicus* contains peridinin (Durand and Berkaloff, 1985).

[c]Absent from *Pelagococcus subviridis* (Lewin et al., 1977), one strain of *Chrysospherella brevistrina*, and from *Giraudyopsis stellifera* (Andersen, private communication).

[d]Absent from *C. brevistrina* (Andersen, private communication).

[e]Absent from seven species of pennate diatoms (Jeffrey and Stauber, 1985).

[f]Originally identified by Ricketts (1966a); Wilhelm (1987) has presented evidence that this pigment is chlorophyll c_1, but Jeffrey and Wright (1987) believe that it is MgDVP.

B. *Chlorophyll b*

This chlorophyll occurs in all classes of the division Chlorophyta (Table 3–1), but, in addition, Lewin (1976, 1977) has erected a new division, the Prochlorophyta Lewin, for prokaryotes containing chlorophyll *b* in place of the usual biliprotein found in the other algal prokaryotes. Burger-Wiersma et al. (1986) have isolated and cultured a second member of this division, a free-living prokaryote they named *Prochlorothrix hollandica,* and Bullerjahn, Matthijs, Mur, and Sherman (1987) have examined its chlorophyll *a/b*–protein complexes (see Table 6–2).

Absence of chlorophyll *b* is a useful taxonomic marker for the green classes of the Chromophyta that in the past tended to be placed in the Chlorophyta, although Fritsch (1935) did not use this characteristic in describing the class Xanthophyceae (Heterokontae), instead describing them as rich in xanthophylls. The genus *Nannochloris* Naumann, previously assigned to the Chlorophyceae, was found to contain some strains such as *N. oculata* Droop (Millport No. 66) lacking chlorophyll *b* (Allen, 1958; Antia, Bisalputra, Cheng, and Kalley, 1975). However, Establier and Lubian (1982) found chlorophyll *b* in the Cambridge strain of *N. oculata* Droop (No. 251/6) and *N. maculata* Butcher (No. 251/3), thus retaining them in the Chlorophyceae. Strain No. 251/6 examined by Establier and Lubian (1982) has now been described as *N. alomus* Butcher in the Cambridge Culture Centre (Turner and Gowen, 1984). Lubian and Establier (1982) confirmed that three strains, originally described as *Nannochloris* sp. (Strain B-3, Instituto de Investigaciones Pesqueras de Cadiz = S.M.B.A. Strain No. 269), *Nannochloris oculata* (Millport No. 66), and *Monallantus salina* (Station Marine d'Endoume, Marsella), all contained no chlorophyll *b*, and these are now transferred to the type genus of Hibberd's new class, the Eustigmatophyceae, as *Nannochloropsis*.

The genus *Vaucheria* is similar in some ways to the Chlorophyceae (e.g., isokont zoospores), and Seybold and Egle (1937) originally claimed that chlorophyll *b* occurred in this genus, though they corrected this in later publications (Seybold, Egle, and Hulsbruch, 1941), a revision later confirmed by Strain (1958) and Sagromsky (1962). Watanabe et al. (1987) have described a dinophyte (Strain Y-100) containing chlorophyll *b*, with a low chlorophyll *a/b* ratio of less than two. The organism can be cultured autotropically and appears to be a result of symbiosis with a chlorophyte. The green chloroplast is contained in a double membrane with lamellae consisting of three appressed thylakoids, as in the Prymnesiophyceae, but as this class does not contain chlorophyll *b* (Table 3–1), this source of a symbiont seems unlikely. Identifying the xanthophylls in this organism should provide evidence for the origin of the chloroplast.

C. *Chlorophyll c*

The chromatographic separation by Jeffrey (1968b, 1969) of the two forms, chlorophyll c_1 and c_2, deduced by Doughtery et al. (1966) using NMR, infrared, and mass spectrophotometry has complicated the use of this pigment as a taxonomic marker. In Table 3–1, one or both of these forms is shown in all classes of the Chromophyta, except the Eustigmatophyceae; Antia et al. (1975) used absence of chlorophyll *c* as part of the evidence assigning *Nannochloris oculata* and *Monallantus salina* to the class Eustigmatophyceae. Using RP-8 TLC plates, Vesk and Jeffrey (1987) discovered a new pigment which they named chlorophyll c_3. In this and subsequent papers, Jeffrey and co-workers have found it in various species, discussed below, of the Chrysophyceae, Prymnesiophyceae, and Bacillariophyceae. Chlorophyll c_3 is not resolved from chlorophyll c_2 on the polyethylene TLC plate separating chlorophyll c_1 and c_2. This new pigment must not be confused with the chlorophyll c_3 designated by Jeffrey (1969), which is apparently a polymerized form of chlorophyll *c* running close to the R_f value of chlorophyll c_3 on RP-8 TLC plates, but can be distinguished from true chlorophyll c_3 by its absorption spectrum (see Sec. 3.3.1). Chlorophyll c_3 could account for some of the anomalies by which some species of a chromophyte class normally containing both chlorophyll c_1 and c_2 appear to contain only chlorophyll c_2 when their pigments are examined with the use of polyethylene plates. The chromophyte classes Cryptophyceae and Dinophyceae are usually found to have only chlorophyll c_2; so far, preliminary studies in our laboratory have shown that eight species of Cryptophyceae appear not to contain chlorophyll c_3 when examined by means of HP-8 TLC plates.

Cryptophyceae. Although only chlorophyll c_2 without c_1 occurred in nine species of this class, Jeffrey (1976a) found both chlorophyll c_1 and c_2 in *Chroomonas mesostigmatica*. However, Andersen, Haxo, Lee, and Kugrens (private communication) could not find chlorophyll c_1 in the same strain of this species, and Hill and Rowan (unpublished) could not find it in a different strain.

Dinophyceae. Withers and Haxo (1975) were the first to find chlorophyll c_1 in addition to c_2 in a member of this class (*Peridinium foliaceum*), subsequently found by Jeffrey et al. (1975) and Jeffrey (1976a) in four of 24 species examined. The four contained fucoxanthin as the major light-harvesting xanthophyll, not peridinin, as in the majority, and this small group is now widely believed to represent a symbiosis between a colorless dinophyte and a chrysophyte (Tomas and Cox, 1873a,b; Tomas, Cox, and Steidinger, 1973; Jeffrey et al., 1975; see also Table 4–2). Although *Gambierdiscus toxicus* contains chlorophyll c_1, it contains peridinin, not fucoxanthin (Durand and Berkaloff, 1985).

Chrysophyceae. Andersen and Mulkey (1983) have listed earlier workers unable to identify chlorophyll *c* in various genera of this class. Of these, Gibbs (1962) did not attempt to look for chlorophyll *c* in *Ochromonas danica,* but apparently accepted the belief common at the time that the pigment did not occur in the Chrysophyceae. Dales (1960) used column chromatography, a method unsuitable for small amounts of pigment, whereas Ricketts (1965) relied on the trichromatic equations of Parsons and Strickland (1963) and the phase separation method of Parsons (1963). These methods are not as satisfactory as detecting pigments on TLC plates by using UV light and measuring the absorption spectrum of the eluted pigment or by examining the fluorescence-emission spectrum of the extracts from the cells at the wavelength specific for chlorophyll *c* in the solvent used.

Andersen and Mulkey (1983) screened 56 strains of Chrysophyceae, finding chlorophyll c_1 and c_2 in all but one strain of *Chrysosphaerella brevispina* (neither c_1 nor c_2) and in *Giraudyopsis stellifera* (no c_1). Lewin et al. (1977) have described an aberrant new alga, *Pelagococcus subviridis,* without chlorophyll c_1 and containing fucoxanthin-like pigments but no fucoxanthin. This is one of the species in which Vesk and Jeffrey (1987) have found the new chlorophyll c_3.

Synurophyceae. Partly because chlorophyll c_2 was absent from 34 strains of marine genera *Synura* and *Mallomonas* (Andersen and Mulkey, 1983; Andersen, private communication), Andersen (1987) has removed them from the Chrysophyceae to this new class containing the two families, Synuraceae and Mallomonadaceae.

Tribophyceae. The amount of chlorophyll *c* in this class is usually much lower than in other chromophytes, with chlorophyll *a*/*c* ratios of 50–100 rather than about 5; this property is used occasionally in separating Tribophyceae and Chrysophyceae (Gibbs, Chu, and Magnussen, 1980). Thus chlorophyll *c* is likely to be overlooked in the Tribophyceae unless specifically looked for (Guillard and Lorenzen, 1972). This low level also makes a physiological role for chlorophyll *c* unlikely; Wiedemann, Wilhelm, and Wild (1983) did not find chlorophyll *c* in pigment–protein complexes from *Tribonema,* though they did so for *Synura* (see Table 6–2). In the past, chlorophyll *c* has been reported absent from *Botrydium granulatum* (Reith and Sagromsky, 1972) and from two strains of *Ophiocytrium majus* (Jeffrey, 1976a; Whittle, 1976) though present in another strain (Jeffrey, 1976a). The failure of Stransky and Hager (1970a) to find chlorophyll *c* in eight genera of the class is hard to explain, as other workers have found the pigment in six of the same genera (Jeffrey 1976a; Sullivan, Entwisle, and Rowan, 1989). The latter found at least traces of chlorophyll *c* in 27 cultures of Tribophyceae by using spectrophotofluorometry.

Prymnesiophyceae. Allen, Goodwin, and Phagpolngarm (1960), Dales (1960), Parsons (1961), Jeffrey (1963, 1969, 1976), and Ricketts (1965) have reported chlorophyll *c* in a number of species of Chrysophyceae now placed in the Prymnesiophyceae (Andersen and Mulkey, 1983). McLean (1967) probably failed to find chlorophyll *c* because he evaporated the solution of pigments to dryness and attempted to redissolve them in hexane, a strongly nonpolar solvent which would not dissolve chlorophyll *c* efficiently. Jeffrey (1969, 1976) found both chlorophyll c_1 and c_2 in five species of this class, and Fawley (1988) found both in *Pavlova gyrans*. The new pigment, chlorophyll c_3, occurs in *Emiliania huxleyi* (Vesk and Jeffrey, 1987; Jeffrey and Wright, 1987), *Phaeocystis pouchetii* (Jeffrey and Wright, 1987), *Pavlova salina,* and two unidentified strains of *Pavlova* and of another prymnesiophyte strain (Stauber and Jeffrey, 1988) in place of the usual chlorophyll c_2. Gieskes and Kraay (1986) found an unknown pigment with spectral properties similar to chlorophyll c_3 in prymnesiophycean phytoplankton.

Bacillariophyceae. Stauber and Jeffrey (1988) found chlorophyll c_2 in 51 species of marine diatoms but, contrary to previous beliefs, chlorophyll c_1 in only 88 percent. Chlorophyll c_3 usually replaced chlorophyll c_1 as a major component when it was absent or present in only very low concentration. *Nitzschia bilobata* contained all three chlorophylls (c_1, c_2, and c_3) in approximately equal amounts, whereas *N. clostridium* contained only c_2, as did *N. frigida* (Andersen and Rowan, unpublished). Of diatoms containing chlorophyll c_2 and c_3, seven species contained true fucoxanthin, whereas *Thalassiothrix heteromorpha* contained a 19′-butanoxyloxyfucoxanthin-like pigment in addition to fucoxanthin.

Prasinophyceae. Some members of this class contain a pigment identified as MgDVP by Ricketts (1966a). However, Wilhelm and colleagues (Wilhelm, Lenarz-Weiler, Wiedemann, and Wild, 1986; Wilhelm, 1987) have presented evidence from HPLC, spectral analysis, and NMR that this pigment is chlorophyll c_1. Brown (1985) has also described the pigment as chlorophyll *c*-like and has shown that it transmits light energy to chlorophyll *a* in thylakoid membranes, whereas Jeffrey and Wright (1987) believe that this pigment is MgDVP. The conflicting evidence will be discussed in Section E, below.

Although the presence of chlorophyll c_3 may account for most of the anomalies in which chromophytes have appeared to contain only one chlorophyll *c* (Table 3–1), the distribution of this new pigment may not be valuable in taxonomy unless it occurs in all members of a taxon. Whether chlorophyll c_3 occurs in the Cryptophyceae and Dinophyceae remains to be seen.

D. Chlorophyll d

Chlorophyll *d* is chemically close to chlorophyll *a* (Fig. 3–1) but has a distinct absorption spectrum. Originally described by Manning and Strain (1943) in the extracts from *Girgartinia gardii* used for large-scale preparations, they detected spectroscopic evidence for it in approximately ten other species. Smith and Benitz (1955) subsequently repeated the procedure using the same species collected from the same site. In a subsequent survey of a wide range of Rhodophyceae, Strain (1958) detected a high concentration of chlorophyll *d* in *Rhodocorton rothii* (Nemalionales), less substantial amounts in four species, and traces in 28 species. Some species considered relatively high in chlorophyll *d* in the 1943 study, including *Erythrophyllum delesseriodes,* one of the major sources, were now classed as negative. These fluctuations possibly account for the lack of interest in this pigment and make a functional role unlikely. The chlorophyll *d*-like pigment reported in extracts from *Chlorella* (Michel-Wolwertz, Sironval, and Goedheer, 1965) seems likely to be an artifact of extraction. Lu, Wang, and Yu (1983) have published maxima of absorption and fluroescence-emission in 20 solvents, some of which are shown in Tables 3–2 and 3–3.

E. *Magnesium 2,4-divinylpheoporphyrin a_5 monomethyl ester (MgDVP)*

Jones (1963a,b) isolated this pigment from the photosynthetic bacterium, *Rhodopseudomonas spheroides,* and it is now known as an important intermediate in the synthesis of chlorophyll (Castelfranco and Beale, 1983; Beale, 1984) (Sec. 3.1.1). Ricketts (1966a) detected a pigment in *Micromonas pusilla* with spectral properties (visible and IR) similar to those of MgDVP (see Sec. 3.1.2C, Prasinophyceae) and in subsequent papers (Ricketts, 1966b, 1967b,c) reported the pigment only in those species with abnormal carotenoids. Wilhelm et al. (1986) and Wilhelm (1987) have identified this pigment as chlorophyll c_1, showing that it cochromatographs by HPLC with authentic chlorophyll *c* from *Ectocarpus* and is separated from MgDVP. Wilhelm (1987), using absorption and fluorescence spectroscopy, has shown that the pigment is closer to chlorophyll c_1 than MgDVP. However, the absorption maxima that he shows for MgDVP in 90% acetone and diethyl ether, and for its pheoporphyrin in diethyl ether, do not agree with those given by Ricketts (1966a) or Jeffrey and Wright (1987). Jeffrey and Wright (1987) have separated this pigment isolated from *M. pusilla* from chlorophyll c_1, c_2, and c_3 by TLC on polyethylene and have shown that the absorption spectrum differs significantly from chlorophyll c_1 and c_2; these authors agree with Ricketts (1966a) that it is MgDVP.

How this conflicting evidence about the identity of this new pigment in some prasinophytes is to be resolved is not clear. The critical step leading to accumulation of the pigment is presumably its incorporation into a light-harvesting pigment–protein complex, and although MgDVP is an intermediate in the synthesis of the other chlorophylls, chlorophyll *a* is an intermediate in the synthesis of chlorophyll *b* (Beale, 1984) and is diverted from this pathway by binding to protein in the thylakoid membrane; there seems to be no reason why MgDVP might not be diverted in a similar way, as there is good evidence for the pigment, which ever it is, in the LHCP fractions from *Micromonas pusilla* (Wilhelm et al., 1986) and from *Mantoniella squamata* (Wilhelm and Lenarz-Weiler, 1987), where Brown (1985) has found that it harvests light for photosynthesis. As all *Micromonoas* cultures used appear to have originated from Plymouth 27, different strains do not appear to cause the variation in pigmentation found; possibly different culture conditions might determine whether MgDVP or chlorophyll c_1 is incorporated into the LHCP.

Table 3–2. *Absorption maxima of the spectra of the chlorophylls and their derivatives and their specific extinction coefficients in various solvents*

Solvent	Absorption maxima (nm)		Ratio	Extinction coefficient	Reference
	Blue	Red	blue/red	$(l \cdot g^{-1} \cdot cm^{-1})$	
Chlorophyll a					
Diethyl ether	427.5	660	1.32	102.1	Zscheile & Comar (1941)
	430	660	1.21	90.1	MacKinney (1941)
	429.0	660.0	–	–	Harris & Zscheile (1943)
	–	661	–	102	Davidson (1954)
	430	662	1.30	100.9	Smith & Benitez (1955)
	428.5	660.5	1.30	96.6	Strain et al. (1960, 1963)
	428.5	660.5	1.295	–	Pennington et al. (1964)
	–	–	1.31–1.32	–	Perkins & Roberts (1964)
	428.8	660.6	–	–	Seely & Jensen (1965)
	–	–	1.29–1.32	–	Wintermans and De Mots (1965)
	430	662	1.3	–	Goedheer (1966)
	428.9	662	1.34	–	S. R. Brown (1968)
	–	660.7	--	98.07	Jeffrey & Humphrey (1975)
Acetone (100%)	430	660	1.26	84.0	MacKinney (1940)
	–	662	–	91.5	Ziegler & Egle (1965)
	–	663	–	92.60	Vernon (1960)
	–	–	1.59	–	Jensen & Aasmundrud (1963)
	430.2	662.0	–	–	Seely & Jensen (1965)
	–	662	–	91.8	Wintermans & De Mots (1965)
	–	663	–	89.7	Ziegler & Egle (1965)
Acetone (90%)	–	665	–	90.8	Vernon (1960)
	–	665	–	89.0	Parsons & Strickland (1963)

	–	664.3	–	87.67	Jeffrey & Humphrey (1975)
	432	664	–	–	Boto & Bunt (1978)
	430	662	–	–	Marker (1972)
Acetone (80%)	430	663	1.11	82.04	MacKinney (1941)
	–	664	–	89.0	Ziegler & Egle (1965)
	433	665	1.21	90.80	Vernon (1960)
	–	664	–	89.5	Wintermans & De Mots (1965)
Acetone (90%)–methanol (5:1)	432	664	–	84.10	Pechar (1987)
Methanol (100%)	432.5	665	1.21	74.5	MacKinney (1941)
	432.5	665.0	–	–	Strain (1949)
	432.0	665.7	–	74.5	Seely & Jensen (1965)
	–	665	–	75.0	Lentz & Zeitzchel (1968)
	–	665	–	76.1	Marker (1972)
	–	665	–	77.9	Reimann (1978b)
Methanol (95%)	430	664	–	–	Marker (1977)
Ethanol (96%)	431.5	664	–	–	Seely & Jensen (1965)
	432	665	1.00	83.4	Wintermans & De Mots (1965)
DMSO	–	666	–	78.7	Seely & Jensen (1965)
Chlorophyll b					
Diethyl ether	452.5	642.5	2.82	56.8	Zscheile & Comar (1941)
	455	642.5	2.74	54.9	MacKinney (1941)
	–	642.5	–	64.5	Davison (1954)
	455	644	2.82	62.0	Smith & Benitez (1955)
	–	–	2.82	–	French (1960)
	452.5	642.0	–	61.8	Strain et al. (1960, 1963)
	–	–	2.82–3.05	–	Wintermans & De Mots (1965)
	453	642	2.92	–	Goedheer (1966)
	455.0	645.0	–	–	Hoffman & Werner (1966)
	453	642.5	2.77	–	S. R. Brown (1968)
	–	643.2	–	62.00	Jeffrey & Humphrey (1975)

Table 3–2. (*Cont.*)

Solvent	Absorption maxima (nm)		Ratio	Extinction coefficient	Reference
	Blue	Red	blue/red	$(1 \cdot g^{-1} \cdot cm^{-1})$	
Acetone (100%)	455	645	2.84	51.8	MacKinney (1940)
	–	644	–	54.3	Ziegler & Egle (1965)
	–	647	–	53.5	Vernon (1960)
Acetone (90%)	–	646	–	53.1	Ziegler & Egle (1965)
	–	648	3.05	52.5	Vernon (1960)
	–	645	–	54.0	Parsons & Strickland (1963)
	–	646.8	–	54.36	Jeffrey & Humphrey (1975)
	459	647	–	–	Boto & Bunt (1978)
Acetone (80%)	460	645	2.66	45.6	MacKinney (1941)
	–	647	–	52.3	Ziegler & Egle (1965)
	460	648.5	3.05	52.5	Vernon (1960)
	–	647.5	–	51.0	Wintermans & De Mots (1965)
Methanol (100%)	475	650	2.81	36.4	MacKinney (1941)
	470.0	650.0	–	–	Strain (1949)
	–	642	–	44.5	Riemann (1978b)
Methanol (90%)	470	650.0	–	–	Marker (1972)
Ethanol (100%)	–	649	–	45.9	Parsons & Strickland (1963)
Ethanol (96%)	464.0	649.0	2.68	44.2	Wintermans & De Mots (1965)
Chlorophyll c_1					
Diethyl ether	444.5–445.8	627.8–628.2	9.93–10.21	–	Jeffrey (1969)
	438	625	8.7	–	Wilhelm (1987)
Acetone (100%)	442.2–443.5	629.8–630.5	6.71	–	Jeffrey (1969)
	–	629.1	8.89	39.2	Jeffrey (1972)

Acetone (90%)	–	630.6	7.11	44.8	Jeffrey (1972)
	442	629	7.3	–	Wilhelm (1987)
Methanol (100%)	443.5–444.9	633.1–634.4	7.20–7.40	–	Jeffrey (1969)
Tetrahydrofuran	453.2	633.0	10.66	36.1	Jeffrey (1972)
Pyridine	461.7	639.7	9.9	35.0	Jeffrey (1972)
Dioxane	450.6	630.2	10.22	–	Jeffrey (1972)
Chlorophyll c_2					
Diethyl ether	448.2–449.4	628.2–629.4	13.4–15.2	–	Jeffrey (1969)
	451.5	624.5	11.25	–	Vesk & Jeffrey (1987)
	445	628	9.3	–	Wilhelm (1987)
Acetone (100%)	444.2–444.7	630.1–630.6	9.64	–	Jeffrey (1969)
	–	629.6	8.69	37.2	Jeffrey (1972)
Acetone (90%)	–	630.9	9.25	40.4	Jeffrey (1972)
	450	629	8.8	–	Wilhelm (1987)
Methanol (100%)	451.1–453.3	634.8–635.0	9.32–9.99	–	Jeffrey (1969)
	451	633	9.4	–	Jeffrey & Shibata (1969)
Tetrahydrofuran	457.2	634.8	14.70	33.2	Jeffrey (1972)
Pyridine	466.0	641.6	14.45	31.6	Jeffrey (1972)
Dioxane	454.2	631.4	14.37	–	Jeffrey (1972)
Chlorophyll c_3					
Diethyl ether	451.3	625.9	32.1	–	Jeffrey & Wright (1987)
	451.3–451.5	623.9–624.5	–	–	Vesk & Jeffrey (1987)
Acetone (100%)	451.7	626.7	30.77	–	Jeffrey & Wright (1987)
Chlorophyll d					
Diethyl ether	445.0	686	–	–	Manning & Strain (1943)
	447	688	0.89	110.4	Smith & Benitez (1955)
	445.0	686.0	0.88	117.8	Holt & Morley (1959)
	–	687	–	–	Lu et al. (1983)

Table 3–2. (*Cont.*)

Solvent	Absorption maxima (nm)		Ratio blue/red	Extinction coefficient	Reference
	Blue	Red		$(l \cdot g^{-1} \cdot cm^{-1})$	
Acetone (100%)	445	693	1.01	–	Jensen & Aasmundrud (1963)
	–	688	–	–	Lu et al. (1983)
Acetone (80%)	–	692	–	–	Lu et al. (1983)
Methanol (100%)	456.0	696	–	–	Manning & Strain (1943)
	–	696	–	–	Lu et al. (1983)
Ethanol (100%)	–	696	–	–	Lu et al. (1983)
Ethanol (80%)	–	698	–	–	Lu et al. (1983)
Petroleum spirit (60–90°C)	–	685	–	–	Lu et al. (1983)
Pyridine	–	696	–	–	Lu et al. (1983)
Dioxane	–	684	–	–	Lu et al. (1983)
Mg 2,4-Divinylpheoporphyrin a_5 *monomethyl ester*					
Acetone (100%)	437.3	624.5	8.23	–	Jeffrey & Wright (1987)
Acetone (90%)	443	630	9.0	–	Ricketts (1966a)
	443	623	8.3	–	Wilhelm (1987)
Diethyl ether	438	624	–	–	Ricketts (1966a)
	437.1	623.5	9.72	–	Jeffrey & Wright (1987)
	431	622	14	–	Wilhelm (1987)
Chlorophyllide a					
Diethyl ether	428	662	–	–	S. R. Brown (1968)
	428.5	660.5	–	–	Strain et al. (1971b)

Pheophytin a					
Diethyl ether	410	665	2.14	59.0	Zscheile & Comar (1941)
	408.5	667	2.07	63.7	Smith & Benitez (1955)
	409	667	2.09	–	Pennington et al. (1964)
	408	667	–	–	Goedheer (1966)
	408	669	2.03	76.7	S. R. Brown (1968)
Acetone (80%)	410	663	2.62	–	Jensen & Aasmundrud (1963)
	409	666.5	2.30	56.6	Vernon (1960)
Pheophorbide a					
Diethyl ether	408.5	667	2.01	–	S. R. Brown (1968)
Chlorophyllide b					
Diethyl ether	452	642	–	–	S. R. Brown (1968)
	451	641.5	–	–	Strain et al. (1971b)
Pheophytin b					
Diethyl ether	433	653	5.35	37.0	Zscheile & Comar (1941)
	434	655	5.08	42.1	Smith & Benitez (1955)
	–	–	5.13	–	French (1960)
	434	655	–	–	Goedheer (1966)
	433	654	–	–	Marker (1972)
Acetone (100%)	433	657	5.88	–	Jensen & Aasmundsrud (1963)
Acetone (80%)	436	655	5.07	34.8	Vernon (1960)
Ethanol (100%)	–	654.5	–	33.7	Bruinsma (1961)
	–	654.5	4.64	33.4	Wintermans & De Mots (1965)
Methanol (95%)	420/435	652	–	–	Marker (1972)
Pheophorbide b					
Diethyl ether	432	654	4.79	–	S. R. Brown (1968)
Pheophytin d					
Diethyl ether	421	692	–	–	Smith & Benitez (1955)
	421	692	–	–	Goedheer (1966)

Table 3–2. (*Cont.*)

Solvent	Absorption maxima (nm)		Ratio	Extinction coefficient	Reference
	Blue	Red	blue/red	$(l \cdot g^{-1} \cdot cm^{-1})$	
Pheoporphyrin c_1					
Diethyl ether	–	574/590	–	–	Jeffrey (1972)
	421	568	11.3	–	Wilhelm (1987)
Pheoporphyrin c_2					
Diethyl ether	–	574/590	–	–	Jeffrey (1972)
	432	572	12.3	–	Wilhelm (1987)
Acetone (100%)	430.5	574/596	–	–	Jeffrey & Wright (1987)
Acetone (90%)	431	574/596	13	–	Jeffrey & Shibata (1969)
Pheoporphyrin c_3					
Acetone (100%)	435	532/602	–	–	Jeffrey & Wright (1987)
Divinylpheoporphyrin a_5 *monomethyl ester*					
Diethyl ether	421	568/590	–	–	Ricketts (1966a)
	416	564	16.1	–	Wilhelm (1987)

Table 3–3. *Maxima of the fluorescence emission of the chlorophylls*

Chlorophyll	Solvent	Excitation (nm)	Emission (nm)	Reference
a	Acetone (100%)	430	668	Jeffrey (1972)
	Acetone (90%)	435	667	Boto & Bunt (1978)
		437	676	Gibbs (1979)
	Acetone (80%)	432	674	White et al. (1972)
	Diethyl ether	—	668	French et al. (1956)
		—	669	French (1960)
		430	668	Goedheer (1966)
		420	669	White et al. (1972)
	Methanol	—	674	Zscheile & Harris (1943)
b	Acetone (100%)	459	652	Jeffrey (1972)
	Acetone (90%)	470	651	Boto & Bunt (1978)
		467	659	Gibbs (1979)
	Acetone (80%)	466	659	White et al. (1972)
	Diethyl ether	—	648	French et al. (1956)
		—	653	French (1960)
		453	648	Goedheer (1966)
		454	649	White et al. (1972)
c_1	Acetone (100%)	450	633	Jeffrey (1972)
	Diethyl ether	443	632	Wilhelm (1987)
c_2	Acetone (100%)	453	635	Jeffrey (1972)
	Diethyl ether	446	632	Wilhelm (1987)
	Ethanol	450	638	Jeffrey (1972)
d	Diethyl ether	—	699	French et al. (1956)
		445	693	Manning & Strain (1943)
		447	696	Smith & Benitez (1955)
		447	695	Goedheer (1966)
		—	695	Lu et al. (1983)
	Acetone (100%)	—	700	Lu et al. (1983)
	Acetone (80%)	—	704	Lu et al. (1983)
	Methanol (100%)	—	714	Lu et al. (1983)
	Ethanol (100%)	—	715	Lu et al. (1983)
	Ethanol (80%)	—	713	Lu et al. (1983)
	Petroleum spirit (60–90°)	—	688	Lu et al. (1983)
	Pyridine	—	708	Lu et al. (1983)
	Dioxane	—	693	Lu et al. (1983)
MgDVP	Diethyl ether	440	627	Chereskin et al. (1983)
		436	628	Wilhelm (1987)
Pheophytin *a*	Acetone (90%)	—	673	French et al. (1956)
		390	667	Boto & Bunt (1978)
		420	672	Gibbs (1979)

Table 3–3. (*Cont.*)

Chlorophyll	Solvent	Excitation (nm)	Emission (nm)	Reference
	Acetone (80%)	398	676	White et al. (1972)
	Diethyl ether	409	673	Smith & Benitez (1955)
		400	673	White et al. (1972)
Pheophytin *b*	Acetone (90%)	435	657	Boto & Bunt (1978)
		442	666	Gibbs (1979)
	Acetone (80%)	398	676	White et al. (1972)
	Diethyl ether	434	661	Smith & Benitez (1955)
		—	661	French et al. (1956)
		416	661	White et al. (1972)
Pheoporphyrin c_1	Acetone (100%)	431	652	Jeffrey (1972)
	Diethyl ether	422	655	Wilhelm (1987)
Pheoporphyrin c_2	Acetone (100%)	431	655	Jeffrey (1972)
	Diethyl ether	432	655	Wilhelm (1987)
Pheophytin *d*	Diethyl ether	—	701	French (1960)
Divinylpheoporphyrin monomethyl ester	Diethyl ether	419	645	Wilhelm (1987)

F. Derivatives of chlorophyll a

Dörnemann and Senger (1981) extracted a derivative, which they named chlorophyll RC I,[1] from a mutant of *Scenedesmus* lacking chlorophyll *b* and carotenoids, with absorption maxima at 672 and 433 nm; it could not be extracted from the cells when formation of PS I was suppressed by chloramphenicol, and always occurred in equal amount to P_{700} determined by reduction by ascorbate. Katoh, Dörnemann, and Senger (1985) extracted chlorophyll RC I from two species of the cyanophyte *Anacystis* and again found that concentrations of P_{700} and chlorophyll RCI were equal; they suggested that P_{700} and RC I were identical. Dörnemann and Senger (1986) determined the structure of chlorophyll RC I as 13′-hydroxy-20-chloro-chlorophyll *a* but were still not certain that it was P_{700}. Rebeiz et al. (1983) and Wu and Rebiez (1988) found monovinyl 10-hydroxy-chlorophyll *a* lactone in extracts from *Anacystis nidulans,* with a fluorescence excitation maximum at 432 nm and emission maximum at 662 nm. Gieskes and Kraay (1983a) found a pigment identified as divinyl chlorophyll *a* (Wu and Rebeiz, 1988), which appears to be the major

[1]Senge, Dörnemann, and Senger (*FEBS Lett.*, 234: 215–17, 1988) now believe that chlorophyll RC I is an artifact of extraction.

chlorophyll species in extracts from phytoplankton collected in the North Sea and the tropical Atlantic Ocean.

3.1.3. Chemical structure

Chlorophylls *a*, *b*, and *d* are derivatives of dihydroporphyrin (chlorin) (Holt, 1961, 1965; Jackson, 1976; Kirk and Tilney-Bassett, 1978), whereas chlorophyll c_1 and c_2 are more oxidized and are derivatives of porphyrin (Dougherty et al., 1966, 1970; Budzikiewicz and Taraz, 1971). All chlorophylls contain an atom of magnesium chelated at the center of the ring system, and all contain a fifth isocyclic ring. Chlorophylls *a*, *b*, and *d* carry a propionic acid residue esterified with the C_{20} monounsaturated alcohol, phytol, whereas chlorophyll c_1 and c_2 contain an unesterified acrylic acid residue in place of the propionic acid residue. MgDVP is an important intermediate in the synthesis of chlorophylls, and Ricketts (1966a) claimed that it was the unknown chlorophyllous pigment isolated from *Micromonas pusilla* (Secs. 3.1.2D and E, above). It differs only from chlorophyll c_2 in that a propionic acid residue replaces the acrylic acid residue. The structures of these chlorophyllous pigments are shown in Fig. 3–1.

Pigments without Mg chelated at the center of the ring system are grayer than pigment molecules containing Mg, though removing the phytyl chain does not alter the spectrum. The molecules without Mg include pheophorbides *a*, *b*, and *d* and pheoporphyrins c_1, c_2, and c_3 (Jeffrey, 1972; Jeffrey and Wright, 1987). Molecules without both Mg and the phytyl chain are pheophorbides *a*, *b*, and *d*. Some authors have referred to chlorophyll c_1 and c_2 as chlorophillides because they lack the phytyl chain, but other differences in structure (i.e., oxidation of C-7 and C-8, and an acrylic rather than a propionic residue) do not justify this nomenclature (Jeffrey, 1972; Kirk and Tilney-Bassett, 1978).

3.1.4 Degradation products

The most common degradation products of the chlorophylls are the Mg-free derivatives, pheophytins or pheoporphyrins, formed rapidly when the pH is lowered. Loss of the phytyl chain by hydrolysis is another common degradation route, forming a chlorophyllide. When both phytyl chain and Mg are lost, the product is a pheophorbide (see Sec. 3.1.3, above). Svec (1978) has constructed a reaction scheme (Fig. 3–2) for the formation of derivatives of chlorophyll *a*, and this scheme applies to the other chlorophylls as well, including the pheophytins. Many of the products shown in this scheme are formed only by treating with chemicals not normally

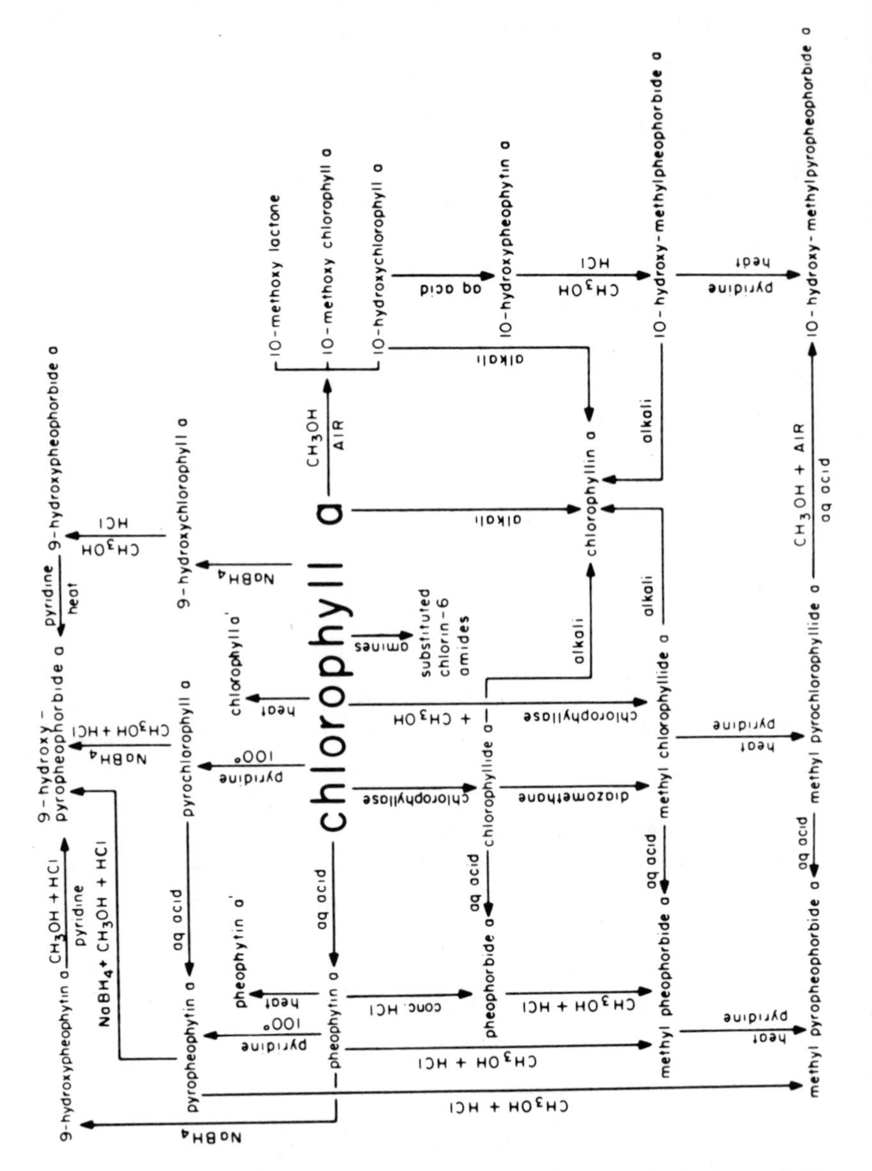

Fig. 3–2. A reaction scheme for the formation of derivatives of chlorophyll *a*. (From Svec, 1978, by permission.)

used during extractions; however, in addition to the two major degradations mentioned above, heating causes isomerization to form an isomer, chlorophyll *a′* (Strain, 1958), whereas extracting with methanol can form methyl chlorophyllides (Holt and Jacobs, 1954; Pennington et al., 1964). Oxidative enzymes in the tissue can replace the H at C-10 of chlorophyll *a* with a hydroxyl group (Pennington et al., 1964). The relevant parts of the scheme are shown in Fig. 3–2 (Svec 1978). Although undoubtedly pheophytin in extracts is mainly from dead cells or an artifact of extraction, Klimov, Klevanik, Shuvalov, and Krasnovsky (1977) proposed that pheophytin equivalent to about 2% of total chlorophyll acts as the first electron acceptor in PS II, and Vermans and Govindjee (1981), Okamura, Feher, and Nelson (1982), and Codgell (1983) have discussed evidence for this function, though Larkum and Barrett (1983) pointed out that selective phytinization of such a small fraction was hard to visualize. More recently, Satoh (1986) and Namba and Satoh (1987) have stated that pheophytin is an essential component of RC II in higher plants. Hallegraeff and Jeffrey (1985) have found up to four unknown chloropigments on TLC plates of extracts from phytoplankton subjected to stress or damage, and they believe that they were similar to those seen by earlier workers such as Gieskes and Kraay (1983a,b) who found high proportions of altered chloropigments in natural samples. Otsuki, Watanabe, and Sugahara (1987) have reported a number of oxidation products of chlorophylls when extracting pigments from diatoms and green algae.

3.1.5. *Methods for preventing degradation during extraction*

The bond between chlorophylls and the proteins to which they are bound in the cell breaks when the pigment is extracted with nonpolar solvents, and thus the extract produced gives no idea of the complexity of the forms of chlorophyll in the cell. Extraction of pigment–protein complexes found in the cell will be considered in Chapter 6.

This section deals with precautions to prevent further changes in the naturally occurring pigment during extraction with nonpolar solvents, specifically the various forms of possible degradation mentioned previously in Section 3.1.4.

A. *Chlorophyllase*

This enzyme catalyzes hydrolysis of the bond linking the phytyl chain attached to the propionic acid residue (Fig. 3–1) of chlorophylls *a* and *b*, thus forming the equivalent chlorophyllide (Fig. 3–2). Barrett and Jeffrey (1964, 1971) found a highly active chlorophyllase when some marine algae from the classes Chlorophyceae and Chrysophyceae were incubated in 50% acetone for 15 min at 20°C, but found the activity to be strongly inhibited by 80–100% acetone. Unless incubated in distilled water for 5 min before incubation in acetone, an anomalous chlorophyllide was produced. Unfortunately, extracting photosynthetic pigments with acetone alone is frequently inefficient, and other solvents must be used. Chromatography of the resulting extract will show whether chlorophyllides have been formed in appreciable amounts because these are clearly separated from the chlorophylls *a* and *b* in all systems. Jeffrey (1968a, 1974, 1976b) has sometimes extracted chlorophylls from phytoplankton with 100% acetone, followed by 90%, whereas others have switched to methanol or another solvent after the 100% acetone (see Chapter 2). Jeffrey (1963, 1972) found that chlorophyllase was not present in some groups of algae, and more recently Jeffrey and Hallegraeff (1987a) have examined 97 species of unicellular algae from ten classes: The enzyme was active only in some but not all diatoms and in two chrysophytes and two chlorophytes. In addition, they found that handling some cultures activated chlorophyllase before extraction, thus posing problems for accurate analysis of the chloropigments in these species.

B. Acidity

Fortunately, few algae have acid sap, but when they do, extracting chlorophyll *a* without degradation to pheophytin is difficult. Holden (1976), Jensen (1978), and Svec (1978) recommended including alkalies such as $NaHCO_3$, diethylamine, or ammonia in the medium, though Strain and Svec (1966) and Svec (1978) warned that $CaCO_3$ or $MgCO_3$ do not neutralize endogenous acids rapidly enough. However, $MgCO_3$ has been widely used in the collection of phytoplankton by filtration (see Chapter 2), both for preventing phytinization and as a filter aid, though the reduction of pheopigments observed with the use of $MgCO_3$ pads could be due to adsorption of existing pheopigments, rather than inhibition of phytinization of chlorophylls (Daley et al., 1973a; Holm-Hansen and Riemann, 1978) or formation of aggregates of chlorophyll and $MgCO_3$ (Sand-Jensen, 1976). Lenz and Fritsche (1980) found no improvement with its use and considered it optional. $MgCO_3$ has also been used for standardizing acidification techniques for phytinization (Moed and Hallegraeff, 1978, table 1), though the cited authors do not recommend it. Storing acetone over $MgCO_3$ and redistillation (Strickland and Parsons, 1968) are unnecessary when analytical-grade reagents are used (Holm-Hansen and Riemann, 1978). Whittle and Casselton (1969) warned that storing microalgae in deep-freeze for too long before extracting the pigments could lead to phytinization. Extraction of pigments from prochlorophyte species symbiotic on the colonial ascidian *Diplosoma virens* was complicated by the acid sap released when the tissues of the ascidian were cut, apparently causing phytinization (Thorne, Newcombe, and Osmond, 1977).

C. *Temperature*

Chlorophylls *a*, *b*, and *d* form an equilibrium mixture of isomers, slowly at room temperature and within 15–20 min at 100°C. The isomers are usually indicated by a prime, for example, chlorophyll *a′*. Their spectra are similar to the parent molecule but run faster on chromatograms (Strain and Svec, 1966). Chlorophyll *c*, with a planar symmetrical structure, does not form an equilibrium mixture of isomers (Strain et al., 1971a). Although some authors recommend heating prior to extracting pigments to inactivate degrading enzymes, Holden (1976) does not recommend it for general use.

D. *Oxidation*

Including reducing agents in the extracting medium such as H_2S (Jensen, 1978) or butylated hydroxytoluene, and extracting in an atmosphere of N_2 gas can prevent oxidation to form 10-hydroxychlorophyll (Fig. 3–2). This derivative can be detected as running slightly more slowly than the parent compound in chromatograms, and it gives a negative phase test (Chapter 2). Jones, Butler, Gibbs, and White (1972), added butylated hydroxytoluene to ethereal solutions of pigments before applying solutions to reverse-phase TLC plates, demonstrating that degradation products did not form as separate spots on the chromatograms. Gillan and Johns (1980) added butylated hydroxytoluene to the solvents used in running TLC plates. Daley and Brown (1973) have shown that oxidation of chlorophyll *a* in senescent cells of *Anacystis nidulans* was greater when cells were aerated in light than in darkness, and that no appreciable degradation occurred in N_2 gas in light.

E. *Aggregation*

Chlorophyll *c*, unlike the other chlorophylls, tends to aggregate after extraction when diluted with water (Jeffrey and Shibata, 1969). A third chlorophyll zone other than chlorophyll c_1 and c_2 appearing on chromatograms on polyethylene powder when aged extracts were run (i.e., chlorophyll c_3) showed decreased band intensities in the absorption spectrum (Jeffrey, 1969) but also showed a new band at 666 nm. Melnikov and Yevstigneyev (1964) found a similar aggregated complex when using paper chromatograms. The implication of Jeffrey's result is that chlorophyll *c* must be estimated as rapidly as possible before aggregation can occur. This chlorophyll c_3 is not the polymerized form recently identified by Vesk and Jeffrey (1987) and Jeffrey and Wright (1987).

3.2. Purification

Analytically pure preparations of the chlorophylls are required for structural analysis and for determining molecular weights and absorption and fluorescence spectra. Smith and Benitez (1955) listed the previous methods used for purifying chlorophylls *a* and *b*; apart from the early purification in Willstätter's laboratory, these have all relied on a preliminary separation of the two pigments and any degradation products by column chromatography. The chlorophyllous pigments thus separated are still contaminated with colorless compounds that must be removed by further chromatography with different solvents and/or by precipitation. The method of Strain and Svec (1966) involved some novel steps, including adsorbing solid chlorophylls on the surface of water droplets in an emulsion with petroleum spirit. This method is essentially that described by Strain, Thomas, and Katz (1963) and appears to be used commercially, for example, by the Sigma Chemical Company. Svec (1978) has reported a method employing a series of chromatographic elutions on sugar columns, followed by removal of colorless contaminants by washing with a series of aqueous methanol solutions from 50% to 85%.

Svec (1978) has also reported a purification using high-speed liquid chromatography, with two preparative columns of polyethylene in series; the preliminary elution with 50% acetone eluted the xanthophylls first; the concentration was then raised to 80%, eluting the chlorophylls and leaving the carotene on the column; the chlorophylls were then transferred to petroleum spirit by phasing with a solution of NaCl, and evaporated to dryness. The chlorophylls were then redissolved in diethyl ether and diluted with petroleum spirit (20–40°C). They were then injected into a column of cellulose, and chlorophyll *a* was eluted successively with petroleum spirit, 1.5% propanol-1-ol in petroleum spirit, and 3% propan-1-ol in petroleum spirit. Chlorophyll *b* was then eluted with ethanol.

Chlorophyll *a* was then purified further by dissolving in ethanol–petroleum spirit (20–40°C) and precipitated by washing with water, chilling, and drying. Chlorophyll *b* required further separation from redissolved chlorophyll *a* by further chromatography on cellulose developing with increasing concentrations of propan-1-ol in petroleum spirit. The chlorophyll *b* was then precipitated and dried in the manner as for chlorophyll *a*.

Chlorophyll *c* has usually been prepared from species of one of the groups rich in it, such as the Bacillariophyceae, Phaeophyceae, or Dinophyceae. Dougherty et al. (1970) boiled the cells of *Nitzschia* (Bacillariophyceae) in culture medium buffered with ammonium acetate to inhibit enzymic alteration, and, after cooling, extracted the pigments with methanol–diethyl ether–petroleum spirit (5:2:1) (see Chapter 2, Sec. 2.1.1). The methanol was separated from the less polar solvents by shaking with NaCl solution, and the combined extracts were dried in vacuo. The dry residue was dissolved in diethyl ether, four volumes of petroleum spirit were added, and the solution was applied to a sucrose chromatography column developed with petroleum spirit containing a gradient of 0–7.5% propan-1-ol. The band of chlorophyll *c* was cut from the column and eluted with ethanol and diethyl ether.

Jeffrey (1972) has described methods for preparing crystalline preparations of chlorophyll c_1 and c_2 from seaweeds, diatoms, and symbiotic dinoflagellates, using a different method of extracting pigments from each type: (1) for seaweeds, boiling, followed by extracting with the mixed solvent (methanol–diethyl ether–petroleum spirit) of Strain and Svec (1966) and a further extraction with acetone; (2) for diatoms, freezing, followed by extraction with 100% acetone; (3) for dinoflagellates, freezing, followed by extraction with methanol. Pigments from all sources were phased into diethyl ether, and the fraction containing chlorophyll *c* was obtained by chromatography on a cellulose column developed first with 0.5% propan-1-ol in petroleum spirit (60–80%C), then with 2% propan-1-ol in petroleum spirit (>40°C) to remove fucoxanthin. Chlorophyll *c* was then concentrated to a narrow band with 5% methanol in diethyl ether. In spite of extracting chlorophyll *c* from diatoms in 100% acetone, a little activity of chlorophyllase remained, and chlorophyllide *a* contaminated the chlorophyll *c* fraction. This could be removed by eluting the column with 2% propan-1-ol in petroleum spirit (>40°C) before concentrating the chlorophyll *c* zone with 5% methanol in diethyl ether. Chlorophyll *c* was precipitated from the eluate from this column after concentrating under a stream of N_2, by adding three to four volumes of petroleum spirit (60–80°C) and holding overnight at 5°C. The resulting black precipitate was collected by centrifugation, dissolved in a minimum volume of pyridine, and crystallized by holding overnight at −20°C. The crystalline chlorophyll *c* was collected by centrifugation and dried. As the dinoflagellate contained only chlorophyll c_2, the purification was taken no further. The chlorophyll *c* from diatoms or seaweed was redissolved in a little pyridine, ten volumes of acetone were added, and the solution was applied to a column of polyethylene (PRX-1025); this was developed with acetone until the two bands of chlorophyll c_1 and c_2 were clearly separated. The column was sucked dry and extruded, and the separate bands were each packed into separate columns from which they were eluted with pyridine. The eluates were dried and then redissolved in pyridine, ten volumes of acetone were added, and the solution was crystallized overnight at −20°C. The crystals were collected by centrifugation, washed with 50% aqueous acetone, and dried in darkness.

Mg 2,4-divinylpheoporphyrin a_5 monomethyl ester extracted from the Prasinophyceae (Ricketts, 1966a) has not been thoroughly purified, but the step of phase separation by the method of Parsons (1963) provides a valuable separation of pigments with and without the phytyl chain attached. In this method, 10 ml of the extract containing the pigment in 90% aqueous acetone was mixed with 3.5 ml of 0.05% NaCl solution and 13.5 ml hexane. On mixing, the phytyl-free pigments moved to the hypophase. The pigments in this phase were then resolved by chromatography with a column of icing sugar (Strain, Manning, and Hardin, 1944) and with 0.5% propan-1-ol in petroleum spirit as solvent. MgDVP could then be eluted with 2% propan-1-ol in petroleum spirit.

3.3. Physical properties

3.3.1. Absorption spectra

The absorption spectra of the chlorophylls as determined by Zscheile and Comar (1941), Jeffrey (1972), and Ricketts (1966a) are shown in Fig. 3–3, and the absorption maxima in various solvents in Table 3–2. Slight variations in the maxima shown in Table 3–2 are no doubt due in part to errors in calibration of the spectrophotometers used. In some recording spectrophotometers, such as the Philips SP 8, variations in the height of the pen in its holder can cause slight variation in readings, and a holmium oxide filter with sharply defined absorption maxima at 453.2 (see Fig. 4–3) and 637.6 nm can be used to check the readings of the peaks shown

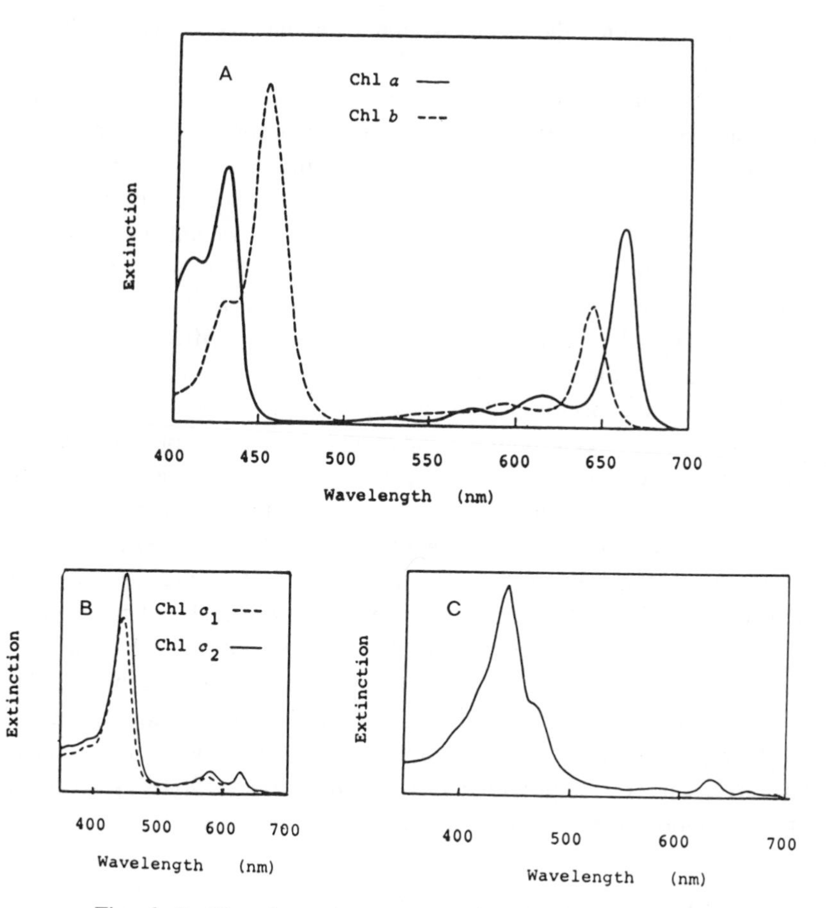

Fig. 3–3. The absorption spectra of chlorophyllous pigments. (A) Chlorophyll *a* and *b* in diethyl ether. (After French, 1960.) (B) Chlorophyll c_1 and c_2 in acetone containing 2% pyridine. (Jeffrey, unpublished data, after Kirk and Tilney-Bassett, 1978.) (C) MgDVP in 90% acetone. (After Ricketts, 1966a.)

in the traces of the spectra. Table 3–2 contains only values from 1940 onward, except for chlorophyll c_1, c_2, and c_3 values, which are only those published after Jeffrey and Shibata (1969) discovered that chlorophyll *c* could polymerize after extraction and occurred in more than one form (Jeffrey 1969, 1972; Vesk and Jeffrey, 1987). Recently, MgDVP has been identified in some Prasinophyceae (Ricketts, 1966a, 1970) and has been shown to function in light harvesting by J. S. Brown (1985). Seely and Jensen (1965) have examined the absorption maxima of chlorophyll *a* in 40 solvents; only the results for the commonly used ones are shown in Table 3–2.

The ratio of the peak heights of the absorption spectra of the chlorophylls at the blue (Soret) and red end of their absorption spectra is one criterion of an undegraded pigment, though this must be taken in conjunction with other properties such as the performance in the phase test and R_f value (Bacon, 1966). Table 3–2 shows that a number of workers have reported this ratio for chlorophyll *a* as 1.3. Perkins and Roberts (1964) have shown that preparations of chlorophyll *a* with a low blue/red ratio claimed by earlier workers as the more purified were due to contamination of the solvent used, diethyl ether, with ethanol or water.

The absorption spectra of the chlorophyll *c* group (including MgDVP) differ from those of chlorophyll *a* and *b* in having two major peaks, rather than one, at the red end of the spectrum, and in having a much larger ratio of absorbance of the blue (Soret) band to the red bands (see Fig. 3–3). The values for these ratios are given in Table 3–2. Jeffrey (1972) designated the three major bands of the chlorophyll *c* group as I, II, and III (Soret) in order of decreasing wavelength. The II/I and III/I ratios of the new chlorophyll c_3 are much larger than those of chlorophyll c_1 and c_2 (Jeffrey and Wright, 1987). This provides a useful means of identifying chlorophyll c_3 and separating it from various alteration products of chlorophyll c_3 such as the original chlorophyll c_3, which has a small but distinct peak at approximately 665 nm (Jeffrey, 1969), as do some other alteration products of chlorophyll c_3 (Jeffrey, 1972).

3.3.2. *Fluorescence spectra*

Much of the light energy absorbed by the chlorophyll molecule in solution is emitted as fluorescence at a longer wavelength than the red absorption maximum (Stokes shift; Stokes, 1852) and is specific for the type of chlorophyll (Table 3–3). This property can be used for quantitative estimation of the chlorophylls and derivatives (Sec. 3.4.1B) and, although not as convenient as spectroscopy, is much more sensitive when modern photomultipliers are used.

Because the Stokes shift is small, fluorescence emission is best measured by activating the molecule at or close to the blue maximum in the absorption spectrum (Soret band). However, in solutions containing a mixture of chlorophyll *a* and *c*, the emission of the chlorophyll *c* in acetone is best measured by activating at about 450 nm (Jeffrey, 1972; Sullivan et al., 1989), a point of minimal absorbance of chlorophyll *a* (Fig. 3–3).

3.4. Methods of estimating chlorophylls

Although pure samples of chlorophyll can be assayed by estimating magnesium, virtually all methods used by biologists involve the properties of the molecule in solution (absorbance or fluorescence), whether direct estimation is made on the extract or the extract is concentrated and analyzed by quantitative chromatography. The published values of the extinction coefficients of the chlorophylls are universally accepted, and standard curves are never prepared by individual workers for spectroscopic analysis. Many of the extinction coefficients determined since 1940 in a range of solvents are shown in Table 3–2. Although some of the earlier extinction coefficients tend to be low owing to impurities in the samples, those reported by MacKinney (1940) have been widely used indirectly, as the dichromatic equations of Arnon (1949) were based on them (Table 3–4). Earlier values for chlorophyll *c* have not been included, because before 1968 the two forms of this chlorophyll (see Secs. 1.2.1 and 3.1.2C) had not been separated and the polymerization of the compounds when purified was undetected (Jeffrey and Shibata, 1969); the extinction coefficients for chlorophyll c_1 and c_2 in Table 3–2 are thus approximately twice the earlier figures. Table 3–2 also includes extinction coefficients for a number of derivatives of the chlorophylls. Concentrations of chlorophyll have been expressed either per cell or per unit weight (fresh or dry), and sometimes the chlorophyll has been expressed as a percentage of the total chlorophyll.

Marker, Nusch, Rai, and Riemann (1980b), in summarizing the opinions expressed at the Workshop held at Plön, West Germany in 1978, have made a number of important recommendations for estimating chlorophylls in phytoplankton. Spectrophotometers with grating monochrometers are preferred to prism instruments with decreasing dispersion at the red end of the spectrum and susceptibility to thermal shifting. A bandwidth of no more than 2 nm is desirable as found by Brown, Dromgoole, and Guest (1980) for estimating chlorophylls *a* and *b* using the equation of Jeffrey and Humphrey (1975). Marker et al. (1980b) also stressed that the solutions of chlorophylls should be free from turbidity with an upper limit of 0.005 absorbance units at 750 nm, and that absorbance units should range from 0.2 to 0.8. They also commented on the accuracy of a number of specific extinction coefficients for the chlorophylls, though they neglected those of Jeffrey and Humphrey (1975).

Using either spectroscopy or fluorometry to measure concentrations of chlorophylls is inaccurate when phytinized derivatives are present (Gieskes and Kraay, 1983a). Many workers have tried to overcome this difficulty

Table 3–4. *Equations for analyzing mixtures of chlorophylls dissolved in various solvents (concentrations in* $mg \cdot l^{-1}$*)*

Chlorophylls	Solvent	Equation	Reference
a & *b*	Acetone (100%)	$a = 9.78E_{662} - 0.99E_{644}$	Holm (1954), Nybom (1955)
		$b = 21.40E_{644} - 4.65E_{662}$	
		$a = 12.3\ E_{663} - 0.86E_{645}$	MacLachlan & Zalik (1963)
		$b = 10.3\ E_{645} - 3.6\ E_{663}$	Jeffrey & Humphrey (1975)
	Acetone (90%)	$a = 11.93E_{664} - 1.93E_{647}$	
		$b = 20.36E_{647} - 5.50E_{664}$	
	Acetone (85%)	$a = 10.3\ E_{663} - 0.91E_{644}$	Hoffman & Werner (1966)
	Acetone (80%)	$a = 12.7\ E_{663} - 2.7\ E_{645}$	
			MacKinney (1941), Arnon (1949)
		$b = 22.9\ E_{645} - 4.7\ E_{663}$	
		$a = 11.78E_{664} - 2.29E_{647}$	Ziegler & Egle (1965)
		$b = 20.05E_{647} - 4.77E_{664}$	
		$a = 11.63E_{663} - 2.39E_{649}$	Vernon (1960)
		$b = 20.11E_{649} - 5.18E_{663}$	
		$a = 11.73E_{664} - 1.97E_{647}$	Jeffrey et al. (1974)
		$b = 20.56E_{647} - 5.42E_{664}$	
		$a + b = 27.8\ E_{652}$	Bruinsma (1961, 1963)
	Diethyl ether	$a = 9.93E_{660} - 0.78E_{642.5}$	Comar & Zscheile (1942)
		$b = 17.6\ E_{642.5} - 2.81\ E_{660}$	

		$a = 9.95E_{661} - 0.95E_{642.5}$ $b = 15.7\ E_{642.5} - 2.31E_{661}$	Davidson (1985)
		$a = 10.10E_{662} - 1.01E_{664}$ $b = 16.40E_{640} - 2.57E_{662}$	Smith & Benitez (1955)
		$a = 9.93E_{660} - 0.78E_{642.5}$ $b = 17.60E_{642.5} - 2.81\ E_{660}$	French (1960)
		$a + b = 100.50E_{600}$	Smith & Benitez (1955)
	Ethanol	$a = 13.70E_{665} - 5.76E_{649}$ $b = 25.80E_{649} - 7.60E_{665}$	Wintermans & De Mots (1965)
	Methanol	$a = 16.50E_{665} - 8.30E_{650}$ $b = 33.80E_{650} - 12.5\ E_{665}$	Hoffman & Werner (1966)
a, b, c_1, & c_2	Acetone (90%)	$a = 11.85E_{664} - 1.54E_{647} - 0.08E_{630}$ $b = -5.45E_{664} + 21.03E_{647} - 2.66E_{630}$ $c_1 + c_2 = -1.67E_{664} - 7.60E_{647} + 24.52E_{630}$	Jeffrey & Humphrey (1975)
a, c_1, & c_2		$a = 11.47E_{664} - 0.46E_{630}$ $c_1 + c_2 = 24.36E_{630} - 3.73E_{664}$	Jeffrey & Humphrey (1975)
a & c_2		$a = 11.43E_{663} - 0.64E_{630}$ $c_2 = 27.09E_{630} - 3.63E_{663}$	Jeffrey & Humphrey (1975)
a, b, & c		$a = 11.63E_{663} - 2.16\ E_{645} + 0.10\ E_{630}$ $b = -3.94E_{663} + 20.97E_{645} - 3.66\ E_{630}$ $c = -5.53E_{663} - 14.81E_{645} + 54.22E_{630}$	SCOR-UNESCO (1966)

by following changes in absorbance after chlorophylls were converted to the equivalent Mg-free derivatives by treating with dilute acid (Sec. 3.4.1A and B).

3.4.1. Direct estimations in extracts from algae

The various methods for estimating chlorophylls in extracts from algae all have their value in suitable circumstances. The technically simple method of using di- or trichromatic equations is accurate when the chlorophylls have not decomposed to Mg-free or phytyl-free derivatives, and Lorenzen and Jeffrey (1980) have made the following recommendations for determining chlorophylls in oceanography:

1. Degradation products of chlorophylls are usually insignificant in samples collected near the surface of the euphotic zone, and spectrophotometric or fluorometric methods can be used.
2. Although the most accurate trichromatic equations are those of Jeffrey and Humphrey (1975), even with these, errors occur when concentrations of chlorophylls *b* and *c* are low as compared to *a*.
3. Although some estimates of chlorophyll *a* are made neglecting concentrations of chlorophylls *b* and *c*, errors will be serious only if chlorophyll *b* is present; however, this is frequently the case.
4. Samples lower in the water column require analysis by a technique involving acidification, since degradation products from senescent cells, detritus, and fecal pellets can be expected.
5. Samples containing sediments could contain a variety of unidentified pigments and chlorophyll derivatives, and therefore, acidification techniques could be inadequate.
6. Precise analysis of any sample requires quantiative chromatography such as TLC or HPLC.

Marker et al. (1980b) came to substantially similar conclusions, though they assumed that chlorophyll *b* is rarely present.

A. *Spectrophotometry*

The simplest method of estimating chlorophyll assumes that chlorophyll *a* predominates, and thus the absorbance is read only at the maximum peak in the red end of the spectrum. This is valid when only chlorophyll *a* is present without degradation products (Cyanophyceae and Rhodophyceae); when chlorophyll *c* is the only other chlorophyll, estimates will be only a few percentage points in error. However, when Chlorophyta are analyzed, the chlorophyll *b* will cause serious errors, because the chlorophyll *a/b* ratio may be as low as 1.5 in this phylum (Keast and Grant, 1976), but the error will be less than 10% if chlorophyll *b* is only 20% of the chlorophyll *a* present (Riemann, 1978b). Hallegraeff (1981) measured chlorophyll *a* in extracts from phytoplankton collected at a coastline station off Sydney, Australia, by measuring extinction at 664 nm only and applying the extinction coefficient for chlorophyll *a* in 90% acetone (Table 3–2). The values differed little when calculated using the trichromatic equations of Jeffrey and Humphrey (1975). Hallegraeff estimated nanoplankton by measuring chlorophyll *a* in organisms passing a 10-μm nylon gauze, finding them relatively constant throughout the year and contributing a substantial fraction of total chlorophyll *a*, except when the larger organisms increased sharply during short-lived diatom blooms.

The absorption spectra and extinction coefficients of MacKinney (1940) were the basis for the first equations, published by Arnon (1949), for the calculation of concentrations of mixtures of chlorophyll *a* and *b* in 80% acetone. The absorption spectrum of chlorophyll *a* (MacKinney, 1940) was measured using a single-beam instrument, and the readings were taken at intervals of 2–3 nm apart. Inspection of the table giving the readings shows that the maximum was at 664 nm, as later reported by Jeffrey (Jeffrey and Humphrey, 1975); although this did not mean that the extinction coefficient at 663 nm was inaccurate, many workers have assumed that 663 nm was the maximum, and, because not many spectrophotometers are calibrated accurately to the nearest nanometer, the wavelength of the observed maximum of chlorophyll *a* has been taken as 663 nm, and the setting of the instruments adjusted accordingly. An error of 1 nm in the setting of the wavelength scale can cause significant error in the concentration of chlorophyll *b* calculated with dichromatic equations (Ogawa and Shibata, 1965; Ziegler and Egle, 1965). The accuracy of the dichromatic equations for calculating concentrations of chlorophyll *a* and *b* in the various solvents shown in Table 3–4 will vary with the accuracy of the extinction coefficients on which they are based, and these have not all been determined recently with modern methods.

Most of the trichromatic equations shown in Table 3–4 for determining simultaneously chlorophylls *a*, *b*, and *c* are based on the new extinction coefficients of Jeffrey (1969, 1972), determined following experiments showing that chlorophyll *c* tends to polymerize in aqueous solution; thus, previous equations gave values for concentrations of chlorophyll *c* double the correct amount. These new equations were calculated assuming that equal amounts of both chlorophyll c_1 and c_2 were present. The dichromatic equations credited to "Humphrey and Jeffrey (in preparation)" in Jeffrey (1968a,b) and Jeffrey and Haxo (1968) when analyzing chlorophyll *a*/*b* and *a*/*c* ratios are not those subsequently published by Jeffrey and Humphrey (1975), and the ratios therein will be low. The earlier equations proposed by Richards and Thompson (1952) and Parsons and Strickland (1963) and used by Ricketts (1965, 1967a,b,c) have not been given in Table 3–4, as these seriously overestimate chlorophyll *a* in addition to the errors due to the old extinction coefficients of chlorophyll *c* (Hallegraeff, 1976; Lorenzen and Jeffrey, 1980), though Chang and Rossmann (1981) used them in their numerical simulation of trichromatic equations for estimating chlorophylls, and they have been used recently by Dustan (1979). Humphrey (1978) and Wartenberg (1978) have attempted to correct earlier estimates made with inaccurate equations. Jeffrey and Hallegraeff (1980a,b), using the trichromatic equations of Jeffrey and Humphrey (1975), measured photosynthetic pigments in phytoplankton from a warm-core eddy of the Eastern Australian current. Spectroscopic curve analysis detected detrital degradation-products of chlorophyll at various depths, and spectroscopic analysis of chlorophylls was accurate only when the degradation products were absent, as judged by examination with TLC, since only 50–70% of chlorophyll measured spectroscopically was chlorophyll *a* in some samples. Hallegraeff and Jeffrey (1984) used trichromatic equations for measuring total chlorophyll *a* in nanoplankton in continental shelf waters of North and Northwestern Australia, though TLC showed that degradation products of chlorophyll were sometimes present.

The trichromatic equations are valuable for measuring chlorophylls in healthy cultures in the laboratory, when chromatography has confirmed that degradation has not occurred (Jeffrey et al., 1975; Jeffrey and Vesk, 1977; Bjørnland, 1982, 1983). As Shoaf and Lium (1976) found that the absorption spectra of both chlorophylls *a* and *b* were identical in 90% acetone and in 90% acetone–DMSO (1 : 1), dichromatic equations may be used with extracts prepared in DMSO. As an alternative method, Ogawa and Shibata (1965) measured the maximum absorbance of the chlorophyll spectrum at 666 nm in 95% methanol at pH 5.8 before and after converting the chlorophyll *b* with hydroxylamine to a derivative having the same absorption maximum as chlorophyll *a*.

Using these equations is known to cause serious error if degradation products of the chlorophylls are present in solution (Jeffrey and Hallegraeff, 1980a,b; Lorenzen and Jeffrey, 1980; Marker et al., 1980b). In attempting to overcome this, many workers have used equations based on absorbance before and after treating solutions of the mixture of pigments with dilute acids, when the chlorophylls are converted to the equivalent pheophytins, thus allowing calculation of the original amount of chlorophylls *a* and *b* in the presence of the degradation products. Most procedures used stem from the paper by Vernon (1960), often as modified by Lorenzen (1967). White, Jones, and Gibbs (1963) determined the concentrations of chlorophylls *a* and *b* and their common derivatives dissolved in diethyl ether (solution 1) by treating one replicate of the solution with HCl to give pheophytins and pheophorbides (solution 2), and fractionating the other replicate by phasing with 0.01 M KOH, to give an epiphase containing the original chlorophylls and pheophytins (solution 3) and a hypophase containing the chlorophyllides and pheophorbides (aqueous solution). Solution 3 was then treated with HCl to convert the chlorophylls to pheophytins. From known extinction coefficients for the chloropigments in diethyl ether, they constructed dichromatic equations enabling them to estimate the eight components, thus improving the method of Vernon (1960), which does not resolve chlorophylls and chlorophyllides or pheophytins and pheophorbides. For algae, the method could be applied not only to green algae, but to those containing only chlorophyll *a* and derivatives. The method of Lorenzen (1967) has been used by Colijn and Dijkima (1981) for estimating chlorophyll *a* in benthic diatoms, and by Wojciechowski et al. (1988) for measuring concentrations of chlorophyll *a* over a ten-year period in a lake undergoing deeutrophication.

The equations of Moss (1967) assumed that pheophytin *b* and pheoporphyrin *c* would not affect estimations of chlorophyll *a* and pheophytin *a* at 665 nm. Acidification has been used by Tett et al. (1975), Holm-Hansen and Riemann (1978), and Marker (1972, 1977). In the second paper, Marker has shown acetone to be preferable to methanol for acidification methods, because in methanol the pheophytins were sensitive to pH. Marker and Jinks (1982) subsequently compared acetone, ethanol, and methanol, using acidification; all gave essentially similar results, but the acidification factors were lower for ethanol and methanol. Whitney and Darley (1979) prevented chlorophyllides and pheophorbides from interfering with determinations of chlorophyll and pheophytin when using acidification by phasing the phytyl-free derivatives into the aqueous acetone, leaving chlorophylls and pheophytins in a hexane-rich epiphase.

Seely et al. (1972) described a method for estimating pigments extracted from brown algae by soaking in DMSO (fraction X1) followed by acetone (fraction X2). Fraction X2 was then fractionated by phasing with hexane and methanol, giving an acetone–hexane phase (fraction X2A) and an acetone–methanol–water phase (fraction X2B). Using extinction coefficients known at that time, they presented the following equations for calculating chlorophyll *a* and *c* concentrations in milligrams per liter ($mg \cdot l^{-1}$):

$$\text{Fraction X1: Chl } a = 13.74\ E_{665}$$

$$\text{Chl } c = 16.18\ (E_{631} + E_{582} - 0.297\ E_{665})$$

$$\text{Fraction X2A: Chl } a = 12.00\ E_{661}$$

$$\text{Fraction X2B: Chl } a = 13.59\ E_{664}$$

$$\text{Chl } c = 16.08\ (E_{631} + E_{581} - 0.300\ E_{664})$$

The principle of measuring chlorophyll *c* at two wavelengths is based on the observation by Wasley, Scott, and Holt (1970) that the sum of E_{630} and E_{580} was more nearly independent of solvent and species than an observation at a single wavelength. Wheeler (1980) used this method for estimating pigments in fronds of *Macrocystis pyrifera*. Riemann (1978a) showed that spectral changes of fucoxanthin interfered with the estimation when extracts were acidified for determining pheopigments. Moed and Hallegraeff (1978) studied the effect of variation in pH on the efficiency of the method, for, in addition to the type of interference seen by Riemann (1978) with carotenoids, the absorption spectrum of the dications of pheophytin formed at the low pH differ from those at the normal pH. They measured values in a narrow range of pH from 2.6 to 2.8, claiming that high pH values sometimes gave incomplete phytinization, though in this laboratory chromatography of the extracts from higher plants treated with oxalic acid (Vernon, 1960) showed no residual chlorophyll.

Chlorophyll *c* is usually present in lower concentrations than chlorophyll *a* in phytoplankton, and its extinction coefficient at the red end of the spectrum is much lower (Table 3–2), thus making accurate analysis using trichromatic equations inaccurate (Lorenzen and Jeffrey, 1980). Parsons (1963), taking advantage of the relatively low solubility of chlorophyll *c* in nonpolar solvents, separated chlorophyll *c* by transferring the chlorophyll *a* present in extracts prepared in 90% acetone into a hexane-rich epiphase, leaving chlorophyll *c* in the hypophase of aqueous acetone. Spectroscopic analysis of the two phases confirmed that the separation was efficient. Through treatment of the hypophase with dilute HCl, the chlorophyll *c* was converted to pheoporphyrin *c*; the concentration of chlorophyll *c* was then estimated from the decrease in absorbance at 450 nm, using appropriate extinction coefficients for chlorophyll *c* and a calibration graph showing the change in absorbance at 450 nm when standard solutions of chlorophyll *c* were treated with HCl. However, chlorophyllide *a* will interfere with this method, and, as has been found in some but not all extracts from marine phytoplankton (Jeffrey, 1974, 1976b; Jeffrey and Hallegraeff, 1980b), this interference limits the value of the method for field work. Ricketts (1966a) used this method for purifying MgDVP, and Loeblich and Smith (1968) used it for estimating chlorophyll *c* in *Gyrodinium*.

B. *Fluorometry*

Early attempts to estimate chlorophylls by fluorescence measurements (Yentsch and Menzel, 1963; Holm-Hansen et al., 1965) relied on the specificity of the narrow-band emission filters used and gave approximate values only, though Holm-Hansen and Riemann (1978) continued to use this method. Loftus and Carpenter (1971) used a Turner III fluorometer with an excitation filter transmitting at 425 nm and three emission filters with maxima at approximately 620, 650, and 720 nm. Using each of the three emission filters, they calculated conditional molecular fluorescence coefficients for chlorophylls *a*, *b*, and *c* from pigments purified by TLC on silica gel using the system of Riley and Wilson (1965), or phase separation using *n*-hexane–90% acetone (Parsons, 1963) and presented a set of six simultaneous equations for the light emitted through the three filters before and after acidification. From these they calculated three simultaneous equations for calculating the concentrations of chlorophylls *a*, *b*, and *c*. Boardman and Thorne (1971) estimated chlorophyll *a*/*b* ratios using a sensitive fluorometer with a narrow bandwidth (±1.5 nm), but the method did not give absolute amounts of the chlorophylls, and the degradation products were presumed absent. Thorne et al. (1977) used this method for determining chlorophyll *a*/*b* ratios in *Prochloron*. White, Jones, Gibbs, and Butler (1972) reported a procedure similar to their earlier spectroscopic method (Sec. 3.4.1A) in that they used acidification, and phase separation with 0.01 M KOH with pigments dissolved in diethyl ether. They fitted a Turner model III fluorometer with a blue fluorescent lamp for greater sensitivity and used three primary and one or two secondary interference filters with narrow bandwidths. They measured relative fluorescence intensities for each of the pigments with the three primary filters and presented simultaneous equations based on fluorescence transmitted with the two appropriate primary filters before and after acidification for each of the chlorophylls (442 and 465 nm) and pheophytins (402 and 442 nm).

Boto and Bunt (1978), using a double monochrometer instrument, have also given equations based on specific coefficients of emission of the chlorophylls and phytinized derivatives at the three wavelengths of the maximum emission of the three chlorophylls *a*, *b*, and *c*; the method agreed well with spectroscopic methods for mixtures of chlorophylls *a* and *b*, but when chlorophyll *c* was present in addition, the estimates of chlorophyll *b* could be in error by 30% when the concentration of chlorophyll *b* was small. In presenting these equations, they assumed that the fluorescence of chlorophyllides *a* and *b* and pheophorbides *a* and *b* were similar to chlorophyll *a* and *b* and pheophytin *a* and *b* and thus would be included with them in the analysis. This method could no doubt be improved by using modern spectrophotofluorometers and the revised extinction coefficients for chlorophyll *c* (Jeffrey and Humphrey, 1975).

Schanz (1982) has used emission filters with maxima at 669.5 and 655.5 nm with a half-bandwidth of 10 nm to estimate chlorophyll *a* and pheophytin *a*, employing a modification of the method of Loftus and Carpenter (1971), in which the interference by fluorescence of chlorophyll *c* at 631 nm was estimated at no more than 5%. Equations were calculated with the use of mean molecular fluorescence coefficients, measured with the filters before and after acidification for estimating the concentrations of chlorophyll *a* and pheophytin *a*. Chlorophyll *b* could also be calculated in the absence of chlorophyll *c*. Tett and Wallis (1978), when studying the annual cycle of chlorophyll in algae in the brackish water (1–8 m) of Lake Ceran between 1970 and 1976, used the acidification technique of Holm-Hansen, Lorenzen, Holms, and Strickland (1965) in most years. They applied a correction factor adjusting estimates for algae passing the filters (Whatman GF/C glass-fiber), loss through grinding, calibration errors, and underestimation by the first fluorometer used. They found a recurring cycle of chlorophyll concentration with a winter low (less than 0.5 mg chlorophyll m^{-3}), rising to 12–37 mg in spring. Gibbs (1979), using a double-monochrometer spectrophotofluorometer, found that when chlorophyll *a* was converted by acid to pheophytin, the acidification factor decreased sharply when the acetone:water ratio was increased above 0.9, and recommended this concentration of acetone for the determinations. Chlorophyll *b* interfered seriously when chlorophyll *a* and pheophytin *a* were determined by this method.

Sakshaug and Holm-Hansen (1986) measured chlorophyll *a* and pheophytin *a* by the fluorometric method of Holm-Hansen et al. (1965) in cultures of phytoplankton collected in the Antarctic, using the results to measure chlorophyll/carbon ratios; these were very variable (0.004–0.018), and they have warned against calculating biomass by multiplying chlorophyll concentrations by ratios found in the literature. Venrick (1987) has found that many filters used to collect phytoplankton increase the background fluorescence when 90% acetone is employed as extractant; he has suggested that controls with an extract from the filter be utilized to standardize the instrument.

3.4.2. Chromatographic separation

Although techniques involving acidification may partially overcome interference by degradation products, increasing numbers of workers believe that only quantitative chromatography gives unambiguous results for estimating all chloroplast pigments, thus overcoming the need for estimating pigments in mixtures by spectroscopy.

A. *Paper chromatography*

Jeffrey (1961) appears to have been the first to have estimated chlorophylls in extracts from algae by quantitative chromatography. Using the system already described (Chapter 2, Sec. 2.2.3B), she cut spots containing chlorophylls from the dried chromatogram, eluted the pigments with acetone, and, using a spectrophotometer, measured the extinction at the appropriate wavelength. She determined the concentrations of chlorophyll *a*, *b*, and *c* in four classes of algae, though at that time the extinction coefficients of chlorophyll were in error. She would regard a subsequent correction of these values (Jeffrey, 1963) with a revised extinction coefficient as also in error (Jeffrey and Humphrey, 1975). She subsequently used paper chromatography for semiquantitative analysis of chlorophylls in algal blooms ranging from the intertidal zones to deep water (Jeffrey, 1965), thus giving an idea of the abundance of different classes of algae in these sites. Jeffrey (1968d) ran chromatograms in one dimension on small strips of kieselguhr-impregnated paper in 2–4% propan-1-ol in petroleum spirit (60–80%) when handling very small amounts of pigments present in samples of water from coral reefs, as the spots were more concentrated than when run by other methods. The levels of chlorophyll ($a + c$) were as low as 0.1 $\mu g \cdot l^{-1}$, and because these levels were calculated using the old extinction coefficients for chlorophyll *c*, the values must have been even lower. Hallegraeff (1976, 1977) used quantitative paper chromatography (Bauer, 1952) to calculate the contribution each pigment made to the absorption of light in an extract from phytoplankton in freshwater lakes, showing that absorption by degradation products of chlorophyll *a* sometimes equaled that by chlorophyll *a*. He found that the absorbance ratio E_{430}/E_{665}, advocated by Margalef (1968) for estimating the relative abundance of carotenoids and chlorophyll *a*, was often subject to serious interference by degradation products of chlorophyll and deviation of the absorbance at 430 nm from Beer's law at high concentrations.

Although most workers now use other forms of chromatography for quantitative analysis, Jensen and co-workers have perfected circular chromatography on paper impregnated with kieselguhr (Jensen, 1978). The circular lines containing each pigment were rapidly cut out after the chromatogram was developed and packed tightly into a glass tube drawn out at the end to a capillary. The pigments could then be eluted with a known volume of acetone or methanol, and the concentration estimated by spectrophotometry. Jensen and Sakshaug (1973) used this method for an ecological study of phytoplankton of the Trondheimsfjord, finding that degradation products of chlorophyll *a* appeared only at the bloom stage of the diatoms when chlorophyll concentration per cell decreased over ten fold. They pointed out that these measurements (i.e., concentration of chlorophyll and degradation products} provide valuable information about the stage of development of the algal populations.

B. *Thin-layer chromatography*

Separation of chlorophylls by TLC was discussed in Chapter 2 (Sec. 2.2.3C). For analyzing the concentrations of chlorophylls by this technique, the pigments from a known amount of tissue or number of cells must be applied to the layer and the resolved spot or line eluted efficiently into a known volume of a solvent in which the extinction coefficient of the chlorophyll is known (Table 3–2). The layer must not cause phytinization of the chlorophyll, as discussed already. Jeffrey (1968a) reviewed previous methods used for quantitative chromatography, dismissing most as causing degradation or giving unsatisfactory recovery. She described a system of separation by TLC on sucrose plates (following the good separations on sucrose columns in the Argonne laboratory under Strain [Chapter 2, Sec. 2.2.3A]), described already in Chapter 2 (Sec. 2.2.3C). She designed a suction device not unlike that of Riley and Wilson (1965) for removing and eluting pigments from the layer and reported recoveries of chlorophylls *a*, *b*, and *c*, chlorophyllide *a*, and pheophytin *a* of 95% or better. As already mentioned, cellulose plates run with 20% acetone in petroleum spirit (60–80°C) gave separations of degradation products of chlorophyll not achieved with sucrose, and chlorophylls c_1 and c_2 were separated only on polyethylene plates. She reported quantitative analysis of pigments from *Dunaliella, Phaeodactylum,* and *Amphidinium*. Jeffrey used the polyethylene plates for measuring ratios of chlorophyll $a/(c_1 + c_2)$ and chlorophyll c_1/c_2 in 24 marine algae.

Jeffrey (1974) used the cellulose plates rather than sucrose for estimating depth profiles of chlorophyll *a* at stations eight miles offshore on the east coast of Australia. Comparing measurements of chlorophyll *a* concentration made by direct spectroscopy on cell extracts in 90% acetone with those made by quantitative TLC showed considerable discrepancies at depths below 30 m at station 1 when pheophorbide *a* and pheophytin *a* could not be distinguished from chlorophyll *a* with the use of chromatic equations. An analysis of fecal pellets from copopods feeding on diatoms showed that almost all the chlorophyll *a* was converted to pheophorbide *a* though chlorophyll *c* remained intact. Chlorophyllide *a* was attributed to senescence of diatoms due to the action of chlorophyllase. She reverted to using sucrose plates (Jeffrey, 1976) for resolving pigments in water sampled in the Central North Pacific area at 5, 100, and 200 m depth. The concentrations of chlorophyll ($\mu g \cdot l^{-1}$) ranged from about 0.1 at 5 m down to 0.01 at 200 m; chlorophyll *b* was present, but the chlorophyll *b*/*a* ratios fell from 0.2–0.3 to less than 0.05 at depth 200 m. Chlorophyll *c* could also be found at all depths. These observations supported the hypothesis that algae were able to adapt to very low light intensities of blue light (Jeffrey and Vesk, 1977; Vesk and Jeffrey, 1977) and confirmed that degradation products of chlorophyll *a* gave spuriously high readings of chlorophyll *b*/*a* ratios (about 0.5) when measured by fluorometry.

Jeffrey and Vesk (1977) used quantitative TLC to measure chlorophyll/carotenoid ratios in their experiments on effects of light quality on pigment concentration. A refined technique using washed cellulose powder (Jeffrey, 1981) was described previously (Chapter 2, Sec. 2.2.3C). This technique is suitable for quantitative analysis of 0.15–1.0 μg of pigment per spot, using the sensitive spectrophotometers now available. If necessary, faint spots of chlorophyll or derivatives can be detected from their red fluorescence in UV light. Lipids from zooplankton interfered with the running of the chromatograms, and thus these animals should be removed before extracting the pigments if possible. Jeffrey and Hallegraeff have used this TLC method in Australian waters for semiquantitative analysis when determining chlorophyll profiles in warm-core eddies off Sydney (Jeffrey and Hallegraeff 1980a, 1987b), when determining relative abundance of chlorophyll and degradation products in water samples at a coastal station off Sydney at intervals down to 100 m throughout the year (Hallegraeff, 1981), and when determining classes of algae present by measuring ratios of chlorophyll *a* to chlorophylls *b* and *c*.

The reverse-phase TLC system of Daley et al. (1973a,b), used in conjunction with direct scanning of the chromatogram with a Turner fluorometer, could detect 1 ng of chlorophyll *a* in a spot. When the method was calibrated with pure compounds by plotting areas under the curves against concentration of pigments, the response was linear up to 0.3 μg of pigment. Calibration coefficients (the slope of area/μg pigment) were calculated for 16 chloropigments. The method required rigorous attention to a number of physical properties of the system, such as thickness of the TLC layer, the impregnating oil, dryness of plates, slit width of the fluorometer, and distance of development. Daley and Brown (1973) used this technique for measuring distribution of chlorophylls in phytoplankton in freshwater.

Although the method of Riley and Wilson (1965) for separating pigments on layers of silica gel G has been criticized as possibly causing phytinization of chlorophylls, the authors claimed that they could detect no degradation products by using this method. Garside and Riley (1969) employed this chromatographic system with a Joyce Loebel Chromoscan in the reflectance mode, using light passed through an Ilford No. 601 filter (maximum transmission, 430 nm), thus suitable for both chlorophylls and carotenoids. The chromoscan was calibrated by eluting the resultant spots into a known volume of solvent and measuring the content by spectroscopy using known extinction coefficients.

The TLC system used by Gowen et al. (1982) did not resolve chlorophillides and pheophorbides. For quantitative analysis, these pigments were eluted from the lower zone of the plates with 90% acetone; with the use of a Turner fluorometer with two filters, simultaneous equations based on measurements of fluorescence before and after acidification were constructed for estimating concentrations of chlorophillide *a*, chlorophyll *c*, pheophorbide *a*, and pheoporphyrin *c* (chlorophyll *b* would interfere with this method). They employed this method for estimating changes in dihydroporphyrin pigment in phytoplankton in two Scottish sea-lochs. Parallel analyses using a simple fluorometric method (Tett and Wallis, 1978), showed that the latter at the time of the spring bloom seriously overestimated chlorophyll *a* through failing to distinguish it from chlorophyllide *a*. Hager and Stransky (1970a–c) and Stransky and Hager (1970a,b), although primarily analyzing carotenoids, also measured chlorophylls by TLC in the ten classes of algae that they examined.

C. *High-performance liquid chromatography*

This method, which has been described in Chapter 2 (Sec.2.2.3D), will be a valuable technique in the future for workers interested in quantitative analysis of chlorophyll and derivatives in phytoplankton. Either peak heights from the traces from the chromatograms (Abaychi and Riley, 1979; L. M. Brown et al., 1981; Gieskes and Kraay, 1983a; Bidigare et al., 1985; Vernet and Lorenzen, 1987) or areas under the curves (Braumann and Grimme, 1981; Falkowski and Sucher, 1981; Gieskes and Kraay, 1983b; Mantoura and Llewellyn, 1983; Paerl et al., 1984; Kleppel and Pieper, 1984; Bowles et al., 1985; Korthals and Steenbergen, 1985; Vernet and Lorenzen, 1987) have been used, most workers using some form of integration unit. Employing fluorometers, several workers have found the limits of detecting chlorophyll *a* at about 10 pg (Mantoura and Llewellyn, 1983; Bidigare et al., 1984; Sartory, 1985). Quantitative methods employing peak heights or area usually require calibration with authentic standards, though in some procedures the chlorophyll has been estimated from the specific extinction coefficient by integration of the absorbance (Korthals and Steenbergen, 1985).

Several comparisons between HPLC and spectrophotometric or fluorometric analysis have been made. Abaychi and Riley (1979) compared quantitative analysis of pigments extracted from three species of algae by HPLC, TLC, and chromatic equations. Results by TLC (the reflectometer method of Garside and Riley, 1969) were similar to those by HPLC, but chromatic equations were inaccurate in that they gave negative values for chlorophyll *b* in *Phaeodactylum* and *Oscillatoria,* and positive values for chlorophyll *c* in *Dunaliella* and *Oscillatoria* where none should be present. Brown et al. (1981) compared analysis by HPLC, spectroscopy with acidification (Holm-Hensen and Riemann, 1978), and fluorometry (Holm-Hansen et al., 1965). The values for chlorophyll *a* by all methods were similar with acetone as extracting solvent but not methanol, whereas for analysis of pheophytin, HPLC differed significantly from the other two methods. Jacobsen (1982) found statistical differences for chlorophyll *a* and pheophytin *a* when comparing results by HPLC and the spectrophotometric and fluorometric methods of Strickland and Parsons (1968).

Gieskes and Kraay (1983a,b) found that concentrations of chlorophyll *a* analyzed in extracts from cultures of *Rhodomonas*, *Isochrysis*, and *Fucus* were similar when analyzed by HPLC, fluorometry with acidification (Holm-Hansen et al., 1965), and spectrophotometry (Lorenzen, 1967), but fluorometric estimates of chlorophyll *a* in extracts from natural phytoplankton were often more than double those made by HPLC, considered due to chlorophyll-like pigments not moving like chlorophyll *a* in HPLC.

Mantoura and Llewellyn (1983) compared analyses of chlorophylls *a*, *b*, and *c* in phytoplankton by HPLC and five spectroscopic methods, finding values by HPLC only about one-tenth of those by the other methods, though large amounts of chlorophyll-like pigments did not appear on the chromatograms, as might be expected. Sartory (1985) found spectroscopic methods gave values for chlorophyll *a* in extracts from cultures little higher than with HPLC, but chlorophyll *b* was often much higher and pheophytin sometimes over tenfold higher. In extracts from natural phytoplankton, values for chlorophyll *a* were similar by both methods, though pheophytin *a* was much lower by HPLC. Trees, Bidigare, and Brooks (1986) measured chlorophylls and degradation products in samples from the northwest Atlantic Ocean, and interpreted them as changes in taxonomic composition, biomass, and grazing pressures by zooplankton. Murray, Gibbs, Longmore, and Flett (1986) have developed a rapid isocratic method in place of the "full" method (Table 2–6) for estimating chlorophyll. They found that results using both HPLC methods agreed well with fluorometric and spectroscopic methods, providing chlorophyllide *a* and allomers of chlorophyll *a* were considered.

HPLC now seems the most commonly used method for ecological work, either for examining algal blooms (Paerl, Tucker, and Bland, 1983; Paerl, 1984) or for distribution of algae by type and quantity at different depths (Gieskes and Kraay, 1983b; Korthals and Steenbergen, 1985), and is also used for identifying chlorophyll *b* and *c* in investigations of phylogeny (Farnham et al., 1985; Burger-Wiersma et al., 1986) and in detecting the types of symbiont present in marine organisms (Gieskes and Kraay, 1983b; Paerl et al., 1984; Knight and Mantoura, 1985).

3.4.3 Conclusions

The methods chosen for estimating chlorophylls in algae will depend in particular on whether the material is a healthy culture growing in the laboratory, when any method described in Section 3.4 may be used, or a natural sample, possibly containing dead and senescent cells and detritus, when some form of chromatography is required so that interference by numerous forms of degradation products of chlorophyll or other pigments may be avoided.

4

The carotenoids

4.1. Distribution and pathways of synthesis

Unlike the chlorophylls, the carotenoids are very numerous, and over 400 are known in both plants, fungi, and animals. In plants, the main interest is in determining their function in the cell; whereas some are concerned with light harvesting and are bound to the pigment–protein complexes in the thylakoid membranes (Chapter 6), some are also believed to protect chlorophylls against photooxidation (Paerl, 1984; Siefermann-Harms, 1987). Unlike those in other classes, the carotenoids of the Cyanophyceae may be numerous in a given species and are not all found in the thylakoid membranes. The carotenoids are of two types, the carotenes and the xanthophylls. The carotenes are hydrocarbons and are few in number; the xanthophylls contain at least one atom of oxygen in the molecule and make up the vast majority of the carotenoids.

4.1.1. Biosynthesis in algae

The early steps of carotenoid biosynthesis from mevalonic acid to lycopene are well known and have been outlined by Goodwin (1971b), Spurgeon and Porter (1980), Schutte (1983), and Bramley (1985); these authors have also summarized work on the mechanisms of cyclization of lycopene to form the β and ϵ rings found in most algal carotenoids. It is also well established that the oxygen atoms of the hydroxy and epoxy groups of the xanthophylls are derived from atmospheric oxygen. Liaaen-Jensen (1978) has proposed hypothetical pathways for the biogenesis of algal carotenoids, including alternate mechanisms possible in triple bond formation. Schütte (1983) has also shown a scheme for formation of the common algal carotenoids from lycopene. Few attempts have been made to demonstrate synthesis of xanthophylls by using enzymic systems from algae. Swift, Milborrow, and Jeffrey (1982) demonstrated incorporation of [^{14}C]zeaxanthin into neoxanthin and then into peridinin and diadinoxanthin; this synthesis of neoxanthin and diadinoxanthin is inconsistent with the schemes proposed for formation of allenic and triple bonds by Liaaen-Jensen (1978), which start with a 3-hydroxy-5,6-epoxide rather than zeaxanthin and neoxanthin, respectively.

4.1.2. Distribution in the algal class

In the past, reviewers have listed the major carotenoids found in the different classes (Goodwin, 1965, 1974, 1976, 1979, 1980; Stransky and Hager, 1970c; Boney, 1975; Kirk and Tilney-Bassett, 1978; Spurgeon and Porter, 1980; Weber and Wettern, 1980; Larkum and Barrett, 1983). Table 4–1 shows the carotenoids found regularly or occasionally in the various classes of algae, and whether they are predominant or minor, or occur only as traces. More recently, Ragan and Chapman (1978) and Liaaen-Jensen (1977, 1978, 1979) have preferred to consider the synthetic pathways leading to synthesis of carotenoids found in the different classes.

4.1.3. Distribution and pathways of synthesis in the algal classes

Of the pathways of carotenoid synthesis in the various classes of algae listed by Liaaen-Jensen (1977), the following seem most important as phylogenetic markers:

Group 1: ϵ-cyclization, 5,6-epoxidation, allene formation, triple bond formation, 19-hydroxylation, 8-keto formation, acetylation, allylic 3-hydroxylation

Group 2 (too general to be useful): β-cyclization, 3-hydroxylation

Group 3 (more restricted than the first group and valuable in assigning a species to one or, perhaps, one of two classes): glycosylation, allylic 4-hydroxylation and 4-keto formation, *trans*-2-hydroxylation, butenolide formation, C_3 expulsion, 5,6-glycol formation, acylation, and 2-hydroxylation

Liaaen-Jensen (1985) has reported rare and unusual xanthophylls requiring other synthetic reactions than those above, but, as they are known so far in few species, they are not yet important phylogenetically.

The following generalizations about the pathways found in the algal classes may be made:

1. The prokaryotic classes (Cyanophyceae and Prochlorophyceae) do not carry out ϵ-cyclization nor do they form epoxides; usually they form glycosides, rarely found in the other classes.

Table 4–1. *The distribution of carotenoids in the algal classes*

Carotenoid (Trivial name)	Cyanophyceae	Prochlorophyceae	Rhodophyceae	Cryptophyceae	Dinophyceae	Chrysophyceae & synurophyceae	Prymnesiophyceae	Bacillariophyceae	Tribophyceae	Eustigmatophyceae	Phaeophyceae	Euglenophyceae	Prasinophyceae	Chlorophyceae & charophyceae
Alloxanthin	–	–	–	H	–	–	–	–	–	–	–	–	–	–
Anhydrodiatoxanthin	–	–	–	–	–	–	–	–	–	–	–	t	–	–
Antheraxanthin	–	–	H	–	–	t	–	t	L	L	L	–	–	H t
Aphanizophyll	L	–	–	–	–	–	–	–	–	–	–	–	–	–
Astaxanthin	–	–	–	–	L	–	–	–	–	L	–	–	–	L
Aurochrome	–	–	L	–	–	–	t	–	t	–	–	–	–	–
Auroxanthin	–	–	L	–	–	–	–	–	–	t	–	–	–	–
Calloxanthin	H L	–	–	–	–	–	–	–	–	–	–	–	–	–
Canthaxanthin	H L	–	–	–	–	–	–	–	–	L	–	–	–	–
α-Carotene	–	–	H	L	–	L	t	–	–	L	L	–	LL t	H

β-Carotene	H L	H	H	–	L	H L	L	H	H	H L	H L	L	H	H L
γ-Carotene	t	–	t	–	L	–	–	–	–	–	t	–	L	L
ϵ-Carotene	–	–	–	t	–	–	–	t	–	–	L	–	–	t
β-Carotene-5,6-epoxide	–	t	–	–	–	–	t	–	–	–	–	–	–	–
Crocoxanthin	–	–	–	L	–	–	–	–	–	–	–	–	–	–
α-Cryptoxanthin	–	–	H L	–	–	–	–	–	t	–	–	–	–	–
β-Cryptoxanthin	L	L	L	–	–	t	–	t	t	–	–	t	–	H t
Cryptoxanthin diepoxide	–	–	–	–	–	–	L	–	L	L	–	–	–	–
Cryptoxanthin epoxide	–	–	–	–	–	t	–	–	L	L	–	t	–	t
Diadinoxanthin	–	–	–	–	H L	LH L		–	H	–	tH		–	–
Diatoxanthin	–	–	–	–	L	–	H L	–	H L	–	HH LL		–	–
Dinoxanthin	–	–	–	–	H L	–	t	–	–	–	–	–	–	–
Echinenone	H L	L	?t	–	–	–	t	–	–	t	–	–	–	–
Eutreptiellanone	–	–	–	–	–	–	–	–	–	–	–	H	–	–
Fucoxanthin	–	–	?H L	–	H	HH	H		–	–	H	–	–	–
Fucoxanthinol	–	–	?L	–	–	L t	t	L t	–	–	H t	–	–	–
Gyroxanthin	–	–	–	–	L	–	–	–	–	–	–	–	–	–
Heteroxanthin	–	–	–	–	–	–	–	–	H	–	–	L	–	–
Hexadehydro-β-caroten-3-ol	–	–	–	–	–	–	–	–	L	–	–	L	–	–

Table 4–1. (*Cont.*)

Carotenoid (Trivial name)	Cyanophyceae	Prochlorophyceae	Rhodophyceae	Cryptophyceae	Dinophyceae	Chrysophyceae & synurophyceae	Prymnesiophyceae	Bacillariophyceae	Tribophyceae	Eustigmatophyceae	Phaeophyceae	Euglenophyceae	Prasinophyceae	Chlorophyceae & charophyceae
19′-Hexanoyloxyfucoxanthin	–	–	–	–	H	–	H	–	–	–	–	–	–	–
3′-Hydroxyechinenone	L t	–	–	–	–	–	–	–	–	–	–	–	–	–
Isocryptoxanthin	L	L	–	–	–	–	–	–	–	–	–	–	–	–
Isozeaxanthin	L t	–	–	–	–	–	–	–	–	–	–	–	–	–
4-Keto-myxol-2′-*O*-methyl-methylpentoside	L	–	–	–	–	–	–	–	–	–	–	–	–	–
Loroxanthin	–	–	–	–	–	–	–	–	–	–	–	–	–	L
Lutein	–	–	H	–	–	–	–	–	–	–	–	–	H L	H
Lycopene	L	–	–	t	–	–	–	–	–	–	–	–	–	a
Monadoxanthin	L	–	–	L	–	–	–	–	–	–	–	–	–	–
Myxol-2′-*O*-methyl-methylpentoside	H	–	–	–	–	–	–	–	–	–	–	–		–

Myxoxanthophyll	H L	–	–	–	–	–	–	–	–	–	–	–	–	–
Neoxanthin	L –	–	?t	–	–	H L	–		t	t	t	L	LH L	L
Nostoxanthin	L	–	–	–	–	–	–	–	–	–	–	–	–	–
Octadehydro-β-carotene	H	–	–	–	–	–	–	–	–	–	–	–	–	–
Oscillaxanthin	L	–	–	–	–	–	–	–	–	–	–	L	–	–
Oscillol	t	–	–	–	–	–	–	–	–	–	–	–	–	–
Peridinin	–	–	?L	–	H	–	–	–	–	–	–	–	–	–
Peridinol	–	–	–	–	L	–	–	–	–	–	–	–	–	–
Prasinoxanthin	–	–	–	–	–	–	–	–	–	–	–	–	–	–
Pyrrhoxanthin	–	–	–	–	L	–	–	–	–	–	–	–	H	–
Siphonaxanthin	–	–	–	–	–	–	–	–	–	–	–	–	H L	H
Siphonein	–	–	–	–	–	–	–	–	–	–	–	H	L	H
Taraxanthin	–	–	?H L	–	–	–	–	–	–	–	–	–	–	H L
Vaucheriaxanthin/ester	–	–	–	–	–	–	–	–	H	H	–	–	–	–
Violaxanthin	–	–	?L	–	–	L	–	L	–	H	H L	–	L	L
β-Zeacarotene	t	–	–	–	–	–	–	–	–	–	–	–	–	–
Zeaxanthin	H L	H	H L	t	–	H L	–	H L	L	H L	H	L	L	L

Abbreviations: H = high concentration; L = low concentration; t = trace amount. Letters to the left of each column: regular occurrence. Letters to the right of each column: occasional occurrence. a = high concentration in antheridia (*Chara*).

2. The Rhodophyceae carry out both β- and ϵ-cyclization but rarely form epoxides.
3. The Cryptophyceae, like the Rhodophyceae, carry out both cyclizations but never form epoxides or allenes; they form xanthophylls with one or two acetylenic bonds.
4. The classes of the Chromophyta (sensu Cavalier-Smith, 1982) can form allenic bonds, and carry out hydroxylation at C-19 and oxidation at C-8.
5. The Chlorophyta (except Euglenophyceae) cannot synthesize acetylenic bonds.

4.1.4. Comments on the pathways found in the algal classes

A. Division Prokaryota

Unlike the Eukaryota, the members of this division rarely carry out ϵ-cyclization or form epoxides, though small amounts of mutatochrome are sometimes found. β-Carotene and echinenone are the only carotenoids common to most species of the division (Goodwin, 1974, 1980; Liaaen-Jensen, 1978).

Cyanophyceae. Members of this class contain a wide range of pigments; 24 are listed by Goodwin (1980) for the whole class, but usually a given species contains no more than seven or eight, though Britton (unpublished) has found 14 in *Chlorogloea fritschii,* and in preliminary experiments in this laboratory we have found at least 12 in *Rivularia firma*. Cyanophyceae usually form glycosidic esters, the most common being myxoxanthophyll, only absent from some strains of *Oscillatoria, Phormidium,* and *Spirulina* in the list published by Goodwin (1980), and from *Oscillatoria bornetti* (Hallenstvet, Liaaen-Jensen, and Skulberg, 1979). β-Carotene and echinenone have been reported in all species; Stransky and Hager (1970b) did not report echinenone in *Anacystis nidulans,* but Halfen and Francis (1972) and Francis and Halfen (1972) found it present, the first paper reporting increases in concentration of most carotenoids with increasing temperature during culture. Thus the glycosidic esters and echinenone are valuable indicators for the Cyanophyceae. Although epoxyxanthophylls are said to be absent from this class, Hertzberg and Liaaen-Jensen (1967) found mutatochrome (= flavacin), a 5,8-epoxide, in small quantities in a few species.

Prochlorophyceae. This small group of prokaryotes, originally described as the genus *Prochloron* Lewin (Lewin, 1976, 1977), is unique in the Prokaryota in containing chlorophyll *b*, and has only recently been cultured in vitro when Patterson and Withers (1982) discovered that it was a tryptophan auxotroph, though Paerl et al. (1984) could not duplicate this culture. Withers et al. (1978a,b) and Foss, Lewin, and Liaaen-Jensen (1986) have found β-carotene to be the major carotene and zeaxanthin the major xanthophyll. Both groups found other xanthophylls typical of the Cyanophyceae (isocryptoxanthin, echinenone, and β-cryptoxanthin), but Foss et al. (1986) found traces of 5,6-epoxy β-carotene, a possible precursor of mutatochrome, no trace of which had been previously reported in the Cyanophyta. The carotenoid glycosides, particularly myxoxanthophyll, have not been found in the Prochlorophyceae. Recently, Burger-Wiersma et al. (1986) described a filamentous prokaryote containing chlorophyll *b* (Chapter 3, Sec. 3.1.2B); zeaxanthin was the only typical prokaryote pigment identified in it so far. Unknowns included one thought to be canthaxanthin (though from the R_f value of HPLC, it might be echinenone) and two said to be derivatives of zeaxanthin.

B. Division Eukaryota

The diversity of important pathways listed in Section 4.1.3 are found mainly in the Eukaryota and account for the distinctive pattern of pigments found in the classes of that division. The 5,6-epoxides found in most classes arise by the activity of a mixed-function oxidase (Hager, 1980); this is one of the few biological reactions in which atmospheric oxygen is incorporated into an organic compound. The cycles described by Hager (1980) (Fig. 4–1) are:

Zeaxanthin ⇌ antheraxanthin ⇌ violaxanthin
Diatoxanthin ⇌ diadinoxanthin

Low light promotes epoxidation and high light deepoxidation. A third cycle, lutein ⇌ lutein 5,6-epoxide (taraxanthin), is less well known but apparently replaces the regular cycle of the Chlorophyta in a strain of *Dunaliella* (see Chlorophyceae, below).

Some of the classes of the Eukaryota contain xanthophylls specific to a single class (peridinin, siphonaxanthin, or alloxanthin) or to a limited number (fucoxanthin, heteroxanthin, vaucheriaxanthin), lending support to classifying species in conjunction with other properties.

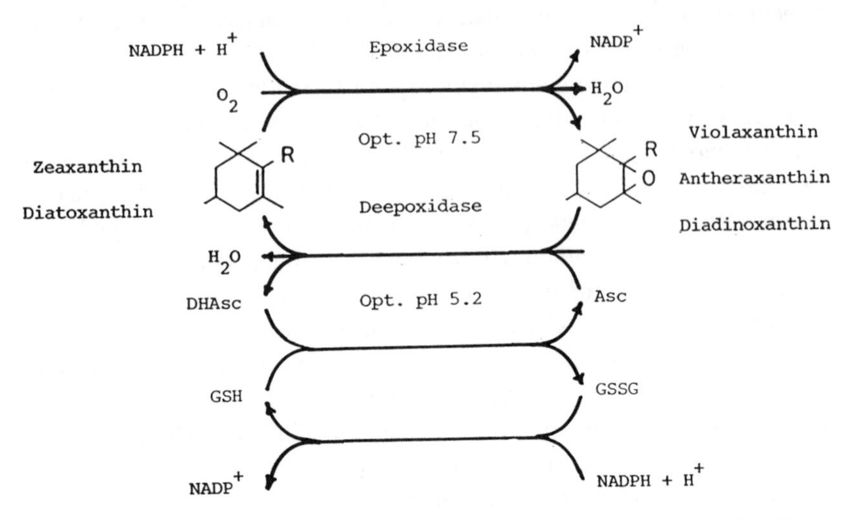

Fig. 4–1. The reactions of the xanthophyll cycle as proposed by Stransky and Hager (1970c) and Hager (1980). DHAsc/Asc, oxidized and reduced ascorbic acid; GSSG/GSH, oxidized and reduced glutathione.

Rhodophyceae. Species of this class, except for six (Goodwin, 1980), usually carry out ϵ-cyclization, and the common pigments are α- and β-carotene and their 3-hydroxy derivatives, lutein, and zeaxanthin (Strain, 1958, 1966). Liaaen-Jensen (1978) listed nine other xanthophylls found occasionally, but, as material from monoalgal culture has rarely been used, warned that fucoxanthin in particular could arise from contaminating diatoms. Bjørnland and Aguilar-Martinez (1976) found no neoxanthin or fucoxanthin in cultured, as opposed to natural material of three species, though so much was found in *Nemalion helmenthoides* that they believed it could not have come from contaminating diatoms. Whether or not members of the nine xanthophylls listed by Liaaen-Jensen (1978) as occurring occasionally are due to contamination requires testing on cultured material. So far, the only identifications from cultured material are for α- and β-cryptoxanthin and antheraxanthin (Bjørnland and Aguilar-Martinez, 1976; Bjørnland, 1983, 1984; Bjørnland et al., 1984a), apart from lutein and zeaxanthin.

Cryptophyceae. In addition to the ϵ-cyclase system, the Cryptophyceae synthesize acetylenic carotenoids, but not epoxy xanthophylls. The diacetylenic dihydroxyxanthophyll, alloxanthin, is a major component and is restricted to this class, thus being a useful taxonomic marker; the monoacetylenic xanthophylls found include the monohydroxy crocoxanthin and the dihydroxy monadoxanthin. These acetylenic pigments were first identified and named by Chapman (1966a), who differentiated them from spectrally similar diatoxanthin, α-cryptoxanthin, and lutein, respectively. Although earlier workers (Haxo and Fork, 1959; Allen, Fries, Goodwin, and Thomas, 1964) could not detect β-carotene, it is now usually found (Chapman and Haxo, 1963; Pennington et al., 1985), though it seemed to be absent from *Chroomonas salina* (Norgard et al., 1974b). The rare carotenes, ϵ- and ψ-carotene, have been reported by Chapman and Haxo (1963). Pennington et al. (1985) also found traces of lycopene and zeaxanthin in *Cryptomonas ovata*. Manixanthin, a cis isomer of alloxanthin, can be induced by controlling the conditions used for culturing the alga (Cheng et al., 1974).

Dinophyceae. In the majority of the Dinophyceae the major carotenoid is the complex C_{37} peridinin, whose structure was derived by Strain et al. (1976; see also Liaaen-Jensen, 1978; Goodwin, 1980). This pigment requires butenolide formation, C_3 expulsion, and formation of allenic, epoxy, and acetaloxy groups. Although traces of peridinin are occasionally reported in other classes, the pigment seems highly specific to the Dinophyceae. Pyrrhoxanthin, an acetylenic C_{37} with a similar butenolide and epoxy group, was reported by Loeblich and Smith (1968) as a minor pigment in *Gyrodinium resplendens,* and the alcohols of the two pigments, peridinol and pyrrhoxanthinol, were detected in low concentration by Johansen, Svec, Liaaen-Jensen, and Haxo (1974), the latter pigment only in *G. dorsum*.

Diadinoxanthin, in common with most of the Chromophyta, is present as the second most abundant xanthophyll, and the amounts of the minor xanthophylls are given by Johansen et al. (1974) and listed by Liaaen-Jensen (1978) and Goodwin (1979, 1980). β-Carotene is the major carotene (but see below). α-Carotene is shown in error in Table 9 of Liaaen-Jensen (1978) for γ-carotene. The distribution of diatoxanthin, the deepoxide of diadinoxanthin, is variable; Johansen et al. (1974) found it only in the species lacking peridinin (see below); however, the amount of the pigment probably reflects the reversible equilibrium between the epoxide, diadinoxanthin, and the reduced form, diatoxanthin (see Sec. 4.1.4B). The recent study of carotenoids in zooxanthellae from clams, corals, and nudibranchs by modern methods (Skjenstad, Haxo, and Liaaen-Jensen, 1984) confirmed that their pigments were similar to those of the majority of the free-living dinoflagellates (Taylor, 1967; Jeffrey and Haxo, 1968; Jeffrey et al., 1975).

The majority of the dinoflagellates are heterotropic, sometimes feeding on smaller autotropic dinophytes whose pigments persist after ingestion (Larsen, 1988). Balch and Haxo (1984) have examined the spectral properties of the heterotroph, *Noctiluca melitaxis,* pointing out that surveys of productivity of oceanic water could be distorted if based on measuring photosynthetic pigments.

Some Dinophyceae contain fucoxanthin (or derivatives) rather than peridinin, widely believed to occur through endosymbiosis between a heterotrophic flagellate and a unicellular alga containing fucoxanthin. Steidinger and Cox (1980) have discussed the origin of the chloroplasts of these abnormal species, originally found with a eukaryotic nucleus in addition to the normal mesokaryotic nucleus typical of the Dinophyceae (Dodge, 1971, 1975; Tomas and Cox, 1973a, b; Tomas et al., 1973; Loeblich, 1976). However, not all Dinophyceae containing fucoxanthin have two nuclei (Loeblich, 1976) and some contain derivatives of fucoxanthin (Table 4–2). *Gyrodinium aureolum* and *Gyrodinium* sp. A contain 19′-hexanoyloxyfucoxanthin (Bjørnland and Tangen, 1979; Tangen and Bjørnland, 1981), thus suggesting affinity with the prymnesiophyte, *Emiliania huxleyi,* in which this derivative is found (Norgard et al., 1974b; Arpin, Svec, and Liaaen-Jensen, 1976). However, Kite and Dodge (1985), on the basis of the type of plastid DNA found, have suggested that whereas the endosymbiont of *Gymnodinium foliaceum* could be a diatom, the endosymbiont of *G. aureolum* could not be of the same type as *E. huxleyi* as both contained different plastid DNA. The Trondheim group (Bjørnland and Tangen, 1979; Liaaen-Jensen, 1985) (see Table 4–2) have recently found four other pigments chemically related to fucoxanthin, in *Gyrodinium* sp. A and *Gymnodinium breve*. *Peridinium balticum* (Withers, Cox, Tomas, and Haxo, 1977) and *Gymnodinium nagasakiensis* (Larsen and Rowan, unpublished) contains fucoxanthin, but, in addition, a pigment spectrally similar to fucoxanthin and 19′-hexanoyloxyfucoxanthin, but running with a higher R_f value on TLC. The dinophyte strain Y-100, with a nucleus apparently derived from a symbiont containing chlorophyll *b*, does not contain either peridinin or fucoxanthin (Watanabe et al., 1987).

Chrysophyceae and Synurophyceae. Christensen (1962) and Hibberd (1976) split off the Prymnesiophyceae (Haptophyceae) from the original members of this class as defined by Pascher in 1914, leaving a small group as the present Chrysophyceae. Few members of the new group of Chrysophyceae had been examined until Withers et al. (1981) analyzed pigments from six genera. Fucoxanthin was invariably present in large

Table 4–2. *Dinophyceae containing fucoxanthin and fucoxanthin-like pigments in place of peridinin*

Pigment	Species	Reference
Fucoxanthin	*Gymnodinium veneficum*	Riley & Wilson (1967)
	G. micrum	Whittle & Casselton (1968)
	G. sp.	
	G. nagasakiensis	Larsen & Rowan (unpublished)
	Peridinium foliaceum	Mandelli (1968)
		Jeffrey et al. (1975)
		Withers & Haxo (1975)
	P. balticum	Tomas & Cox (1973a,b)
		Jeffrey et al. (1975)
		Withers et al. (1977)
Fucoxanthin-like	*P. balticum*	Withers et al. (1977)
	G. nagasakiensis	Larsen & Rowan (unpublished)
19′-Hexanoyloxyfucoxanthin	*Gyrodinium* sp. A	Bjørnland & Tangen (1979)
	Gymnodinium breve	Liaaen-Jensen (1985)
Fucoxanthin acetate	*Gyrodinium* sp. A	Liaaen-Jensen (1985)
19′-Butanoyloxyfucoxanthin	*Gymnodinium breve*	Liaaen-Jensen (1985)
Gyroxanthin	*Gyrodinium* sp. A	Liaaen-Jensen (1985)

amounts and β-carotene was always found, though usually under 5% of total carotenoids. The carotenoids occasionally found included five epoxides, with neoxanthin as 16% in one species of *Ochromonas*. Violaxanthin, one of the occasional epoxides, never occurs in the Prymnesiophyceae, and Craigie, Leigh, Chen, and McLachlan (1971) cited this finding to justify placing *Phaeosacceon callinsii* in the Chrysophyceae.

A small number of marine species, *Pelagococcus subviridis* (Lewin et al., 1977), *Rhizochromalina marina* (Hibberd and Chretiennot-Dinet, 1979), and *Sarcinochrysis marina* (Withers et al., 1981), contain the acetylenic diadinoxanthin, suggesting links to other classes of the Chromophyta, but these cannot be considered Prymnesiophyceae on anatomical evidence. *P. subviridis,* an aberrant member of this class (Chapter 3, Sec. 3.1.2C, Chrysophyceae), contains, in addition, two pigments chemically similar to fucoxanthin, 19′-hexanoyloxyfucoxanthin, and a 19′-butanoyloxyfucoxanthin-like pigment (Lewin et al., 1977; Vesk and Jeffrey, 1987; Wright and Jeffrey, 1987); the latter was not fully characterized and, from its R_f value on TLC plates, appeared to be too polar to be 19′-butanoyloxyfucoxanthin (Wright and Jeffrey, 1987).

Gibbs et al. (1980) summarized evidence based on xanthophyll content that *Olisthodiscus luteus* should be placed in the Chrysophyceae, not the Chloromonadophyceae or Tribophyceae, because the latter classes do not contain fucoxanthin (Chapman and Haxo, 1966), whereas it is the major carotenoid in the Chrysophyceae (Hager and Stransky, 1970c); Withers et al. (1981) have confirmed that fucoxanthin occurs as a major xanthophyll in two strains of this species. Fiksdahl et al. (1984c) have found peridinin rather than fucoxanthin when reexamining *Heterosigma akashiwo,* thus relocating it in the Dinophyceae, though Loeblich and Loeblich (1973) reported that this species contained fucoxanthin rather than peridinin. The carotenoids of the Synurophyceae are similar to those of the Crysophyceae, from which this class has recently been split (Andersen, 1987).

Prymnesiophyceae (Haptophyceae). As mentioned above, this class was recently split from the Chrysophyceae, but, as Withers et al. (1981) showed, cannot be separated by identifying carotenoids, because overlap of the acetylenic xanthophylls occurs in the two classes. However, violaxanthin has never been reported in the Prymnesiophyceae and, as it occurs occasionally in the Chrysophyceae, may be used as evidence for inclusion in the latter class (Craigie et al., 1971). Liaaen-Jensen (1978) gives β-carotene, diatoxanthin, diadinoxanthin, and fucoxanthin as the invariable carotenoids, with ten others as occasional or minor. These include the derivatives of fucoxanthin, 19′-hexanoyloxyfucoxanthin and 19′-butanoyloxyfucoxanthin, discussed in the previous section. The former has been reported several times in *Emiliania huxleyi* (Norgard et al., 1947b; Arpin et al., 1976; Hertzberg et al., 1977; Wright and Jeffrey, 1987); the latter authors have found it in *Phaeocystis pouchetii,* but not in *Pavlova lutheri*. A 19′-butanoyloxyfucoxanthin-like pigment occurs in *P. pouchetii*. Other members of the Isochrysidiales are yet to be examined to determine whether these pigments are common to the order. Although Tangen and Bjørnland (1981) could not separate fucoxanthin from 19′-hexanoyloxyfucoxanthin by TLC, Wright and Jeffrey (1987) separated fucoxanthin from both 19′-hexanoyloxyfucoxanthin and the 19′-butanoyloxyfucoxanthin-like pigment, using the new RP-8 TLC plates (Table 2–5) and the HPLC system of Wright and Shearer (1984). This class differs from the Chrysophyceae in its invariable synthesis of acetylenic linkages and occasional ϵ-cyclization.

Bacillariophyceae (diatoms). The pathways important in this class are epoxidation, and 8-keto and acetylene formation; thus the important pigments are similar to those of the Prymnesiophyceae (see above) – that is, β-carotene, fucoxanthin, diadinoxanthin, and diatoxanthin. Goodwin (1971, 1980) has tabulated earlier reports describing this distribution, and Jeffrey and Stauber (1985) have found it in 51 species, along with minor carotenoids. Strain, Manning, and Hardin (1944) found ε-carotene in *Nitzschia clorobacterium,* and Hager and Stransky (1970c), neoxanthin. The latter pigment is a probable precursor of fucoxanthin. *Thalassiothrix heteromorpha* is unusual in containing 19′-butanoloxyfucoxanthin in addition to fucoxanthin (Stauber and Jeffrey, 1988).

Tribophyceae (Xanthophyceae). A group of genera have been removed from the Tribophyceae into a new class, the Eustigmatophyceae, by Hibberd and Leedale (1971, 1972), and the distribution of chloroplast pigments also falls into these two groups (Whittle and Casselton, 1969, 1975a,b). The redistribution involved separation of two species of *Pleurochloris*. Heteroxanthin is a major pigment in all species of the Tribophyceae and is unusual in being an acetylenic tetrol containing a 5,6-glycol (Strain et al., 1970; Buchecker and Liaaen-Jensen, 1977), found only outside this class in small amounts in the Euglenophyceae. Another unusual pigment is vaucheriaxanthin, the 19′-OH of neoxanthin, shared with the Eustigmatophyceae, and usually found only as an ester, probably a diester (Stransky and Hager 1970b). In some earlier reports, this pigment was not reported (Goodwin, 1980). The other major pigment is diadinoxanthin, accompanied by the reduced form, diatoxanthin, found in variable amounts within a given species in our experience in this laboratory. Stransky and Hager (1970b) found that in light, diatoxanthin increases through deepoxidation of diadinoxanthin (Sec. 4.1.4); thus the environment before extraction may determine whether or not diatoxanthin is present in detectable amount. Other xanthophylls present occasionally or in small quantity are all epoxides (Liaaen-Jensen, 1978). The identification of zeaxanthin in *Botrydium granulatum* by Egger, Nitsche, and Kleinig (1969) was not confirmed by Hager and Stransky (1970b). Thus the Tribophyceae synthesize ε-rings, epoxides in high proportion, triple bonds, allenes, acetates, 5,6-glycols, and 19-OH groups. In this laboratory, Sullivan, Entwisle, and Rowan (1989) have found this distribution of carotenoids described above in 27 species of the Tribophyceae from six genera.

Eustigmatophyceae. As mentioned in the previous section, pigments clearly separate this class from the Tribophyceae (see Chapter 3, Sec. 3.1.2B and C), split from it by Hibberd and Leedale (1971, 1972) on the basis of differences in cytology. Although xanthophyll pigments cannot be used alone to transfer species to this class, pigment content of *Nannochloris oculata* and *Monallantus salina* prompted reexamination of their ultrastructure by Antia et al. (1975), who recommended their transfer to the Eustigmatophyceae from the Chlorophyceae. Hibberd (1981) subsequently transferred both species to the new genus, *Nannochloropsis*. The Eustigmatophyceae can contain the pigments of the xanthophyll cycle – zeaxanthin, antheraxanthin, and violaxanthin (Stransky and Hager, 1970a) – and, in common with the Tribophyceae, contain vaucheriaxanthin, usually as an ester, and, sometimes, the chemically related neoxanthin, but do not contain chlorophyll *c* (Chapter 3, Sec. 3.1.2C). Norgard et al. (1974a) found the keto carotenoid, canthaxanthin, in three species of the class, but Whittle and Casselton (1975b) did not report it. Antia and Cheng (1982) found minor amounts of keto carotenoids, including astaxanthin ester, echinenone, and canthaxanthin in cultured *Nannochloropsis oculata* and *N. salina,* and the canthaxanthin and astaxanthin increased in proportion during the aging of the cultures. Lubian and Establier (1982) found canthaxanthin and traces of astaxanthin in one strain of *Nannochloropsis* (B-3) but not in *N. oculata* or *salina*.

Phaeophyceae. The major pigment is always fucoxanthin, followed by β-carotene and violaxanthin; this distribution is invariable, shown in the wide survey of the class by Strain (1958, 1966) and Jensen (1966). Goodwin (1971a, 1980) listed other earlier investigations, and Liaaen-Jensen (1978) shows the class as containing seven minor carotenoids. Zeaxanthin, reported by Jensen (1966) and Hager and Stransky (1970c), could be an artifact of extraction (Jensen, 1966), arising by reduction of violaxanthin. Traces of the acetylenic pigments, diadinoxanthin and diatoxanthin, occur more frequently than in the Chrysophyceae; otherwise the pigments are similar.

Euglenophyceae. Although the Euglenophyceae are included in the Chlorophyta (sensu Christensen), their xanthophylls are rather those of the Chromophyta because they contain diadinoxanthin as the major pigment and, usually, diatoxanthin and heteroxanthin. Some confusion exists in the earlier literature where antheraxanthin and other typical chlorophyte pigments with somewhat similar visible spectra were identified in place of diadinoxanthin (see Bjørnland, 1982, for references). Using IR spectroscopy, MS, and ^{1}H-NMR, Aitzetmuller, Svec, Katz, and Strain (1968) resolved this by identifying diadinoxanthin in *Euglena gracilis,* where Johannes, Brezezinka, and Budzeikiewicz (1971) and Buchecker and

Liaaen-Jensen (1977) also identified it. Hager and Stransky (1970b) and Bjørnland (1982) found neoxanthin also in this class. Nitsche (1973) and Buchecker and Liaaen-Jensen (1977) identified heteroxanthin in *E. gracilis,* but this finding is not reported by others (Bjørnland, 1982; Hager and Stransky, 1970b). Fiksdahl et al. (1984a) examined three unidentified xanthophylls found in the marine euglenophyte, *Eutreptiella gymnastica,* by Bjørnland (1982). One was siphonein, the ester of siphonaxanthin, usually associated with the siphonous green algae, the others being eutreptiellanone and anhydrodiatoxanthin. Eutreptiellanone is unique among algal carotenoids in having a 3,6-oxa-bicycloheptane ring system, and made up 20% of the total carotenoid. The third, anhydrodiatoxanthin, present in only a small amount, has a unique end-group, the trivial name being 3′,4′-anhydrodiatoxanthin. Traces of ketocarotenoids (Bjørnland et al., 1985) are believed to be in the eyespots (Liaaen-Jensen, 1978; Goodwin, 1971a, 1980) because these pigments persist in chlorotic strains of *Euglena gracilis* (Goodwin and Gross, 1958). In *E. viridis,* Fiksdahl and Liaaen-Jensen (1988) found not only diadinoxanthin, diatoxanthin, heteroxanthin, siphonein, and neoxanthin, but also two diacetylenic carotenoids, octadehydro-β,β-carotene and esterified hexadehydro-β-caroten-3-ol. This is the first report of diacetylenic compounds outside the Cryptophyceae.

Chlorophyceae. The Chlorophyceae are generally believed to have the same major pigments as the higher plants (Goodwin, 1965, 1976; Liaaen-Jensen, 1977, 1978), and Goodwin (1971a, 1980) has listed a number of species of this class where this is true. Hager and Stransky (1970b) always found lutein in highest concentration, and β-carotene, neoxanthin, and the components of the xanthophyll cycle – zeaxanthin, antheraxanthin, and violaxanthin (Hager, 1980) – always present, though Antia and Cheng (1983) found the components of the usual cycle absent in a strain of *Dunaliella tertiolecta* and replaced by the lutein $\rightleftharpoons$ lutein 5,6-epoxide (taraxanthin) cycle (Sec. 4.1.4B). Over half the species they examined contained α-carotene, whereas β-cryptoxanthin, β-cryptoxanthin epoxide, taraxanthin, and what is now believed to be loroxanthin were present occasionally. Liaaen-Jensen (1978) also listed ψ- and ϵ-carotenes as minor occasional carotenoids. *Trentepolia iolithus* contains 2-hydroxylated β- and ϵ-rings (Liaaen-Jensen, 1978) and their epoxides (Nybraaten and Liaaen-Jensen, 1974).

Other carotenoids are present occasionally in high concentration. The best known is siphonaxanthin and its ester, siphonein (Strain, 1965; Kleinig, 1969), esterified with various fatty acids. Goodwin (1980) tabulated the distribution of siphonaxanthin and siphonein based on the surveys by Strain (1958, 1965, 1966) and Kleinig (1969). All Derbesiales, Codiales, Caulerpales, and Dichotomosiphonales contained siphonein, but siphonaxanthin was not detected in the Dichotomosiphonales. The two pigments are also found in some Cladophorales and Siphonocladales. Yokohama (1981a,b), in surveying 47 species of green algae, also found this distribution but, in addition, found siphonaxanthin and siphonein in two species of *Ulva* from an order where this pigment and its ester had not previously been detected. Usually lutein occurred with siphonaxanthin. Yokohama et al. (1977) considered siphonaxanthin to be important as a light-harvesting pigment in green algae growing in shade or deep water, though realizing that it also occurred in *Codium* sp. growing in high light intensities. O'Kelly (1982) studied eight marine species from the Cladophorales growing between 3 m above to 2 m below lower low water level. The three highest species contained lutein but no siphonaxanthin, whereas two of the lowest contained siphonaxanthin but no lutein, with three intermediates with both pigments. A further eight species examined (six Cladophorales and two Chaetosiphonales) had similar variations in lutein and siphonaxanthin. O'Kelly (1982) cited this distribution of pigments as evidence for placing Chaetophorales in the Ulvales. Jeffrey (1968c) found siphonein and a trace of siphonaxanthin but no lutein in the green symbiotic *Ostreobium* of the brain coral, *Favia,* and suggested that this placed the alga in the order Siphonales as it was then described.

In many of the species containing siphonaxanthin and/or siphonein in the Chlorophyceae, α-carotene is usually much more abundant than β-carotene (Strain, 1965), and Benson and Cobb (1981) could not find β-carotene in *Codium fragile*. Their quantitative analyses showed that approximately half the carotenoid was siphonaxanthin.

Loroxanthin also occurs in more than trace amounts in some algae (Aitzetmuller et al., 1969; Francis, Knutsen, and Lien, 1973) but has previously been called pyrenoxanthin (Yamamoto, Yokohama, and Boettger, 1969) or trihydroxy α-carotene (Hager and Stransky, 1970b). Yokohama (1982) examined the distribution of xanthophylls in seven orders of green algae. As a general rule, he found large concentrations of siphonaxanthin and loroxanthin in samples from deep water but large concentrations of lutein in those from shallow water. However, siphonous algae from shallow water contained less lutein than did the nonsiphonous orders. He also surveyed 18 species from the Ulvales, Cladophorales, and Siphonocladales (Yokohama 1983), finding siphonaxanthin or loroxanthin in all but one species of the seven growing in shade or deep water, though loroxanthin was not restricted to species growing in these ecological niches. Jeffrey and Hallegraeff (1980a) found a xanthophyll with spectral properties consistent with loroxanthin in samples of marine phytoplankton taken at depths of 60–75 m. We have found a pigment with spectral and chromatographic properties of loroxanthin in *Chaetomorpha linum* growing in a salt swamp.

Prasinophyceae. This class, whose members are known as the scaly flagellates, contain chlorophylls *a* and *b*, and most have the xanthophylls of the Chlorophyceae, usually with zeaxanthin as well as lutein. The members of the major group within the class are listed in Goodwin (1980, Table 7–4) based on the work of Ricketts (1970, 1971b). The taxonomy of this class requires revision (Ricketts, 1970; Norris, 1980), and the genus *Pyramimonas* is currently under investigation in this laboratory (McFadden, Hill, and Wetherbee, 1986). In Ricketts (1970, table 6), three other groups of prasinophytes with pigments differing from those normal in the Chlorophyta are shown. The first group, represented by *Astromonas,* simply contains siphonein in addition to the normal pigments of the Chlorophyceae.

The second group contains siphonein but zeaxanthin in place of lutein. We have found that *Pyramimonas olivacea,* a member of this group, contains siphonaxanthin also. The third group contains a pigment which was designated "xanthophyll K" by Ricketts (1970) but was renamed prasinoxanthin when extracted from a marine coccoid alga by Foss et al. (1984), who determined the structure. The pigment, micronone, described by Ricketts (1966b), is now considered a degradation product of prasinoxanthin caused by attempted saponification (Foss et al., 1984). Foss et al. (1984) extracted pigments from a further three clones of ultraplankton, finding all with prasinoxanthin but no lutein, whereas one clone contained a dihydro derivative of prasinoxanthin and a lactonic xanthophyll they named uriolide after the clone serial URI-2669. A species of *Pterosperma* cultured in this laboratory contains pigments with R_f values and chromatographic and spectral properties similar to both prasinoxanthin (class 3) and siphonaxanthin (class 2) and could be a representative of yet a fourth group with abnormal pigments.

Charophyceae. This class is essentially similar to the Chlorophyceae except that α-carotene has not been detected in it (Strain, 1958; Hager and Stransky, 1970b). Strain (1958) found traces of γ-carotene and lycopene in *Chara foetida,* but as these occur in the antheridia (Karrer, Fatzer, Favarger, and Jucker, 1943), the material extracted may have included fertile male plants. Hager and Stransky (1970b) also found γ-carotene, and traces of antheraxanthin. Rowan and Ducker (unpublished) estimated that the concentration of total carotenoids in the shield cells of the antheridia of *C. australis* was approximately 25 mM. They found γ-carotene and lycopene in extracts from a fertile male colony prepared from fertile fronds from which the antheridia were cut, and also from sterile fronds. However, in an extract from an asexual colony, γ-carotene and lycopene were not found.

4.2. Structure

Goodwin (1971b, 1980) has provided reasons for believing that the C_{40} carotene, lycopene, is the main precursor of the cyclic carotenoids, though this molecule is rarely found in significant quantities in algae. However, he states that there is also some evidence for cyclization of neurosporene as an alternative pathway. The majority of the carotenoids in algae contain two cyclic end-groups linked by a C_{18} chain, formally derived from lycopene, whose structure is shown in Fig. 4–2 from Liaaen-Jensen (1978, 1980). The cyclization to form cyclohexane end-groups occurs between C-1 and C-6 and between C-1′ and C-6′. The structures of most of the carotenoids found in algae are also shown in Fig. 4–2. About half the carotenoids fully characterized have up to six centers of chirality (Liaaen-Jensen, 1980), and since the mid-1970s the absolute configuration of chiral C atoms in many of them has been determined, as shown in Table 4–3. The configuration at a chiral center is designated using the *R*,*S* convention (Goodwin, 1973, 1980). Liaaen-Jensen (1980) has discussed the various methods available for determining configuration of carotenoids: ORD and CD and, more recently, ^{13}C-NMR, with verification by partial synthesis, have proved to be the most powerful methods.

4.3. Degradation

Oxidation, low or high pH, light, and heat or a combination of these can cause degradation of carotenoids during extraction (Britton and Goodwin, 1971; Liaaen-Jensen and Jensen, 1971; Moss and Weedon, 1976; Liaaen-Jensen, 1978; Liaaen-Jensen and Andrewes, 1985). As will be seen in Section 4.5.1, the ratio %III/II measures the degree of fine structure of

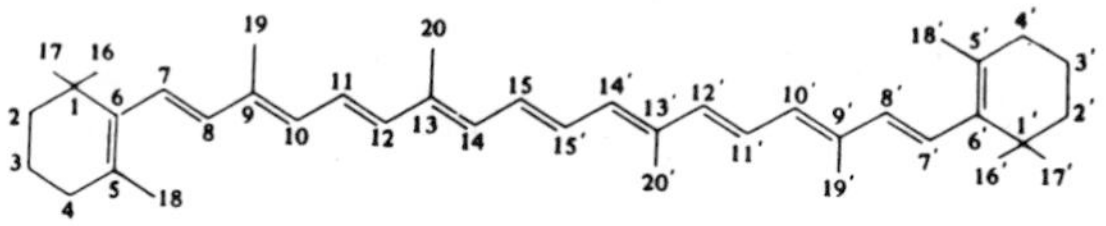

β-Carotene, showing numbering system

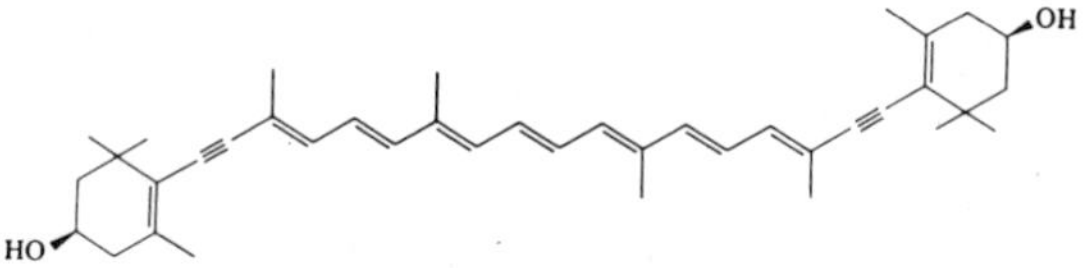

Alloxanthin

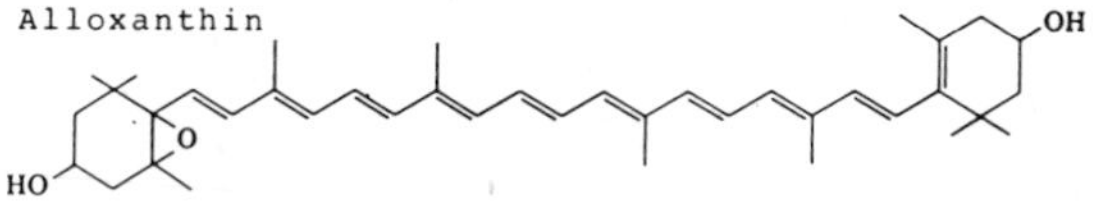

Antheraxanthin

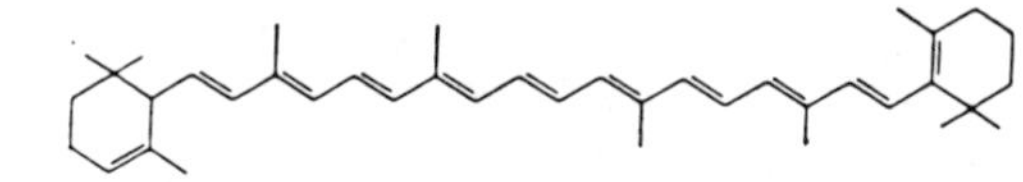

α-Carotene

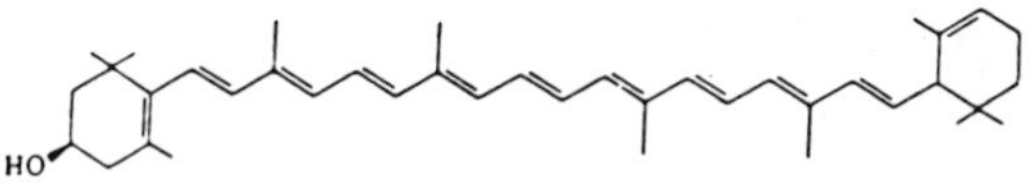

α-Cryptoxanthin

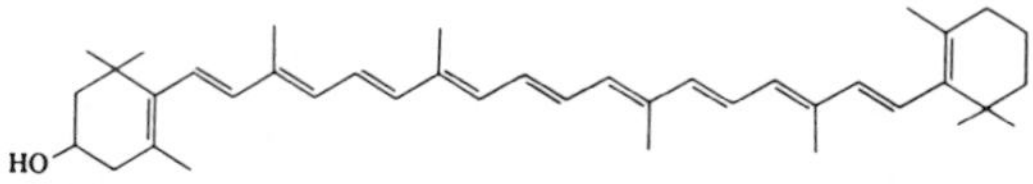

β-Cryptoxanthin

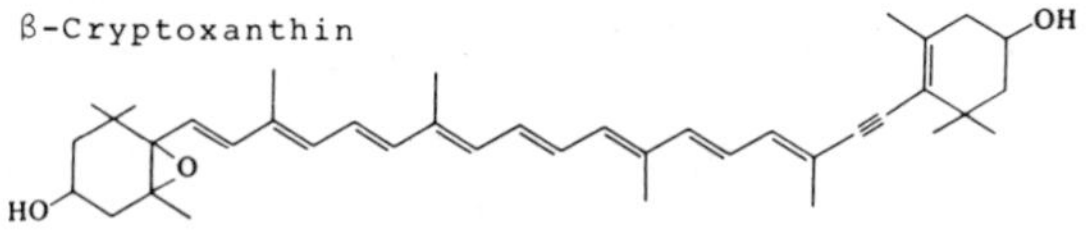

Diadinoxanthin

Fig. 4–2. The chemical structure of carotenoids found in algae. The numbering system used is shown at the head of the table using β-carotene as example.

a carotenoid (Ke, Imsgard, Kjøsen, and Liaaen-Jensen, 1970) and can provide evidence for degradation occurring during extraction (Davies, 1976). The published values of %III/II in Table 4–4 sometimes show variation for a given pigment, and, as a fall in the ratio usually indicates degradation, the highest value should be taken as correct. Astaxanthin, fucoxanthin, 19′-hexanoyloxyfucoxanthin, and peridinin are unstable at high pH, and thus saponification cannot be used during purification. Bases slowly convert fucoxanthin to a number of degradation products with increasing fine structure and lower optical maxima (Liaaen and Sørensen,

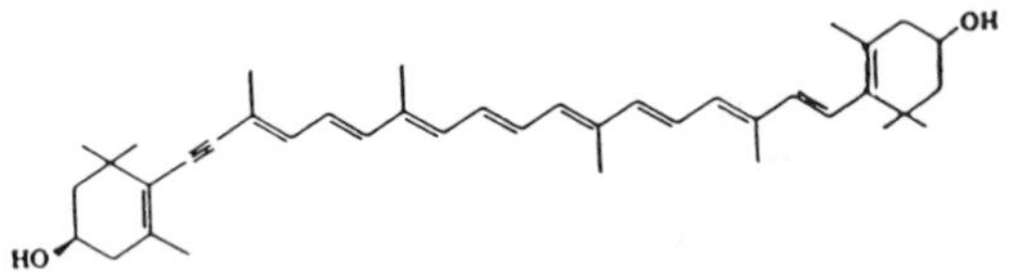

Diatoxanthin

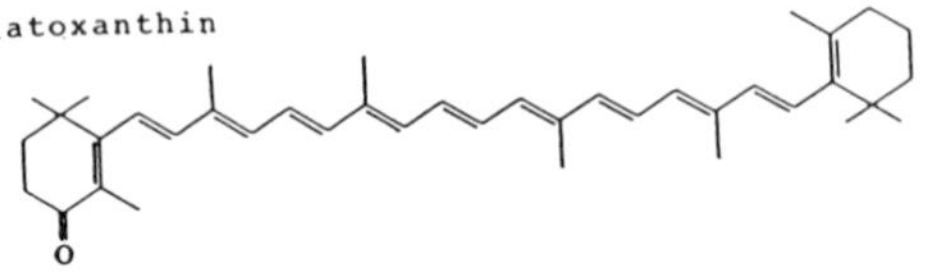

Echinenone

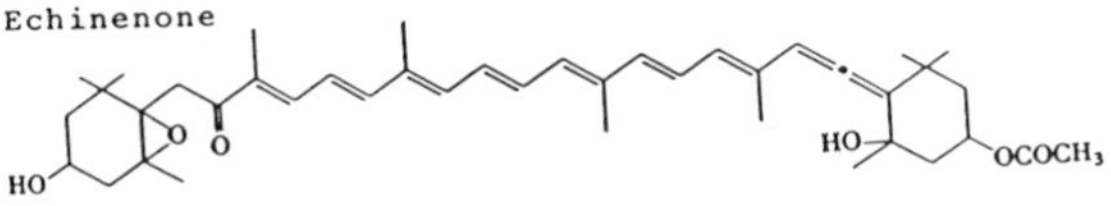

Fucoxanthin

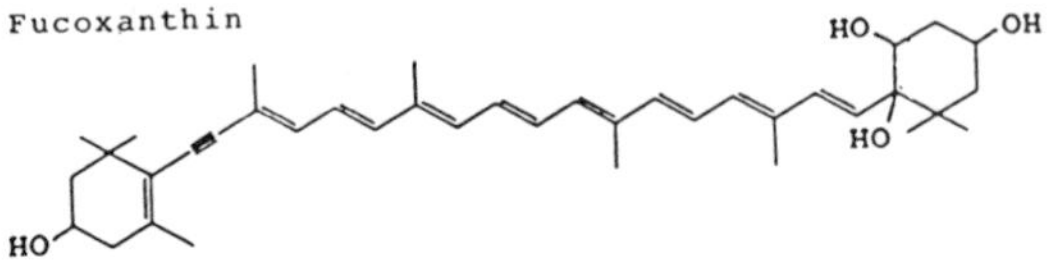

Heteroxanthin

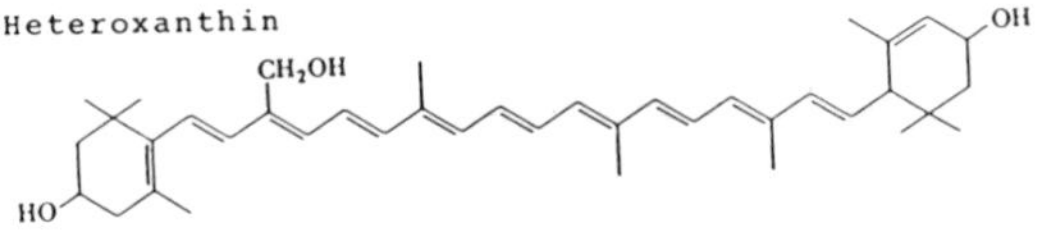

Loroxanthin

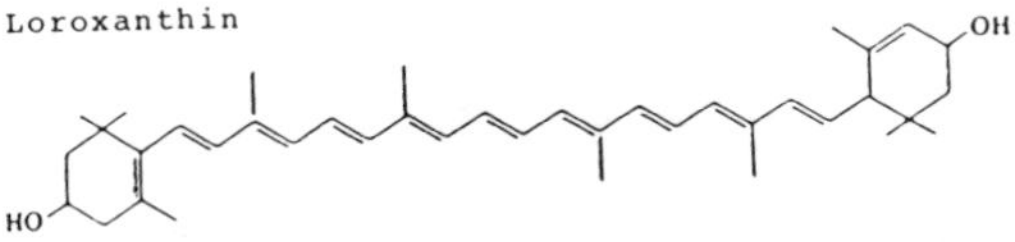

Lutein

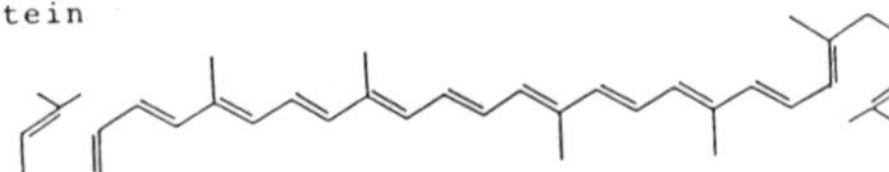

Lycopene

Fig. 4–2. Cont.

1956), whereas peridinin is decolorized rapidly by 4% ethanolic KOH (Loeblich and Smith, 1968). Treating 19′-hexanoyloxyfucoxanthin for 1 h with 2% methanolic KOH or for 3 h with 0.1% methanolic K_2CO_3 caused a hypsochromic shift of 19 nm, to a compound now having fine structure (Arpin et al., 1976).

4.3.1. Oxidation products

Carotenoids are frequently described as easily oxidized during extraction (Spoehr, Smith, Strain, and Milner, 1936; Liaaen-Jensen and Jensen, 1971; Davies, 1976; Goodwin, 1980; Britton, 1983), but the oxidation products

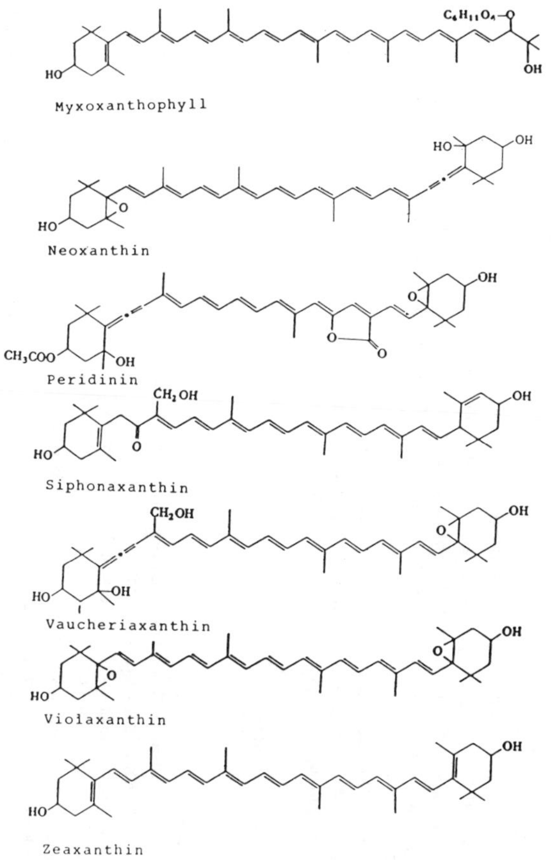

Myxoxanthophyll

Neoxanthin

Peridinin

Siphonaxanthin

Vaucheriaxanthin

Violaxanthin

Zeaxanthin

Fig. 4–2. Cont.

have not often been examined. Thomas and Goodwin (1965) found that the monoepoxide, antheraxanthin, exposed to air for 24 h at 50°C, was partly oxidized to the diepoxide, violaxanthin. Xanthophylls with 3-hydroxy-4-oxo end-groups, such as astaxanthin, are rapidly oxidized by bases in air (Moss and Weedon, 1976; Britton, 1983), and Rodriguez et al. (1976) found substantial oxidation of β-carotene during chromatography on Micro-Cel C.

4.3.2 Isomers

Acids, heat, and light may cause cis–trans isomerization, or isomerization of 5,6-epoxides to the furanoid 5,8-epoxides.

Table 4–3. *Trivial and semisystematic names of the algal carotenoids after Straub (1976), unless otherwise stated*

Trivial name	Semisystematic name
Alloxanthin	(3*R*,3′*R*)-7,8,7′,8′-Tetradehydro-β,β-carotene-3,3′-diol
Anhydrodiatoxanthin	(3*R*)-3′,4′,7′,8′-Tetradehydro-β,β-caroten-3-ol (Fiksdahl et al., 1984a)
Antheraxanthin	(3*S*,5*R*,6*S*,3′*R*)-5,6-Epoxy-5,6-dihydro-β,β-carotene-3,3′-diol (Bjørnland et al., 1984)
Aphanizophyl	2′-(β,L-Rhamnopyranosyloxy)-3′,4′-didehydro-1′,2′-dihydro-β,β-carotene-3,4,1′-triol
Astaxanthin	(3*S*,3′*S*)-3,3′-Dihydroxy-β,β-carotene-4,4′-dione
Aurochrome	5,8,5′,8′-Diepoxy-5,8,5′,8′-tetrahydro-β,β-carotene
Auroxanthin	5,8,5′,8′-Diepoxy-5,8,5′,8′-tetrahydro-β,β-carotene-3,3′-diol
Caloxanthin	6,7-Didehydro-5,6-dihydro-β,β-carotene-3,3′-diol
Canthaxanthin	β,β-Carotene-4,4′-dione
α-Carotene	(6′*R*)-β,ε-Carotene
β-Carotene	β,β-Carotene
γ-Carotene	β,ψ-Carotene
ε-Carotene	(6*S*,6′*S*)-ε,ε-carotene
β-Carotene-5,6-epoxide	(5*S*,6*R*)-5,6-Epoxy-5,6-dihydro-β,β-carotene (Eschenmoser & Eugster, 1978)
Crocoxanthin	(3*R*,6′*R*)-7,8-Didehydro-β,ε-caroten-3-ol
α-Cryptoxanthin	(3′*R*,6′*R*)-β,ε-caroten-3-ol (Bjørnland et al., 1984)
β-Cryptoxanthin	(3*R*)-β,β-Caroten-3-ol
Cryptoxanthin diepoxide	5,6,5′,6′-Diepoxy-5,6,5′,6′-tetrahydro-β,β-caroten-3-ol
Cryptoxanthin epoxide	5,6-Epoxy-5,6-dihydro-β,β-caroten-3-ol
Diadinoxanthin	(3*S*,5*R*,6*S*,3′*R*)-5,6-Epoxy-7′,8′-didehydro-5,6-dihydro-β,β-carotene-3,3′-diol (Buchecker & Liaaen-Jensen, 1977)
Diatoxanthin	(3*R*,3′*R*)-7,8-Didehydro-β,β-carotene-3,3′-diol

Dihydroprasinoxanthin epoxide	4′,5′-Epoxy-3,6,3′-trihydroxy-7,8,4′,5′,7′,8′-hexahydro-γ,ϵ-7,8,4′,5′,7′,8′-caroten-8-one (Foss et al., 1986)
Dinoxanthin	(3*R*,5*R*,6*R*,3′*S*,5′*R*,6′*S*)-5′,6′-Epoxy-6,7-didehydro-5,6,5′,6′-tetrahydro-β,β-carotene-3,5,3′-triol-3-acetate
Echinenone	β,β-Caroten-4-one
Eutreptiellanone	(3*S*,5*R*,6*S*?)-3,6-Epoxy-3′,4′,7′,8′-tetradehydro-5,6-dihydro-β,β-caroten-4-one (Fiksdahl et al., 1984a)
Fucoxanthin	(3*S*,5*R*,6*S*,3′*S*,5′*R*,6′*R*)-5,6-Epoxy-3,3′,5′-trihydroxy-6′,7′-didehydro-5,6,7,8,5′,6′-hexahydroxy-β,β-caroten-8-one acetate
Fucoxanthinol	(3*S*,5*R*,6*S*,3′*S*,5′*R*,6′*R*)-5,6-Epoxy-3,3′,5′-trihydroxy-6′,7′-didehydro-5,6,7,8,5′,6′-hexahydroxy-β,β-caroten-8-one
Heteroxanthin	(3*S*,5*R*,6*R*,3′*R*)-7′,8′-Didehydro-5,6-dihydro-β,β-carotene-3,5,6,3′-tetrol (Buchecker et al., 1984)
Hexadehydro-β-caroten-3-ol	7,8,3′,4′,7′,8′-Hexadehydro-β,β-caroten-3-ol (Fiksdahl & Liaaen-Jensen, 1988)
19′-Hexanoyloxyfucoxanthin	(3*S*,5*R*,6*S*,3′*S*,5′*R*,6′*R*)5,6-Epoxy-3,3′,5′,19′-tetrahydroxy-6′,7′-didehydro-5,6,7,8,5′,6′-hexahydro-β,β-caroten-8-one-3′-acetate-19′-hexanoate (Hertzberg et al., 1977)
3′-Hydroxyechinenone	3′-Hydroxy-β,β-caroten-4-one
Isocryptoxanthin	β,β-Caroten-4-ol
Isozeaxanthin	β,β-Carotene-4,4′-diol
4-Ketomyxol-2′-(methylpentoside)	3,1′-Dihydroxy-2′-rhamnosyloxy-3′,4′-didehydro-β,ϵ-caroten-4-one
Loroxanthin	(3*R*,3′*R*,6′*R*)-β,ϵ-Carotene-3,19,3′-triol (Marki-Fischer et al., 1983)
Lutein	(3*R*,3′*R*,6′*R*)-β,ϵ-Carotene-3,3′-diol
Lycopene	ψ,ψ-Carotene
Monadoxanthin	(3′*R*,6′*R*)-7,8-Didehydro-β,ϵ-carotene-3,3′-diol (Pennington et al., 1985)
Myxol-2′-*O*-methyl-methyl pentoside	2, (*O*-Methyl-5*C*-methylpentosyloxy)-3′,4′-didehydro-1′,2′-dihydro-β,ψ-carotene-3,1′-diol
Mutatochrome	5,8-Epoxy-5,8-dihydro-β,β-carotene
Myxoxanthophyll	(3*R*,2′*S*)-2′-(β-L-Rhamnopyranosyloxy)-3′,4′-didehydro-1′,2′-dihydro-β,ψ-carotene-3,1′-diol
Neoxanthin	(3*S*,5*R*,6*R*,3′*S*,5′*R*,6′*S*)-5′,6′-Epoxy-6,7-didehydro-5,6,5′,6′-tetrahydro-β,β-carotene-3,5,3′-triol
Nostoxanthin	6,7,6′,7′-Tetradehydro-5,6,5′,6′-tetrahydro-β,β-carotene-3,3′-diol

Table 4–3 (*Cont.*).

Trivial name	Semisystematic name
Octadehydro-β-carotene	3,4,7,8,3′,4′,7′,8′-Octadehydro-β,β-carotene (Fiksdahl & Liaaen-Jensen, 1988)
Oscillaxanthin	(2*R*,2′*R*)-2,2′-Bis(β-L-rhamnopyranosyloxy)-3,4,3′,4′-tetradihydro-1,2,1′,2′-tetrahydro-ψ,ψ-carotene-1,1′-diol
Oscillol	2,2′-Bis(*O*-methyl-5-*C*-methyl-pentosyloxy)-3,4,3′,4′-tetradehydro-1,2,1′,2′-tetrahydro-ψ,ψ-carotene-1,1′-diol
Peridinin	(6*R*)-5′,6′-Epoxy-3,5,3′-trihydroxy-6,7-didehydro-5,6,5′,6′-tetrahydro-12,13,20-trinor-β,β-caroten-19′,11′-olide-3-acetate (Strain et al., 1976)
Peridinol	5′,6′-Epoxy-3,5,3′-trihydroxy-6,7-didehydro-5,6,5′,6′-tetrahydro-12,13,20-trinor-β,β-caroten-19′,11′-olide (Foss et al., 1984)
Prasinoxanthin	(3′*R*,6′*R*)-3,6,3′-trihydroxy-7,8-dihydro-γ,ϵ-caroten-8-one (Foss et al. 1984)
Pyrrhoxanthin	5′,6′-Epoxy-3,3′-dihydroxy-7,8-didehydro-5′,6′-dihydro-10,11,20-trinor-β,β-caroten-19′11′-olide-3-acetate
Pyrrhoxanthinol	5′,6′-Epoxy-3,3′-dihydroxy-7,8-didehydro-5′,6′-dihydro-10,11,20-trinor-β,β-caroten-19′11′-olide
Siphonaxanthin	3,19,3′-Trihydroxy-7,8-dihydro-β,ϵ-caroten-8-one
Siphonein	(3*R*?,3′*R*,6*R*)-3,19,3′-Trihydroxy-7,8-dihydro-β,ϵ-caroten-8-one 19-ester (Fiksdahl et al., 1984a)
Taraxanthin	(3*S*,5*R*,6*S*,3′*R*,6′*R*)-5,6-Epoxy-5,6-dihydro-β,ϵ-carotene-3,3′-diol
Uriolide	(3*S*,5*R*,6*S*,3′*R*,6′*R*)-5,6-Epoxy-3,3′-dihydroxy-5,6,7′,8′-tetrahydro-β,ϵ-caroten-11′,19′-olide (Foss et al., 1986)
Vaucheriaxanthin	5′,6′-Epoxy-6,7-didehydro-5,6,5′,6′-tetrahydro-β,β-carotene-3,5,19′,3′-tetrol
Violaxanthin	(3*R*,5*R*,6*S*,3′*S*,5′*R*,6′*S*)-5,6,5′,6′-Diepoxy-5,6,5′,6′-tetradehydro-β,β-carotene-3,3′-diol
β-Zeacarotene	7′,8′-Dihydro-β,ψ-carotene
Zeaxanthin	(3*R*,3′*R*)-β,β-carotene-3,3′-diol

Table 4–4. *Absorption maxima of the carotenoids, their specific extinction coefficients ($E^{1\%}_{1cm}$), and %III/II ratios in nine solvents (shoulders shown in parentheses)*

Carotenoid	Absorption maxima (nm)			Reference	$E^{1\%}_{1cm}$	Ref.	%III/II	Ref.
	I	II	III					
Acetone								
Alloxanthin	(430)[a]	453	483	23	–	–	–	–
Anhydrodiatoxanthin	(445)	467	495	12	–	–	14	12
Aphanizophyll	450	476	507	35	–	–	–	–
Astaxanthin	–	472–480	–	14, 35, 37, 82	–	–	–	–
Calloxanthin	(430)	454	481	19	–	–	–	–
Canthaxanthin	–	466	–	50	–	–	–	–
α-Carotene	420–425	442–448	469–476	10, 28, 35, 37	–	–	55	37
β-Carotene	420–1432	449–454	475–480	10, 14, 28, 35, 37–39, 50, 68, 82, 101	–	–	7–12	16, 88
γ-Carotene	(439)	461	491	37	–	–	–	–
β-Carotene 5,6-epoxide	(420–427)	441–448	470–475	10	–	–	–	–
β-Cryptoxanthin	(427–429)	450–453	475–478	10, 38	–	–	–	
Cryptoxanthin diepoxide	419	443	472	10	–	–	80	10
Diadinoxanthin	(426–430)	448–449	478–479	10, 12, 16, 68, 101	2230	68	61–75	12, 16, 68
Diatoxanthin	(429–434)	448–454	477–482	10, 12, 16, 68, 101	–	–	35	12
Dihydroprasinoxanthin epoxide	–	430	485	39	–	–	–	–
Dinoxanthin	418	442	470	68	–	–	–	–
Echinenone	(475)	459–460	–	39, 50	–	–	–	–
Eutreptiellanone	(441)	461	491	12	–	–	–	–
Fucoxanthin	(425–428)	444–449	467–473	10, 16, 42, 95a, 101, 101a	1060 1600	42 66	9 3	16 95a, 101a

Table 4–4 (*Cont.*)

Carotenoid	Absorption maxima (nm)			Reference	$E^{1\%}_{1cm}$	Ref.	%III/II	Ref.
	I	II	III					
19′-Hexanoyloxyfucoxanthin	423	445–450	471–478	9, 95a, 101a	–	–	20 44	79 95a, 101a
Isocryptoxanthin	(425)	449	472–275	38, 39	–	–	–	–
4-Keto-myxol-2′-*O*-methyl-methylpentoside	–	483	(510)	28	–	–	–	–
Loroxanthin	(423)	446	473	41	–	–	–	–
Lutein	(420)	445	473	82	–	–	–	–
Lycopene	447–448	472–475	504–506	1, 35, 39, 67	–	–	–	–
Mutatochrome	405	426	452	38	–	–	–	–
Myxol-2′-*O*-methyl-methylpentoside	450	478	508	28, 35, 40	–	–	–	–
Myxoxanthophyll	450–452	475–478	508–510	35, 39, 50, 52, 56	2160	56	57–60	52, 56
Neoxanthin	413–423	436–445	463–473	12, 82, 101	–	–	87	12
Nostoxanthin	(431)	453	481	19	–	–	–	–
Oscillol	469	496	530	32, 35, 40	–	–	–	–
Oscillaxanthin	466–470	490–499	522–534	39, 40, 52, 57	–	–	60	52, 57
Peridinin	–	465–471	–	16, 62, 68, 88	1340	62	–	–
Peridinol	–	466	–	68	–	–	–	–
Pyrrhoxanthin	–	458	–	68	–	–	–	–
Pyrrhoxanthinol	–	458	–	68	–	–	–	–
Uriolide	–	448	472	37	–	–	11	37
Vaucheriaxanthin	420	441	467	78	–	–	–	–
Violaxanthin	401–417	440–442	469–470	78, 82	2400	66	–	–
Zeaxanthin	(424–425)	449–452	474–479	28, 35, 38, 37, 101	2340	1, 28	–	–
Benzene								
Alloxanthin	(438)	464	494–495	23, 26	–	–	–	–

Astaxanthin	–	478	–	28	–	–	–	–
Aurochrome	387	409	434	28	2042	28	–	–
Canthaxanthin	–	480–485	–	27, 28, 35	2092	28	–	–
α-Carotene	(432)	456	485	46, 76, 96	–	–	–	–
β-Carotene	(435)	462–464	478–493	28, 46, 76, 96	2337	28	–	–
γ-Carotene	447–450	474–477	510	25, 28, 44, 46	–	–	–	–
ϵ-Carotene	425	451	481–482	21, 35	–	–	–	–
β-Carotene 5,6-Epoxide	–	460	482	28	–	–	–	–
Crocoxanthin	432	458	488	26	–	–	–	–
α-Cryptoxanthin	432–435	459	485–489	35, 46	2355	28	–	–
β-Cryptoxanthin	435	463	489	72	–	–	–	–
Cryptoxanthin diepoxide	426	451–455	481–486	28, 35	–	–	–	–
Cryptoxanthin epoxide	–	461	494	28, 35	–	–	–	–
Diatoxanthin	–	463	492	26	–	–	–	–
Echinenone	–	470	(490)	28	2091	28	–	–
Fucoxanthin	443	460–461	485	11, 28, 36	–	–	–	–
Fucoxanthinol	(436)	460	487	28	–	–	–	–
3′-Hydroxyechinenone	–	472	–	28, 35	–	–	–	–
Lutein	(432–433)	458	487	28, 46, 76	2236	28	–	–
Lycopene	455–456	485–487	517–522	25, 28, 35, 44, 46	3370	28	–	–
Monadoxanthin	430	456	487	23, 26	–	–	–	–
Mutatochrome	(416)	439–440	463–470	27, 28, 35, 44	1989	28	–	–
Myxoxanthophyll	462	488	522	28, 35, 56	–	–	51	56
Neoxanthin	423–424	445–453	478–483	28, 35, 46	2245	35, 46	–	–
Peridinin	–	465–467	494–502	28, 35, 62	1290	62	–	–
Peridinol	–	468	493	68	–	–	–	–
Pyrrhoxanthinol	–	468	494	68	–	–	–	–
Taraxanthin	426–429	452–455	482–485	28, 35, 44, 46, 76	2373	28	–	–
Violaxanthin	426–428	451–454	483–484	28, 44, 46, 76	2240	28	–	–
					2216	27		
Zeaxanthin	(440)	463	491–493	27, 28, 35, 46	–	–	–	–

Table 4–4 (*Cont.*)

Carotenoid	Absorption maxima (nm) I	II	III	Reference	$E^{1\%}_{1cm}$	Ref.	%III/II	Ref.
Carbon bisulfide								
Antheraxanthin	–	476–478	505–510	27, 28, 46	–	–	–	–
Astaxanthin	–	503	–	24, 28, 35	–	–	–	–
Auroxanthin	401	423	451	35	–	–	–	–
Canthaxanthin	–	500	–	35	–	–	–	–
α-Carotene	448–450	475–484	504–510	28, 35, 46, 84	2180	44	–	–
β-Carotene	450	475–485	505–509	35, 46, 84	2008	28	–	–
γ-Carotene	463–468	493–496	525–534	25, 28, 35, 44, 46, 84	–	–	–	–
β-Carotene 5,6-Epoxide	–	479	511	28	–	–	–	–
α-Cryptoxanthin	446–453	477	508–509	35, 46	–	–	–	–
β-Cryptoxanthin	452–453	483	516–518	28, 35	–	–	–	–
Cryptoxanthin diepoxide	–	473	503	28, 35	–	–	–	–
Cryptoxanthin epoxide	–	479	512	28, 35	–	–	–	–
Diadinoxanthin	(449–452)	474	505–506	28, 35, 73, 74	–	–	–	–
4,4′-Diketo,3′-hydroxy-β-carotene	–	494	–	35	–	–	–	–
Dinoxanthin	(441)	467	498	73	–	–	–	–
Echinenone	–	488–494	–	44	–	–	–	–
Fucoxanthin	450	478	508	28, 35	2036 2025	28 35	–	–
Isocryptoxanthin	–	482	509	28, 35	–	–	–	–
Isozeaxanthin	–	479	505	28, 35	–	–	–	–
Lutein	(443–450)	472–475	503–508	28, 35, 46, 84	2160	28, 35	–	–
Lycopene	471–476	505–508	542–547	1, 25, 28, 35, 44, 46, 84	–	–	–	–

Mutatochrome	–	459–460	488–490	27, 28, 35, 44	–	–	–	–
Neoxanthin	(439)	463–466	493–497	28, 35, 46	–	–	–	–
Oscillaxanthin	494	528	568	28	–	–	–	–
Peridinin	450–454	480	512–516	28, 35, 73	–	–	–	–
Pyrrhoxanthin	–	512	586	28, 35	–	–	–	–
Taraxanthin	441–442	468–470	498–501	28, 35, 46	–	–	–	–
Violaxanthin	440–441	468–470	500–501	28, 46, 84	–	–	–	–
β-Zeacarotene	425	450	480	35	–	–	–	–
Zeaxanthin	(450)	481–483	513–518	26, 28, 35, 46	–	–	–	–
Chloroform								
Alloxanthin	·436–438	460–461	489–491	23, 49	–	–	–	–
Antheraxanthin	430	456	484	93	–	–	–	–
Aphanizophyll	453–457	482–488	510–523	35, 94	–	–	–	–
Caloxanthin	(432)	458	484	35, 94	–	–	–	–
Canthaxanthin	–	482	–	28, 35, 94	–	–	–	–
α-Carotene	427–435	455–457	484–485	28, 35, 46, 78, 83, 84, 96	2420	45	–	
β-Carotene	(431–438)	460–465	485–493	28, 35, 46, 73, 76, 83, 84, 92, 93, 96, 99	2200 2396	92 28	–	–
γ-Carotene	443–450	470–475	502–509	25, 28, 35, 46, 84	–	–	–	–
β-Carotene 5,6-Epoxide	–	459	492	28	–	–	–	–
Crocoxanthin	430	454	482	35, 94	–	–	–	–
α-Cryptoxanthin	435	459	485–488	35, 46	–	–	–	–
β-Cryptoxanthin	(434–435)	459–464	485–495	35, 48	–	–	–	–
Cryptoxanthin diepoxide	432	452–453	480–482	35, 88, 93	–	–	–	–
Cryptoxanthin epoxide	434	456	483–488	28, 35, 93	–	–	–	–
Diadinoxanthin	432–433	455	482	35, 93	–	–	–	–
Diatoxanthin	433	458	486	35, 93	–	–	–	–
4,4′-Diketo-3-hydroxy-β-carotene	–	483	–	35	–	–	–	–

Table 4–4 (*Cont.*)

Carotenoid	Absorption maxima (nm) I	II	III	Reference	$E^{1\%}_{1cm}$	Ref.	%III/II	Ref.
Echinenone	–	471	–	94	–	–	–	–
Fucoxanthin	(450–460)	478	508	22, 28, 35	–	–	–	–
Heteroxanthin	426	448	477	35, 93	–	–	–	–
3′-Hydroxyechinenone	–	472	–	28, 35, 87	–	–	–	–
Loroxanthin	431	455	482	35, 93	–	–	–	–
Lutein	428–435	454–458	483–486	28, 35, 46, 76, 84	–	–	–	–
Lycopene	455–458	483–484	515–520	1, 25, 35, 44, 46, 84	–	–	–	–
Mutatochrome	–	438	469	28, 63	–	–	–	–
Myxoxanthophyll	457–460	485–488	512–522	56, 94	–	–	50	56
Neoxanthin	420–423	445–449	475–478	27, 35, 46, 48, 84, 93, 96	–	–	–	–
Nostoxanthin	(432)	457	485	94	–	–	–	–
Oscillaxanthin	476–480	501–510	534–548	57, 94	–	–	60	54
Peridinin	–	470	490	62	1290	62	–	–
Prasinoxanthin	–	461	–	85	–	–	–	–
Siphonaxanthin	–	466	–	85, 86, 102	–	–	–	–
Siphonein	–	463–471	–	12, 85, 86, 102	–	–	–	–
Taraxanthin	424–433	450–452	480–489	35, 46, 48, 76, 96	–	–	–	–
Vaucheriaxanthin	426	451	480	93	–	–	–	–
Violaxanthin	423–426	449–452	478–482	35, 46, 76, 83, 84, 93	–	–	–	–
β-Zeacarotene	(414)	439	465	28	–	–	–	–
Zeaxanthin	(430–434)	458–461	488–491	46, 76, 83, 93, 94, 96	–	–	–	–

Diethyl ether								
Alloxanthin	(430)	451	481	81	–	–	–	–
Anhydrodiatoxanthin	(447)	461	488	32	–	–	–	–
Canthaxanthin	–	456	–	53	–	–	–	–
α-Carotene	417–429	443–445	470–475	2, 35, 60, 81	–	–	57	81
β-Carotene	(421–432)	447–450	472–476	2, 32, 33, 34, 62, 63	2080	44	5	81
γ-Carotene	427	438	468	81, 88	–	–	91	81
Crocoxanthin	428	445	475	81	–	–	58	81
β-Cryptoxanthin	(420)	446–447	472–475	2	–	–	–	–
Diadinoxanthin	424	448	477	33	–	–	–	–
Echinenone	–	455	–	53	–	–	–	–
Eutreptiellanone	(434)	457	483	32	–	–	48	32
Fucoxanthin	(420)	444–446	468–470	33	–	–	3	95a, 101a
Fucoxanthinol	(420–423)	445	470	33	–	–	–	–
Heteroxanthin	(419)	442–445	469–474	33	–	–	–	–
19′-Hexanoyloxyfucoxanthin	–	444	470	95a	–	–	45	95a, 101a
Loroxanthin	–	442	(465)	33	–	–	–	–
Lutein	(422)	443–448	472–476	46, 58, 59	2480	44	67	48
					2600			
Lycopene	456	469	500	81	–	–	50	81
Monadoxanthin	428	446	476	81	–	–	60	81
Mutatochrome	415–418	428	436–452	53, 55	2260	55	–	–
Neoxanthin	413–420	437–444	464–470	33, 59, 60	–	–	–	–
Peridinin	–	453–454	475	34, 62	1450	62	–	–
Prasinoxanthin	–	446	(466)	36	–	–	–	–
Pyrrhoxanthin	439	453	480	34	–	–	–	–
Siphonaxanthin	–	441	(464)	32	–	–	–	–
Siphonein	–	448	–	32	–	–	–	–
Violaxanthin	418–421	441	471–472	58, 59	–	–	–	–
β-Zeacarotene	(405)	427–428	450–455	28, 35	–	–	–	–
Zeaxanthin	(423–428)	447–450	474–477	33, 81	–	–	22	81

Table 4–4 (*Cont.*)

Carotenoid	Absorption maxima (nm) I	II	III	Reference	$E^{1\%}_{1cm}$	Ref.	%III/II	Ref.
Ethanol								
Alloxanthin	(427–430)	450–454	478–483	23, 29, 43, 49	–	–	29	49
Antheraxanthin	422–423	444–447	472–477	7, 22, 46, 48, 49, 92, 93	2350	92, 45	54	47, 50, 93
Aphanizophyll	445	472	542	35, 94	–	–	51	94
Astaxanthin	–	475–476	–	7, 28, 35	–	–	–	–
Auroxanthin	381	402	427	28	1850	28	–	–
Caloxanthin	426	449–450	475–478	89, 94	–	–	25	94
Canthaxanthin	–	474–477	–	7, 35, 48, 94	–	–	–	–
α-Carotene	420–425	442–446	474–480	36, 46, 48, 49, 76, 94	–	–	55–61	36, 48, 49, 50, 94
β-Carotene	(423–427)	447–451	474–480	7, 22, 28, 35, 36, 48, 49, 51, 81, 93, 94, 98	2620	28	27	48, 49, 50, 93, 94
γ-Carotene	438–440	460–462	489–495	25, 28, 46, 48	–	–	23	48
ϵ-Carotene	417	440	469–470	21, 28, 49	–	–	59	49
Crocoxanthin	421	443–447	472–477	26, 35, 49	–	–	62	49
α-Cryptoxanthin	426–428	449	473	35, 46	–	–	–	–
β-Cryptoxanthin	(420–428)	449–452	473–486	27, 38, 48, 87, 94	–	–	28–30	48, 49, 94
Cryptoxanthin diepoxide	423	442	472	35, 93	–	–	54	93
Cryptoxanthin epoxide	(422–424)	445–448	475–477	35, 46, 48, 49, 93, 98, 99	–	–	43	48, 49, 93
Diadinoxanthin	(424–429)	445–448	472–79	3, 29, 31, 31a, 35, 49, 58, 59, 62, 64, 68, 71, 73, 74, 91, 93, 99	2500	28	30 65	68 50

Diatoxanthin	(425–429)	449–453	475–483	49, 58, 59, 71, 74, 93	–	–	25	49
Dinoxanthin	(416–418)	439–443	468–472	35, 59, 62, 64, 73	–	–	–	–
Echinenone	–	453–461	–	7, 94	–	–	–	–
Fucoxanthin	(426)	447–451	465–470	6, 11, 35, 49, 59, 63, 64, 74, 85, 101a	1140	6	–	–
Fucoxanthinol	–	452	–	28	1453	28	–	–
Heteroxanthin	(418)	442–448	470–478	31a, 91	–	–	–	–
Hexadehydro-β-caroten-3-ol	–	457	–	31a	–	–	–	–
19′-Hexanoloxyfucoxantin	–	447	471	95a, 101a	–	–	25	95a, 101a
3′-Hydroxyechinenone	–	460–462	–	24, 94	–	–	–	–
Loroxanthin	(425)	446	473–474	4, 28	–	–	–	–
Lutein	420–424	445–446	473–476	8, 36, 44, 46, 48, 49, 60, 76, 84, 85, 94	2540 2550	92 28	62	49, 50, 94
Lycopene	443–448	471–472	500–503	25, 28, 35, 46	–	–	–	–
Monadoxanthin	424–425	447–448	476–477	23, 26, 28	–	–	–	–
Myxoxanthophyll	448–450	471–474	503–504	35, 51, 94	–	–	64	94
Neoxanthin	413–416	436–438	463–467	28, 35, 36, 46, 48, 49, 83, 84, 96	2243 2470	28 27, 92	89	48, 49, 93
Nostoxanthin	(426)	448	475	94	–	–	25	94
Oscillaxanthin	468	492	526	94	–	–	59	94
Peridinin	–	472–475		59, 62, 64, 73	1325	62	–	–
Prasinoxanthin	–	450	(466)	85	–	–	–	–
Pyrrhoxanthin	–	471	–	35, 73	–	–	–	–
Siphonaxanthin	–	448–452	–	59, 61, 85, 86, 90, 102	–	–	–	–
Siphonein	(427–434)	448–467	471–480	28, 31a, 59, 60, 85, 86, 90, 102	–	–	–	–

Table 4–4 (*Cont.*)

Carotenoid	Absorption maxima (nm) I	II	III	Reference	$E^{1\%}_{1cm}$	Ref.	%III/II	Ref.
Taraxanthin	416–424	439–445	469–474	6, 22, 28, 46, 48, 69, 70, 76, 83, 84, 96	2400 2800	44 28	79	48
Vaucheriaxanthin	418–420	440–444	468–472	7, 93, 98, 99	–	–	78–80	93
Violaxanthin	415–423	439–441	468–471	7, 18, 35, 45, 46, 48, 49, 60, 76, 83, 93, 98	2500	44, 92, 93	48, 49, 50, 93	
Zeaxanthin	(424–428)	449–454	474–482	7, 29, 35, 36, 46, 48, 49, 51, 76, 83, 84, 93, 96	2550 2480 2540	28 9226 28		44, 49, 93, 98
Hexane								
Alloxanthin	(427)	451	480	20, 23, 26	–	–	50	20
Antheraxanthin	(419–425)	443–445	470–474	18, 27, 28, 35, 47, 76, 96, 97	–	–	50–64	18, 47, 97
Astaxanthin	–	466–468	–	24, 28, 100	2100	100	–	–
Aurohrome	381–387	401–409	426–434	28, 35	2035	35	–	–
Auroxanthin	380	400	425	28, 35	–	–	–	–
Canthaxanthin	–	467	–	28, 35	2200	35	–	–
α-Carotene	420–422	442–445	472–474	28, 35, 46, 72, 76, 96	2710 2735 2700	44	65	47

β-Carotene	423–429	447–451	473–479	6, 15, 35, 46, 62, 64, 73, 76, 88, 96	2592	28, 35	8–27	18, 47
γ-Carotene	431–437	460–462	489–494	25, 28, 35, 46	2760	28	31	47
					3100	35		
ϵ-Carotene	414	439	470	21, 35	2900	35	–	–
β-Carotene 5,6-epoxide	423	444	473	28	–	–	–	–
Crocoxanthin	422	445	475	26, 35	–	–	79	26
α-Cryptoxanthin	421–424	444–448	472–478	20, 35, 38, 46, 72	2636	28	69–71	20, 47
					2625	35		
β-Cryptoxanthin	(422–425)	446–452	475–480	28, 35, 72, 80	2460	35	–	–
Cryptoxanthin diepoxide	416	439	470	28	–	–	–	–
Cryptoxanthin epoxide	419	443	472	28	–	–	–	–
Diadinoxanthin	(421–424)	445–448	474–475	28, 35, 68, 73, 74, 100	2110	68	57	68
					2250	100		
Diatoxanthin	–	447–450	475–479	20, 26, 35, 68, 74	–	–	48	50
4,4′-Diketo,3-hydroxy-β-carotene	–	462–478	–	35	–	–	–	–
Dinoxanthin	416	439–442	470–471	35, 73	–	–	–	–
Echinenone	(432)	459	483	28, 35	–	–	–	–
Fucoxanthin	425–427	449–450	476–478	28, 31, 35, 44, 100	1600	100	–	–
3′-Hydroxyechinenone	–	456	–	28, 35	–	–	–	–
Isocryptoxanthin	(427)	451	479	28, 35	–	–	–	–
Isozeaxanthin	(427–428)	450–451	478–479	28, 35	2400	27	–	–
Lutein	420	443–445	472–475	20, 35, 76	–	–	60–74	20, 47
Lycopene	443–448	471–476	501–507	25, 28, 35, 46, 75	3450	35	54	47
Monadoxanthin	422	445	474–475	2, 20, 23, 35	–	–	72	20
Mutatochrome	397–404	422–427	450–456	27, 28, 35, 44	–	–	–	–
Neoxanthin	412–416	435–439	462–468	27, 28, 35, 46, 76, 96	–	–	87	47
Octadehydro-β-carotene	(445)	472	501	31a			14	31a
Peridinin	(431)	454–457	484–490	28, 35, 62, 73	–	–	–	–

Table 4–4 (*Cont.*)

Carotenoid	Absorption maxima (nm)			Reference	$E^{1\%}_{1cm}$	Ref.	%III/II	Ref.
	I	II	III					
Pyrrhoxanthin	–	459	487–489	28, 35, 68	–	–	32	68
Pyrrhoxanthinol	–	457	484	68	–	–	–	–
Taraxanthin	416–420	439–445	469–471	28, 35, 46, 96	–	–	86	47
Ureolide	(427)	448	472	37	–	–	38	37
Violaxanthin	416–418	439–443	469–473	28, 46, 76, 96	–	–	97	48
β-Zeacarotene	400–407	425–428	451–454	28, 35, 80	1940	28, 35	–	–
Zeaxanthin	(425–429)	447–450	447–480	18, 20, 27, 28, 35, 76, 96	2340	35	25–46	20, 48
Methanol								
Aphanizophyll	444	475	505	35	–	–	–	–
Astaxanthin	–	472	–	24, 28, 35	–	–	–	–
Diadinoxanthin	–	445	474	68	2250	68	–	–
19′-Hexanoyloxyfucoxanthin	–	443	–	9	–	–	–	–
Lutein	418	444	474	28	–	–	–	–
Oscillaxanthin	464	496	531	28	–	–	–	–
Oscillol	464	492	524	87	–	–	–	–
Peridinin	–	462–467	–	28, 35	–	–	–	–
Peridinol	–	464	–	68	–	–	–	–
Pyrrhoxanthin	–	457	–	68	–	–	–	–
Pyrrhoxanthinol	–	459	–	68	–	–	–	–
Ureolide	–	448	470	37	–	–	2	37
Violaxanthin	415	440	469	28	–	–	–	–
Zeaxanthin	422	450	481	28	–	–	–	–

Petroleum spirit								
Alloxanthin	–	450	479	28	–	–	–	–
Anhydrodiatoxanthin	(440)	462	490	12	–	–	26	12
Antheraxanthin	421–424	443–445	472–473	14, 27, 28	–	–	67	14
Aphanizophyll	445	472	502	35, 52, 53	–	–	64	53
Aurochrome	380	400	425	35	–	–	–	–
Auroxanthin	379–382	399–402	423–427	28, 35	–	–	–	–
Canthaxanthin	–	461–463	–	28, 79	2200	28	–	–
α-Carotene	418–424	443–448	473–476	12, 13, 14, 15, 28, 35, 46, 60, 83, 84, 93	2800	28, 35	60–73	12, 13, 14, 15
β-Carotene	(421–430)	448–451	474–482	12, 13, 14, 15, 28, 35, 52, 53, 55, 68, 83	2592	28	20–40	12, 13, 14, 15, 95
γ-Carotene	431–437	456–462	486–494	8, 20, 35, 46, 54, 84	3100	28	–	–
ϵ-Carotene	413–416	437–440	467–470	28, 30, 35	3010	35	–	–
					3120	28		
β-Carotene 5,6-epoxide	(420)	442–447	470–476	28, 55, 69	–	–	–	–
α-Cryptoxanthin	420–421	441–446	472–473	13, 15, 17	–	–	57–73	11, 13, 15
β-Cryptoxanthin	(425–426)	446–449	472–476	13, 14, 15, 28, 46, 52	2386	28	12–33	13, 14, 47
Cryptoxanthin diepoxide	417	439–443	470–471	28, 35, 55	–	–	–	–
Cryptoxanthin epoxide	–	447	478	35	–	–	–	–
Diadinoxanthin	(424)	444–445	475	68, 95	2110	68	57–65	68, 95
Diatoxanthin	(428–430)	448–451	476–482	5, 35, 95	–	–	38	95
Echinenone	–	455–458	(482)	28, 36, 52	2158	27	–	–
Eutreptiellanone	(438)	458	487	12	–	–	50	12

Table 4–4 (*Cont.*)

Carotenoid	Absorption maxima (nm) I	II	III	Reference	$E^{1\%}_{1cm}$	Ref.	%III/II	Ref.
Fucoxanthin	(425–427)	446–449	473–478	28, 35, 46, 47, 65, 66, 85	1650	66	–	–
Fucoxanthinol	425	448	476	28	–	–	–	–
19′-Hexanoyloxyfucoxanthin	(427)	445	474	95	–	–	33	95
3′-Hydroxyechinenone	–	452–457	–	25, 28, 35, 52	–	–	–	–
Isocryptoxanthin	(425)	447–448	474–475	28, 35	–	–	–	–
Isozeaxanthin	(428)	446–451	473–379	28, 35	2400	28	–	–
Lutein	418–421	443–445	472–474	13, 15, 28, 55, 84	–	–	67–75	13, 15
Lycopene	440–446	468–472	500–505	1, 28, 30, 35	3450	28	–	–
Mutatochrome	402–409	425–428	452–453	28, 36, 55, 69	–	–	–	–
Neoxanthin	410–418	435–442	465–468	5, 28, 35, 83	–	–	–	–
Prasinoxanthin	(429)	452	480	85	–	–	–	–
Siphonaxanthin	(427)	446–450	468–478	28, 60, 85, 86	–	–	–	–
Siphonein	(427–434)	450–457	473–481	12, 28, 60, 85, 86	–	–	10	12
Taraxanthin	(419)	442–443	470–472	28, 35, 83	–	–	–	–
Violaxanthin	415–420	438–443	466–472	5, 28, 35, 83	–	–	–	–
β-Zeacarotene	406	427–429	454	28, 35	2520	21, 35	–	–
Zeaxanthin	424–428	446–449	475–476	14, 15, 28, 35, 47	2350	28	32–38	14, 15, 48

Sources: 1, Aasen & Liaaen-Jensen (1966); 2, Aihara & Yamamoto (1968); 3, Aitzetmuller et al. (1968); 4, Aitzetmuller et al. (1969); 5, Allen et al. (1964); 6, Antia (1965); 7, Antia & Cheng (1982); 8, Antia & Cheng (1983); 9, Arpin et al. (1976); 10, Berger et al. (1977); 11, Bernhard et al. (1974); 12, Bjørnland (1982); 13, Bjørnland (1983); 14, Bjørnland (1984); 15, Bjørnland & Aguilar-Martinez (1976); 16, Bjørnland & Tagen (1979); 17, Bjørnland et al. (1984); 18, Brown & McLachlan (1982); 19, Buchecker et al. (1976); 20, Chapman (1966a); 21, Chapman & Haxo (1963); 22, Chapman & Haxo (1966); 23, Cheng et al. (1974); 24, Cooper et al. (1975); 25, Czygan & Heumann (1967); 26, A. J. Davies et al. (1984); 27, B. H. Davies (1965); 28, B. H. Davies (1976); 29, Egger et al. (1969); 30, Eidem & Liaaen-Jensen (1974); 31, Eskins et al. (1977); 31a, Fiksdahl & Liaaen-Jensen (1988); 32, Fiksdahl et al. (1984a); 33, Fiksdahl et al. (1984b); 34, Fiksdahl et al. (1984c); 35, Foppen (1971); 36, Foss et al. (1984); 37, Foss et al. (1986); 38, Foss et al. (1987); 39, Francis & Halfen (1972); 40, Francis et al. (1970); 41, Francis et al. (1973); 42, Garside & Riley (1968); 43, Gieskes & Kraay (1983b); 43a, Gieskes & Kraay (1986); 44, Goodwin (1955); 45, Hager & Meyer-Bertenrath (1966); 46, Hager & Meyer-Bertenrath (1967a); 47, Hager & Meyer-Bertenrath (1967b); 48, Hager & Stransky (1970b); 49, Hager & Stransky (1970c); 50, Halfen & Francis (1972); 51, Healey (1968); 52, Hertzberg & Liaaen-Jensen (1966a); 53, Hertzberg & Liaaen-Jensen (1966b); 54, Hertzberg & Liaaen-Jensen (1966c); 55, Hertzberg & Liaaen-Jensen (1967); 56, Hertzberg & Liaaen-Jensen (1969a); 57, Hertzberg & Liaaen-Jensen (1969b); 58, Jeffrey (1961); 59, Jeffrey (1968a); 60, Jeffrey (1968b); 61, Jeffrey (1968c); 62, Jeffrey & Haxo (1968); 63, Jeffrey & Vesk (1977); 64, Jeffrey et al. (1975); 65, Jensen (1961); 66, Jensen (1966); 67, Johansen & Liaaen-Jensen (1974); 68, Johansen et al. (1974); 69, Jungalwala & Cama (1962); 70, Kleinig & Egger (1967); 71, Lewin et al. (1977); 72, Loeber et al. (1971); 73, Loeblich & Smith (1968); 74, Mandelli (1968); 75, Marchand et al. (1965); 76, Mues et al. (1973); 77, Nitsche (1973); 78, Norgard et al. (1974a); 79, Norgard et al. (1974b); 80, Palla et al. (1970); 81, Pennington et al. (1985); 82, Renstrom et al. (1981); 83, Ricketts (1966b); 84, Ricketts (1967a); 85, Ricketts (1967c); 86, Ricketts (1971a); 87, Ronneberg et al. (1980); 88, Skjenstad et al. (1984); 89, Smallidge & Quackenbush (1973); 90, Strain (1958); 91, Strain et al. (1970); 92, Stransky (1978); 93, Stransky & Hager (1970a); 94, Stransky & Hager (1970b); 95, Tangen & Bjørnland (1981); 95a, Vesk & Jeffrey (1987); 96, Weber (1969); 97, Whitfield & Rowan (1974); 98, Whittle & Casselton (1975a); 99, Whittle & Casselton (1975b); 100, Withers et al. (1977); 101, Withers et al. (1981); 101a, Wright & Jeffrey (1987); 102, Yokohama et al. (1977).

A. Cis–trans isomerization

Natural marine carotenoids are almost entirely in the trans configuration, but traces of acids during extraction can cause a slight but characteristic cis-peak in the absorption spectrum about 142 nm below the wavelength of the maximum with longest wavelength in the spectrum of the unaltered pigment. Thus inspecting this part of the absorption spectrum can show whether this isomerization has occurred to any extent. As a downward shift of 2–5 nm also occurs in the absorption spectrum, this can affect the value of spectroscopy as a diagnostic tool. Cis-isomerization also decreases the %III/II value (Davies, 1976). The cis and trans isomers can be separated by chromatography, some adsorbing more strongly than the parent compound and some more weakly. Davies (1976) discussed the conventions for nomenclature of these isomers; those less polar form the series neo A, neo B, etc., with decreasing polarity, and the more polar are neo U, neo V etc. Davies (1976) discussed the chromatographic separation of isomers of fucoxanthin, peridinin, and diadinoxanthin, reported also by Strain (1958), Strain and Svec (1969), Stransky and Hager (1970a,b), and Hallegraeff (1976). Cheng et al. (1974) found that the cis isomer of alloxanthin, manixanthin, was a postmortem isomerization and not a naturally occurring pigment in the healthy cells of *Chroomonas salina*.

B. Epoxide formation

Most 5,6-epoxides are rapidly converted by dilute acid to the furanoid 5,8-epoxides (Davies, 1965, 1976; Goodwin, 1980; Rowan, 1981; Britton, 1983) (Fig. 4–3), though fucoxanthin is stable to acid (Jensen, 1966), and peridinin reacts more slowly than usual with other epoxides (Loeblich and Smith, 1968; Kjøsen et al., 1976). Davies, Matthews, and Kirk (1970) exposed developed TLC plates to fumes of HCl, distinguishing the rich-blue reaction product of diepoxides from the blue-green of the mono-epoxides; similar color changes occur on unbuffered TLC plates of silica gel left exposed to light after development. Mutatochrome, a 5,8-epoxide, occurs naturally in some Cyanophyceae (Liaaen-Jensen, 1978).

4.4. Methods of preventing degradation

As partial degradation distorts the visible spectrum and will affect other methods of chemical analysis, all types of it should be avoided during extraction of carotenoids.

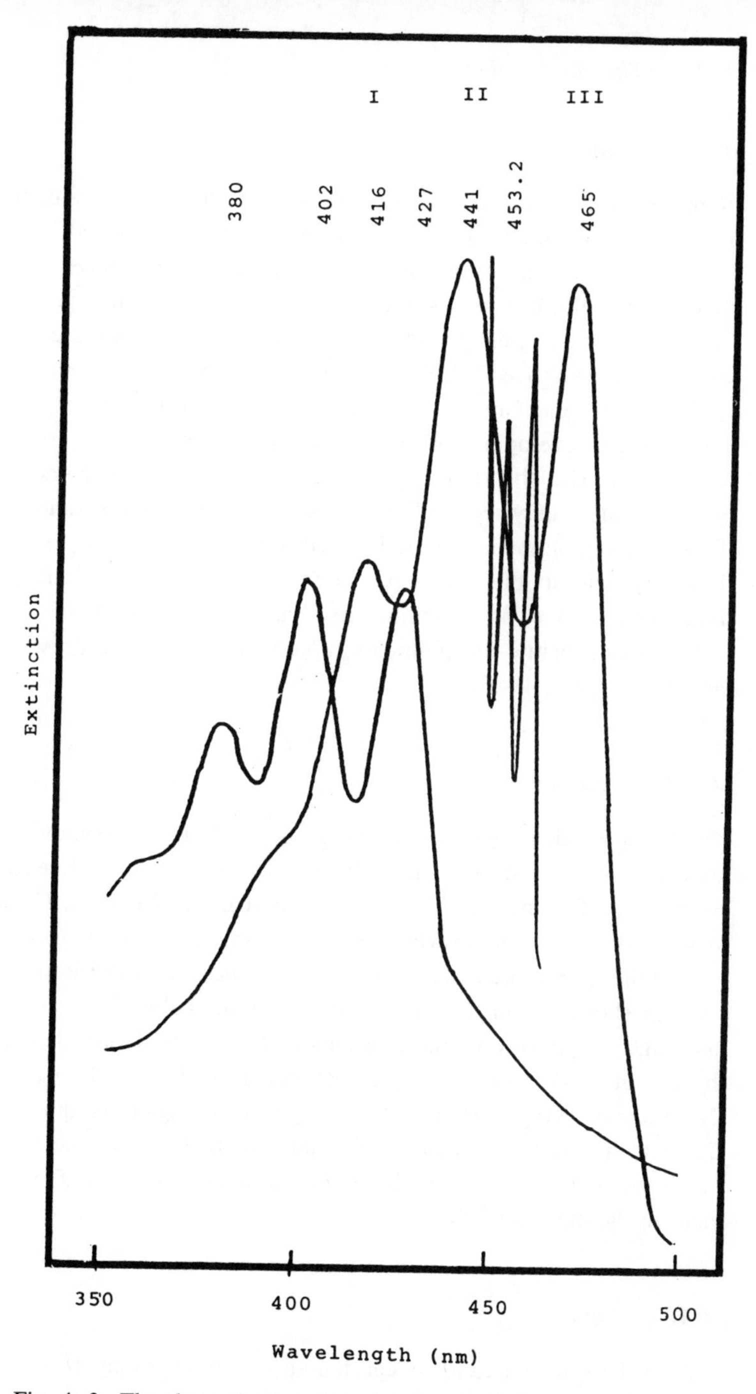

Fig. 4–3. The absorption spectrum in ethanol of violaxanthin extracted from *Nitella* (peaks I, II, and III), with a calibration peak from the spectrum of a holmium filter (453.2 nm). The hypsochromic shift following conversion of violaxanthin to auroxanthin by dilute HCl (ca. 40 nm) is shown. %III/II = 93 (cf. Table 4–4).

4.4.1. Acidity

Some organic solvents, such as chloroform, contain HCl as a degradation product and should be used when fresh or freshly redistilled if used as a component of a chromatographic solvent (Jeffrey, 1961; Hager and Meyer-Bertenrath, 1966) or as a solvent for spectroscopy. Silica gel is widely reported to be sufficiently acidic to cause degradation when used for TLC, but in Chapter 2 (Sec. 2.2.3C), we saw that suspending the gel in alkaline buffer (Keast and Grant, 1976) and including an alkaline organic compound in the developing solvent (Riley and Wilson, 1965) overcame these difficulties. The absorption spectrum of violaxanthin prepared by silica gel chromatography seen in Fig. 4–3 shows a high %III/II ratio. Solvents for extracting pigments should be used in sufficient volume to dilute rapidly any acid in the cell sap (Strain and Svec, 1969), and a water-immiscible solvent is sometimes used as a component of the extracting solution to separate the pigments rapidly from any acid present (Strain and Svec, 1966, 1969).

4.4.2. Oxidation

Oxidation at all stages of extracting and purifying carotenoids must be avoided, and Garside and Riley (1969), Liaaen-Jensen and Jensen (1971), Britton and Goodwin (1971), Davies (1976), and Goodwin (1980) have described precautions preventing it. Flushing atmospheric oxygen away with cylinder nitrogen and handling in darkness are most important, and storage should be in darkness at temperatures below 0°C, as both light and high temperature promote oxidation. Jensen and Liaaen-Jensen (1959) have cautioned against drying kieselguhr paper before eluting pigments. Nelson and Livingstone (1967) found that including ethoxyquin (6-epoxy-2,2,3-trimethyl-1,2-dihydroxyquinoline) up to 0.7% in the chromatography solvent delayed degradation of carotenoids when TLC plates were held in darkness for 2 h.

4.4.3. Light

All workers in the field of carotenoid chromatography recommend reducing light to a minimum when handling extracts of carotenoids. Jensen (1966) has described how a very mild exposure was sufficient to form cis isomers of fucoxanthin during chromatography, but this could be prevented by rigorously excluding light. As carotenoids are particularly sensitive to light when adsorbed on chromatographic materials, chromatograms of all types should be shaded during development (Davies, 1976).

4.4.4. *Heat*

Heating is sometimes used during evaporation or saponification, but is usually not necessary for either. Rotary evaporation at low pressure should not be at more than 40°C (Davies, 1976), and solvents with low boiling points are to be preferred to reduce the time of heating, though the temperature does not rise above the boiling point of the solution at the pressure used. The advantage of heating during saponification is in reducing the time taken, but longer treatment at room temperature is probably less liable to cause degradation.

4.5. Physical properties

4.5.1. *Absorption spectra*

The number of conjugated double bonds, the number of carbonyl groups, the number and type of epoxide groups, and the solvent in which the carotenoid is dissolved will affect the position of their main absorption maxima (Moss and Weedon, 1976). The majority of algal carotenoids (all trans) show three peaks of absorption in the visible spectrum (Fig. 4–3), designated I, II, and III in order of ascending wavelength. Peak I is sometimes present only as a shoulder in molecules containing the β end-group, making accurate measurement of the maximum difficult. In those molecules containing an ϵ end-group such as α-carotene, peak I is distinct. The structure of the visible spectrum of the carotenoids is due to the polyene chromophore seen in simple form in lycopene (Fig. 4–4). Peaks I, II, and III are distinct and the fine structure sharp, shown by the large value of the ratio of heights of peaks III to II relative to the trough between them, expressed as %III/II (Ke et al., 1970). Cyclization to a β end-group leads to steric hindrance between the methyl group at C-5 and the polyene chain and thus a shortening of the chromophore; fine structure decreases with each cyclization and the absorption maxima fall by about 12 nm (Fig. 4–4), seen in the series lycopene–γ-carotene–β-carotene (Goodwin, 1980; Britton, 1983). The effect of ϵ-cyclization of one end-group only in α-carotene leads to less steric hindrance than in

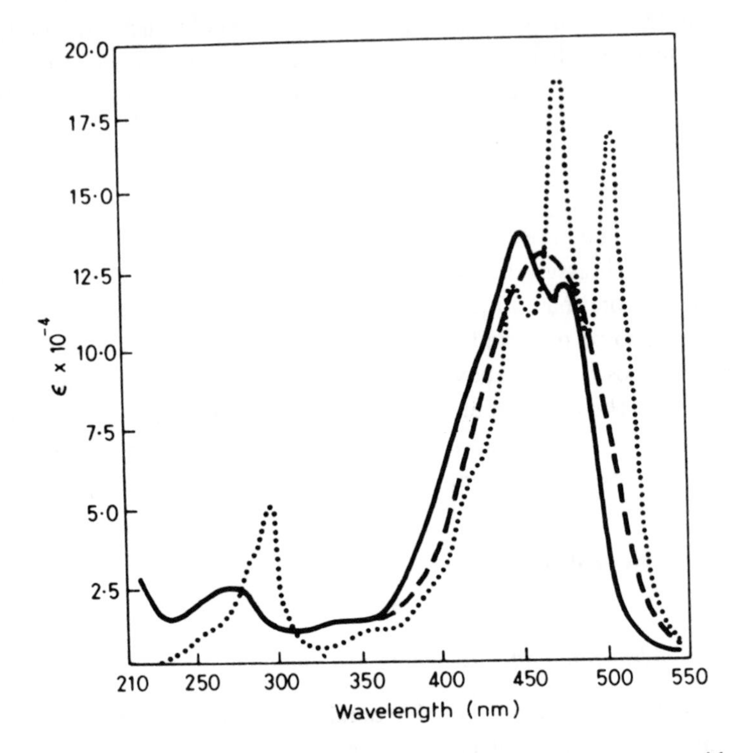

Fig. 4–4. The absorption spectra of lycopene (. . .), a carotene without cyclic end-groups, and of β-carotene (—) and echinenone (----), showing the effect of β-cyclization and insertion of a carbonyl group, respectively (see Sec. 4.5.1). (From Goodwin, 1980, by permission.)

β-carotene, giving less loss of fine structure and less change in wavelength of the absorption maxima. Moss and Weedon (1976) have discussed other effects of modifying the polyene chromophore by conjugation with another chromophore. Changes in the structure of the polyene chromophore may be complex; steric hindrance does not occur when an acetylenic bond replaces the 7,8 double bond, as in alloxanthin: Steric hindrance between the methyl group at C-5 and the polyene chain no longer occurs, but the expected increase in wavelength of the spectrum does not occur, because replacing the 7,8 double bond by the acetylenic bond lowers the spectrum by an equal amount.

Insertion of carbonyls in the end-groups, as in echinenone (one) and canthaxanthin (two) raises the maximum absorption and causes loss of fine structure (Fig. 4–4). Similarly, loss of structure occurs with a similar substitution at C-8 on the polyene chain, as in fucoxanthin and siphonaxanthin, and with butenolide formation in peridinin and related compounds. 5,6-Epoxide groups lower the maximum by about 8 nm, but hydroxyls cause negligible changes, whether on the end-groups or polyene chain (compare β-carotene with zeaxanthin and vaucheriaxanthin).

4.5.2. *Absorption spectra of algal carotenoids in nine solvents*

The absorption maxima shown in Table 4–4 have been taken from more recent reports. The maxima for peak II are most important, as the spectra of peaks I and III are sometimes not as sharp, and, in some pigments containing β end-groups, peak I is only a shoulder. Listing each report of the maxima is impractical, and the table shows the range in which the values for each peak fall. The maxima reported for peak II usually vary by 2–5 nm; this may reflect the degree of isomerization and oxidation during extraction or the inaccurate calibration of the spectrophotometer used. The peak at 453.2 nm in the absorption spectrum of a holmium filter provides a valuable calibration, as this falls close to the value of peak II of many carotenoids. The holmium absorption spectrum is shown in Fig. 4–3, superimposed on a trace made with a Philips SP 8 spectrophotometer.

4.5.3. *Chemical examination by spectroscopy*

Although complete elucidation of chemical structure requires modern methods of physical chemistry, such as NMR, MS, and ORD, chemical reactions causing change in the absorption spectra provide valuable preliminary information.

A. *Treatment with acids: the hypsochromic shift*

Dilute acid (i.e., a drop of 0.1 M HCl added to the cuvette of a spectrophotometer) rapidly converts a 5,6-epoxide into a 5,8-furanoid oxide, decreasing the absorption maxima by about 20 nm per epoxide group (Davies, 1965; Schimmer and Krinsky, 1966; Goodwin, 1980; Krinsky and Welankiwar, 1984). Figure 4–3 shows this change for violaxanthin, when it is converted into auroxanthin. The test thus distinguishes the diepoxide, violaxanthin, from the monoepoxide xanthophylls such as diadinoxanthin, neoxanthin, antheraxanthin, and taraxanthin. As mentioned in Section 4.3.2B, peridinin reacts more slowly with mineral acid than do the epoxides mentioned above, and fucoxanthin is stable.

B. *Iodine isomerization*

The all-trans configuration of the algal carotenoids in solution can be converted to a quasi-equilibrium mixture of the cis and trans isomers by several means. Although heating and lighting alone bring about the conversion, the usual method used is to illuminate a solution of the carotenoid to which catalytic amounts of iodine have been added. Zechmeister (1962) illuminated solutions of carotenoids in hexane or benzene containing 0.1–1.0 mg of carotenoid per milliliter and iodine to give 1–2% of the mass of the carotenoid; a mixture containing one-third to one-half of cis isomers is rapidly formed, and a peak appears about 142 nm below the peak of the absorption spectrum with the longest wavelength main peak of the visible spectrum. The peaks in the visible spectrum also fall by about 2–5 nm. As light and heat alone cause some isomerization, they must be avoided as far as possible during extraction and handling of carotenoids in solution, and the size of a cis peak provides a test for the extent of isomerization during extraction. Ricketts (1967a) found a cis peak and a fall in optical maxima of about 2 nm in aged samples of γ-carotene. A loss of fine structure also occurs during isomerization, and a decrease in the value of %III/II also provides evidence for degradation during extraction. Table 4–4 shows the variations of %III/II reported in the literature for some pigments.

4.6. Estimation

The techniques for estimating carotenoids have been reviewed by Davies (1965, 1976), Liaaen-Jensen and Jensen (1971), Goodwin (1974, 1976, 1979, 1980), Jensen (1978), Weber and Wettern (1980), Krinsky and Welankiwar (1984), and Liaaen-Jensen and Andrewes (1985).

Carotenoids are usually assayed by workers interested in taxonomy, where certain carotenoids are more or less specific to different classes (Table 4–1), or in ecophysiology of standing crops of algae, marine or freshwater. Hallegraeff (1976) has discussed Margalef's ideas about succession in developing populations; he claimed that an increase in carotenoid/chlorophyll ratios paralleled a change from an eutropic to an oligotropic population in freshwater lakes. Although, as mentioned below, the ratio E_{430}/E_{665} used by Margalef does not give values as accurate as those using chromatography, the concept is valuable; however, measuring the concentrations of pigments specific for the different classes of algae gives more precise information about the changes in the populations with time.

4.6.1. By spectroscopy

As with chlorophylls, carotenoids may be assayed by measuring their absorption in unpurified extracts, but these assays will not be as accurate as those for the chlorophylls. Four or more carotenoids usually occur in any class of algae; their absorption maxima are clustered around 450 nm, and, in addition, the extinction coefficients of peridinin and fucoxanthin are much lower than those of the other common carotenoids, thus making equations based on a single extinction coefficient at 450 nm inaccurate. However, approximate concentrations of the carotenoids can be measured by use of equations based on the optical properties of β-carotene. The equation below, proposed by Jaspers (1965), was intended for analysis with higher plants using a 1-cm light path, but would be suitable for extracts from algae unless they contained chlorophyll *c*, fucoxanthin, or peridinin.

$$\text{Total carotenoids } (\text{mg}\cdot\text{l}^{-1}) = 4.1\, E_{450} - 0.0435\, C_a - 0.367\, C_b$$

where C_a and C_b are the concentrations of chlorophyll *a* and *b* determined with a chromatic equation (Arnon, 1949; Jeffrey and Humphrey, 1975). The factor for multiplying E_{450} is based on the specific extinction coefficient of β-carotene in 90% acetone and will not be seriously in error for the carotenoids of the Chlorophyta. Richards and Thompson (1952) proposed equations correcting for chlorophyll *a*, *b*, and *c*, reading absorbance at 480 nm rather than at 450 nm, thus minimizing absorbance by the chlorophylls. They made a correction for the animal pigment astacin, found in zooplankton, but Parsons and Strickland (1963) considered that this correction could give rise to artifacts. Strickland and Parsons (1972) proposed using two equations (light path, 10 cm), depending on the crop analyzed:

$$\text{mg}\cdot\text{m}^{-3}\text{ Carotenoid} = 4.0\, E_{480} \text{ (Chlorophyta or Cyanophyceae)}$$
$$\text{mg}\cdot\text{m}^{-3}\text{ Carotenoid} = 10.0\, E_{480} \text{ (Chrysophyceae or Dinophyceae)}$$

Haxo (Prézelin, 1976) used the following equations for determining peridinin mixed with chlorophyll *a*:

In 90% acetone: $\text{Peridinin } (\text{mg}\cdot\text{l}^{-1}) = 7.52\, E_{469} - 0.124\, E_{663}$

In diethyl ether: $\text{Peridinin } (\text{mg}\cdot\text{l}^{-1}) = 7.35\, E_{475} - 0.095\, E_{661}$

The error involved in using the ratio E_{430}/E_{665} proposed by Margalef (1968) has been discussed in Chapter 3 (Sec. 3.4.2A) (Hallegraeff, 1976, 1977), and the chromatic equations are to be preferred, though these have not been corrected for the revised extinction coefficients for chlorophyll *c* determined by Jeffrey (1969, 1972). The equations of Parsons and Strickland (1963) have been used by Hallegraeff (1976, 1977) for measuring carotenoid/chlorophyll ratios in the summer-succession in freshwater lakes, and by Dustan (1979) for measuring changes in carotenoids per cell in zooxanthellae down to 60 m. The equations calculated by Seely et al. (1972) for measuring concentrations of chlorophylls, fucoxanthin, and β-carotene in brown algae extracted with DMSO have been discussed in Chapter 3 (Sec. 3.4.1). Although complicated by requiring three fractions in different solvents, they would be valuable when both DMSO and acetone must be used for extracting pigments. Wheeler (1980) employed this method for assaying fucoxanthin in *Macrocystis*.

4.6.2. By chromatography

As chromatic equations rarely (Seely et al., 1972) provide accurate assays of carotenoid concentrations, chromatography has been widely used for this purpose. The techniques are generally similar to those described for chlorophylls in Chapter 3 (Sec. 3.4.2). Saponification of the acidic pigments is sometimes used to remove chlorophylls and degradation products, thus simplifying subsequent chromatography (Geisert et al., 1987). When the specific extinction coefficient of the eluted pigments is not known, taking a value of 2500 gives a good approximation (see Table 4–4), as recommended by Liaaen-Jensen and Jensen (1971).

A. Column chromatography

This method is used for separating carotenoids, particularly in large-scale preparations, but has not been widely utilized for quantitative analysis, probably because the fractions containing the pigments are usually eluted in a mixture of solvents in which their extinction coefficients are not known accurately. Ricketts (1966b) calculated approximate percentage composition of carotenoids extracted from *Micromonas*, and Establier and Lubian (1982) and Lubian and Establier (1982) calculated the percentage composition of carotenoids in *Nannochloropsis* using this method.

B. Paper chromatography

Assay of carotenoids by using circular paper chromatography was described by Jensen and Liaaen-Jensen (1959) and Jensen (1959), using kieselguhr-filled paper, S & S 287, as described in Chapter 2 (Sec. 2.2.3B, Factors regulating R_f values), for chlorophylls. This technique was applied in the extensive study of carotenoids in Norwegian Phaeophyceae by Jensen (1966), and by Jensen and Sakshaug (1973) in a study of the seasonal changes in abundance and physiological state of phytoplankton in the Trondheimsfjord. Norgard et al. (1974a,b) used it for measuring the percentage composition of carotenoids. Although S & S paper gives good resolution of carotenoids, few workers other than Jensen and associates have used it for quantitative analysis. Jeffrey (1961) used her two-dimensional method (Chapter 2, Sec. 2.2.3B, Two-dimensional paper chromatography) for measuring the concentration of carotenoids in four classes of algae, and it has been used for calculating the percentage composition of carotenoids in two chrysophytes (Jeffrey and Allen, 1964), in zooxanthellae (Jeffrey and Haxo, 1968), and in calculating the approximate proportions of carotenoids in phytoflagellates (Ricketts, 1970). Hallegraeff (1976, 1977) employed a modification of the method of Bauer (1952) in calculating total carotenoid/chlorophyll absorption ratios and percentage absorption of individual carotenoids.

C. Thin-layer chromatography

Although TLC has been used in some physioecological studies, assay of carotenoids by TLC has been employed mainly for determining the percentage composition of carotenoids in a particular species, principally in the schools of Hager (Munich), Jeffrey (Australia), Liaaen-Jensen (Trondheim), Riley (Liverpool), and Ricketts (Leeds). Table 2–4 shows the most common TLC systems used and the R_f values of the algal carotenoids found with these systems. The thin layer is scraped or sucked (Riley and Wilson, 1965; Jeffrey, 1968a) from the plates and eluted with a small volume of a suitable solvent, when it is used for spectroscopy, or it is concentrated for further chromatography on another layer. The solvent must dissolve the pigment efficiently (the more polar carotenoids are not soluble in nonpolar solvents such as hexane or petroleum spirit), and, if quantitative analysis is required, the extinction coefficient of the pigment in that solvent must be known (Table 4–4). Hager and Meyer-Bertenrath (1966) have described an apparatus for applying extracts to plates for quantitative analysis.

Dubinsky and Polna (1976) used Merck 5720 silica gel plates for quantitative chromatography of pigments extracted from a *Peridinium* bloom, showing the changes in concentration of β-carotene, diadinoxanthin, and peridinin with time. Using commercial plates of silica gel, Sardana and Mehkotra (1979) found that diadinoxanthin increased in parallel with biomass during a bloom of *Gymnodinium*. Jeffrey and Hallegraeff (1980b) calculated ratios of chlorophyll *a* to fucoxanthin and peridinin, using quantitative chromatography on cellulose plates (Jeffrey, 1981) as indicators of the type of phytoplankton present at two sites in the warm-core eddy of the Eastern Australian current; they used this method also for analyzing pigments from continental shelf waters of Northern and Northwestern Australia (Hallegraeff and Jeffrey, 1984) and in a second warm-core eddy (Mario) (Jeffrey and Hallegraeff, 1987b). These workers believed that the dinoflagellates found in eddy Mario were heterotropic, since peridinin was rarely seen on the TLC plates. Hallegraeff (1981) used this method also for estimating approximate concentrations of fucoxanthin and peridinin at a coastal station off Sydney, Australia. Jeffrey and Vesk (1977) have performed quantitative TLC on sucrose plates (Jeffrey, 1968a) to examine the effect of light quality on the ratios of chlorophyll *a* to fucoxanthin, and diadinoxanthin–diatoxanthin in *Stephanopyxis turnis*. Guillard, Murphy, Foss, and Liaaen-Jensen (1985) have used TLC to measure a high proportion of zeaxanthin in the carotenoid extracts from several clones of the blue-green alga, *Synechococcus*, taking this as evidence in support of the suggestion by Gieskes and Kraay (1983a) that the particle passing in 1-μm filter was in fact a blue-green alga.

D. *High-performance liquid chromatography*

Quantitative analysis of carotenoids by HPLC, as with the chlorophylls (Chapter 3, Sec. 3.4.2C) requires measuring peak-heights or areas under the curves on the traces of the chromatograms. Paerl et al. (1983) used HPLC analysis for studying the role of carotenoids in blooms of the blue-green alga, *Microcystis*, finding that the ratio of myxoxanthophyll, zeaxanthin, and β-carotene to chlorophyll *a* increases two- to fourfold during the proliferation of the blooms, and concluded that the increased carotenoid provides protection from UV damage and enhances light utilization in the lower and middle regions of the PAR spectrum. Paerl (1984) has carried out laboratory experiments examining the role of carotenoids in enhancing photosynthesis in bloom species of the Cyanophyta. Gieskes and Kraay (1983b) used HPLC analysis to correct microscopic cell counts of the composition of spring blooms in the central North Sea, presumed to be diatoms. No Chrysophyceae had been detected in the fixed material used for counting, but carotenoid analysis showed that alloxanthin, specific to the Cryptophyceae (Table 4–1), was more abundant than fucoxanthin during the peak of the bloom. Reexamination of fixed material by spectroscopy showed that part of the alloxanthin was derived from a cryptophyte endosymbiont in a cillate, *Mesodinium rubrum,* which would not have been considered as an alga in the original count. Kleppel and Pieper (1984) looked for fucoxanthin, peridinin, and neoxanthin in the gut contents of copopods harvested off the California coast as indicators of diatoms, dinoflagellates, or Chlorophyta in their diet. By this means, they found that dinoflagellates were the principal food; fucoxanthin appeared only in one species at one of the five sampling sites, and neoxanthin was not found. Thus diatoms were rarely, and nanoplankton never ingested, though both were detected in the area studied.

Korthals and Steenbergen (1985) have examined 27 pigments from phytoplankton (bacterial and algal) sampled from the metalimnion of Lake Vechten (Netherlands) during autumn. They were able to follow the change from a population dominated by bacteria (bacteriochlorophylls *d* and *e*, isorenieratene, and okenon) and Cyanophyceae (zeaxanthin) at 6–7 m depth, to one dominated by Cryptophyceae (alloxanthin), Dinophyceae (peridinin), and Bacillariophyceae (fucoxanthin). Paerl et al. (1984) analyzed carotenoids in *Prochloron* strains isolated from five hosts using HPLC, finding results essentially similar to those of earlier workers (Withers et al., 1978a,b), with high levels of β-carotene. Their tentative reference to cryptoxanthin has been confirmed by Foss, Lewin, and Liaaen-Jensen (1987), using TLC. Peak 6 on their trace from the chromatogram may be the β-carotene monoepoxide identified by Foss et al. (1987) as mutatochrome.

Burger-Wiersma et al. (1986) used HPLC when examining a newly discovered prochlorophyte. Their analyses did not detect several of the carotenoids commonly found in *Prochloron* (echinenone, cryptoxanthin, mutatochrome). Again, an unknown peak appeared between chlorophyll *a* and β-carotene, as observed by Paerl et al. (1983) with *Prochloron*. Abaychi and Riley (1979) found that assays of the major carotenoids from a species from each of three classes by HPLC agreed well with TLC, and, using polychromatic equations, found that total carotenoids agreed well for two of the three species. Braumann and Grimme (1981) tested their method of HPLC by comparing spectral data for the carotenoids in *Chlorella fusca* with results published elsewhere. Alberte and Andersen (1986) identified antheraxanthin by HPLC in their study of its role in light-harvesting in *Chrysosphaera magna*.

Table 4–5. *Partition coefficients of carotenoids determined in a two-phase system of hexane (epiphase)–aqueous methanol (hypophase)*

Carotenoid	% Methanol in hypophase					Reference
	95	85	75	60	50	
α-Carotene	100	–	–	–	–	Petracek & Zechmeister (1956)
β-Carotene	100	–	–	–	–	
γ-Carotene	100	–	–	–	–	
Lycopene	100	–	–	–	–	
Echinenone	95	–	–	–	–	
Isocryptoxanthin	86	–	–	–	–	
β-Cryptoxanthin	82	–	–	–	–	
Canthaxanthin	50	–	–	–	–	
Lutein	12	43	–	–	–	
Zeaxanthin	11	40	–	–	–	
Neoxanthin	–	–	–	9	38	Goldsmith & Krinsky (1960)
	–	–	–	8	37	Krinsky & Goldsmith (1960)
Zeaxanthin	10	34	81	–	–	Chapman (1966a)
	–	39	85	–	–	
3-Hydroxyechinenone	25	–	–	–	–	Hertzberg & Liaaen-Jensen (1966a)
Echinenone	6	–	–	–	–	Hertzberg & Liaaen-Jensen (1966b)
β-Carotene	100	–	–	–	–	Loeblich & Smith (1968)
Pyrrhoxanthin	77	–	–	–	–	
Dinoxanthin	7	–	–	–	–	
Diadinoxanthin	6	–	–	–	–	
Peridinin	2	–	–	–	–	

Note: The coefficient is the percentage of carotenoid in the epiphase.

4.7. Partition coefficients and relative polarities

In two-phase systems, usually hexane or petroleum spirit (epiphase) and aqueous methanol (hypophase), the carotenoids are partitioned according to the degree of the polarity of their molecules. Fractionation into epiphasic and hypophasic carotenoids was described in Chapter 2 (Sec. 2.2.2) as a preparative procedure, but Petrachek and Zechmeister (1956) proposed that the partition observed when a single carotenoid was introduced to a two-phase system consisting of equal parts of hexane and 85% or 95% methanol provided a quantitative physical property characteristic of the polarity of the molecule. Krinsky (1963) used this method in two systems: (1) petroleum spirit / aqueous methanol, and (2) petroleum spirit–diethyl ether (3:1) / aqueous methanol. Published partition distributions (percent carotenoid in the epiphase) are shown in Table 4–5. Krinsky

Table 4–6. *M_{50} values and relative polarities (RP) of xanthophylls*

Xanthophyll	M_{50}	RP	Reference
Zeaxanthin	85.5	2.0	Schimmer & Krinsky (1966)
Antheraxanthin	78.3	2.24	
Violaxanthin	74.0	2.48	
Neoxanthin	59.0	3.24	
Violaxanthin	65.6	2.51	Ricketts (1966b)
Zeaxanthin	84	—	Cheng et al. (1974)
Alloxanthin	80	—	
Canthaxanthin	95	—	Petrachek & Zechmeister (1956)

Note: The M_{50} value is the percentage of methanol to give a partition coefficient of 50:50 against hexane or petroleum spirit.

(1963) also calculated M_{50} values as that percentage of methanol giving a partition distribution of 50:50 in the two phases, and plotted this against a relative polarity value calculated by giving each functional group of the molecule a polarity, taking a nonallylic hydroxyl as one, and other groups as less than one. The allotted polarities of molecules of known structure were adjusted by trial and error so that the plot of M_{50} against relative polarity gave a good fit to a straight line with a negative slope. The M_{50} value of a molecule of unknown structure could then be entered on the graph to give a relative polarity. Ricketts (1966b) used this method to help identify the structure of unknown xanthophylls in *Micromonas pusilla*. However, Liaaen-Jensen (1971) does not feel that relative polarity calculated in this way has general applicability. M_{50} values reported for xanthophylls from algae are shown in Table 4–6. The unit is impractical for the carotenes, owing to their very low polarity.

5

The biliproteins

5.1. Introduction

In Chapter 1, we saw that four major types of biliproteins – phycocyanin, phycoerythrin, allophycocyanin, and phycoerythrocyanin – are light-harvesting pigments in the Cyanophyceae, Rhodophyceae, and Cryptophyceae. In the cell, the lipid-soluble pigments, the chlorophylls and carotenoids, are bound to proteins, but this bond is broken when these pigments are extracted with nonpolar solvents, and the spectrum of a given pigment is identical from any source. The covalent bond between the phycobilins and the apoprotein is not broken when the pigments are extracted from the cells in water, and, owing to interaction between the chromophore and the protein, the spectrum of a biliprotein will vary between species. Numerous reviews have described the structure and properties of the biliproteins (O'hEocha, 1962, 1965a,b, 1966; Stanier, Kunisawa, Mandel, and Cohen-Bazire, 1971; Bogorad, 1975; Glazer, 1976, 1977, 1981, 1982, 1983, 1984, 1985; O'Carra and O'hEocha, 1976; Gantt, 1977, 1979, 1981, 1986; Stanier and Cohen-Bazire, 1977; Rüdiger, 1980; Scheer, 1981, 1982; Bryant, 1982; Cohen-Bazire and Bryant, 1982; Gysi and Chapman, 1982; MacColl, 1982; MacColl and Guard-Friar, 1987).

Gantt and Conti (1966a,b) reported arrays of particles associated with thylakoid membranes seen in Rhodophyceae that consisted mainly of biliproteins; accordingly, they named them the phycobilisomes. They were subsequently found in virtually all Cyanophyceae and Rhodophyceae, though they do not occur in the Cryptophyceae. Scheer (1981) has pointed out that the absence from the Cryptophyceae is consistent with the failure of cryptophycean biliproteins to form aggregates in vitro. The structure of the phycobilisomes will be considered in Section 5.9.1.

5.2. Biosynthesis

As with the chlorophylls, the biliproteins are formed from protoporphyrin, but the mechanism has been examined mainly in animals. The

A

B

Peptide-linked PHYCOCYANOBILINS

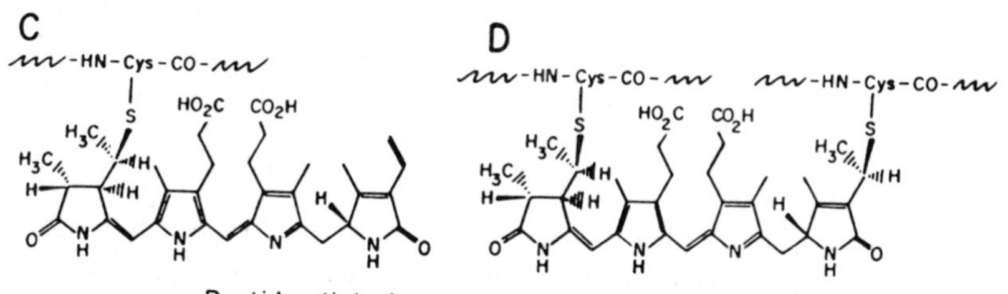

E

F

Peptide-linked PHYCOUROBILINS

Fig. 5–1. Structure of the common phycobilins (PCB, PEB, and PUB), showing peptide linkages to the apoproteins, and the numbering and lettering system used. (Reproduced by permission from the *Annual Review of Biophysics and Biophysical Chemistry,* Vol. 14, ©1985 by Annual Reviews, Inc.)

final steps involve oxidative ring opening of the porphyrin ring with production of carbon monoxide (Troxler and Dokos, 1973). Although the synthesis from [5-^{14}C]ALA gives the expected distribution of ^{14}C in carbon monoxide and phycobilins, no specific labeling could be observed by Troxler and Brown (1975) in Cyanophyceae. The biliprotein subunits require addition of cysteine to the ethylidene group on ring A (Fig. 5–1), (Rüdiger, 1980).

5.3. Distribution in the algal classes

The classes containing the biliproteins are divided into two parts: those with phycobilisomes (Cyanophyceae and Rhodophyceae), and those without phycobilisomes (Cryptophyceae) with biliproteins contained in the intrathylakoid spaces. This leads to differences in the mechanisms of light harvesting in the two groups; in the Cyanophyceae and Rhodophyceae, harvested light-energy passes along the gradient PE → PC → APC → chlorophyll *a*, and the pigments are associated in trimeric or hexameric units. In the Cryptophyceae, the pigments in the intrathylakoid spaces react directly with chlorophyll. The major reviewers in this field have stressed that all phycobilisomes contain APC and PC (Glazer, 1977, 1981, 1984, 1985; Gantt, 1980, 1981; Scheer, 1981, 1982; Bryant, 1982; Zilinskas and Greenwald, 1986; MacColl and Guard-Friar, 1987), though some workers have failed to find APC and/or PC in extracts prepared directly from tissue (Hirose, Kumano, and Madono, 1969; Honsell, Kosovel, and Talarico, 1984) owing to the low concentration of PC in many Rhodophyceae; the pigments should be found more readily by isolating phycobilisomes first. The long table presented by Honsell et al. (1984) is deceptive since few workers were isolating all pigments.

The prefixes C-, R-, and B- were used originally for PC and PE extracted from Cyanophyceae, Floridian Rhodophyceae, and Bangiales, respectively. However, we now know that pigments from these groups do not always fall within the class designated by the prefixes; the Cyanophyceae sometimes contain R-PE (O'hEocha, 1962) or R-PC II (Ong and Glazer, 1987); the Floridian Rhodophyceae, B-PE (Glazer, West, and Chan, 1982) and C-PC (O'hEocha, 1962); and the Bangiales, R-PE (O'hEocha, 1962; Glazer et al., 1982; Jiang et al., 1983; Pan and Tseng, 1985) and C-PC (Pan and Tseng, 1985; Fujiwara-Arasaki, Yamamoto, and Kakiuchi, 1985), but the system has proved convenient for dividing pigments into spectral types (Table 5–1). MacColl (1982) has suggested CU- as a prefix for R-PE-like pigments from the Cyanophyceae; the majority of the Cyanophyceae containing CU-PE (Table 5–1) are marine, and Glazer (1984) has pointed out that this is the dominant biliprotein in marine species of this class, with effective light absorption as low as 500 nm, coinciding with the maximum wavelength of light transmitted by certain types of seawater (Jerlov, 1976). Recently, Olson et al. (1988) have confirmed this finding by analyzing phytoplankton in the sea by using single- and dual-beam flow cytometry. They point out that determinations made by using fluorescence excitation at 540–570 nm rather

Table 5–1. *Maxima of the absorption and fluorescence-emission spectra of the biliproteins of the Cyanophyceae and the Rhodophyceae*

Biliprotein	pH	Absorbance Maxima (nm) 1st peak	2nd peak	3rd peak	Fluorescence maximum (nm)	Reference
APC	5.5	–	651	–	–	Glazer & Cohen-Bazire (1971)
		–	651	660	–	Kursar et al. (1983a)
	6.0	–	–	–	662	Adams et al. (1979)
		–	620	650	–	Murakami et al. (1981)
	6.8		630[a]	650	660	Gantt & Lipschultz (1973, 1974)
		–	–	648	660	Bryant et al. (1976)
		–	620[a]	650	660	Trench & Ronzio (1978)
		605	–	651	–	Grabowski & Gantt (1978)
		–	–	652	662	Bryant & Cohen-Bazire (1981)
	6.85	–	–	650	659	Ley et al. (1977)
	7.0	618[a]	–	650	–	A. S. Brown et al. (1975)
		600[a]	–	650	–	Gray & Gantt (1975)
		618[a]	–	650	–	A. S. Brown & Troxler (1977)
		626[a]	–	647	660	Tyagi et al. (1980)
		–	–	646–653	660–662	Czeczuga (1987)
	8.0	605[a]	–	650	661	Nies & Wehrmeyer (1980)
Type I	7.0	–	–	654	678	Zilinskas et al. (1978)
		–	–	653	–	Zilinskas et al. (1980)
		–	–	654	678	Troxler et al. (1980)
		–	–	654	681	Canaani & Gantt (1980)
Type II	7.0	590[a]	620[a]	650	660	Zilinskas et al. (1978)
		–	–	650	662	Gysi & Zuber (1979)
		–	625[a]	650	660	Canaani & Gantt (1980)
		–	–	650	661	MacColl et al. (1980)
		590[a]	620[a]	648	–	Zilinskas et al. (1980)
Type III	7.0	–	–	650	660	Zilinskas et al. (1978)
		–	625[a]	650	660	Canaani & Gantt (1980)
		–	–	650	660	Troxler et al. (1980)
		592[a]	–	650	–	Zilinskas et al. (1980)
Type A	7.0	595	623	654	–	Leclerc (1983)
Type B	7.0	–	618	671	–	Glazer & Bryant (1975)
		619	–	669	673	Ley et al. (1977)
		618	–	673	681	Canaani & Gantt (1980)
		622	645	671	–	Troxler et al. (1980)
		622	640	671	680	Zilinskas et al. (1980)
		–	640	660	676	Leclerc (1983)
C-PC						
	3.0	617	–	–	647	Goedheer & Birnie (1965)
	3.9	612	–	–	649	MacColl et al. (1980)

Table 5–1 (*Cont.*)

Biliprotein	pH	Absorbance Maxima (nm) 1st peak	2nd peak	3rd peak	Fluorescence maximum (nm)	Reference
	5.5	621	–	–	–	Glazer & Cohen-Bazire (1971)
		621	–	–	–	Glazer et al. (1973)
	6.0	615	–	–	646	Adams et al. (1979)
		616	–	–	652	Seibert and Connolly (1984)
	6.8	614	–	–	637	Bryant et al. (1976)
		618	–	–	–	Grabowski & Gantt (1978)
		620	–	–	645	Trench & Ronzio (1978)
	7.0	620	–	–	648	Dale & Teale (1970)
		625	–	–	–	Glazer et al. (1973)
		620	–	–	–	Gray & Gantt (1975)
		625	–	–	–	Troxler et al. (1975)
		622	–	–	–	Gardner et al. (1980)
		625	–	–	–	MacColl et al. (1980)
		610	–	–	643	Tyagi et al. (1980)
		620	–	–	–	Fuglistaller et al. (1981)
		635				
		623–625	–	–	646	Priestle et al. (1982)
		615	–	–	632	Fujiwara-Arasaki et al. (1985)
		612–618	–	–	637–643	Czeczuga (1987)
	8.0	617	–	–	651	Goedheer & Birnie (1965)
		628	–	–	641–650	Nies & Wehrmeyer (1980)
PC 623	7.0	623	–	–	642	Gray et al. (1976)
PC 637	7.0	637	–	–	653	Gray et al. (1976)
		630	–	–	652	Gantt et al. (1986)
Type II	7.0	618	–	–	–	Brown et al. (1975)
		617	–	–	642	Bryant et al. (1981)
		618	–	–	648	Gantt et al. (1986)
R-PC						
Type I	5.5	555[a]	614	–	641	Kursar et al. (1983a)
	6.8	555	617	–	636	Gantt & Lipschultz (1973)
		555	617	–	637	Gantt & Lipschultz (1974)
		555	620	–	637	Grabowski & Gantt (1978)
		550	618	–	650	Zeng et al. (1984)
	7.0	555	618	–	–	Glazer & Hixson (1977)
Type II	7.0	533	554	615	646	Ong & Glazer (1987)
B-PE	5.5	498[a]	545	563	–	Glazer & Hixson (1977)
	6.8		545	–	575	Gantt & Lipschultz (1973)
		498[a]	545	563	575	Gantt & Lipschultz (1974)
		498[a]	545	563	575	Grabowski & Gantt (1978)
		–	–	–	575	Priestle et al. (1982)
Type I	6.3	490–498	533–539	562–565	574–575	Van der Velde (1973a,b)
	7.0	500[a]	542	568	575	Koller & Wehrmeyer (1975)
		496–503	538–550	564–570	574–577	Glazer et al. (1982)

Table 5–1 (*Cont.*)

Biliprotein	pH	Absorbance Maxima (nm) 1st peak	2nd peak	3rd peak	Fluorescence maximum (nm)	Reference
Type II	6.3	490–498	533–539	558–560	570–572	Van der Velde (1973a,b)
	7.0	500[a]	542	568	578	Koller & Wehrmeyer (1975)
		498–499	529–537	566–568	574–575	Glazer et al. (1982)
b-PE	5.5	–	545	563	–	Glazer & Hixson (1977)
	6.8	–	545	–	570	Gantt & Lipschultz (1973)
		–	545	563[a]	575	Gantt & Lipschultz (1974)
		–	545	563[a]	575	Grabowski & Grantt (1978)
C-PE	5.5	–	565	–	–	Glazer & Cohen-Bazire (1971)
		–	559	–	578	Bryant & Cohen-Bazire (1981)
		–	550–558	568	–	Bryant (1982)
	6.8	–	551	–	570	Alberte et al. (1984)
	7.0	–	565	–	–	Teale & Dale (1970)
		–	562	–	–	Glazer & Hixson (1975)
			551	566[a]	–	Gray & Gantt (1975)
		–	558	–	–	Rüdiger et al. (1980)
		–	557	572	581	Leclerc (1983)
		–	556–562	–	576–577	Czeczuga (1987)
Type I	6.8	540[a]	560	–	–	Grabowski & Gantt (1978)
Type II	6.8	540[a]	563	–	–	Grabowski & Gantt (1978)
	7.0	–	564	–	–	Bennett & Bogorad (1973)
CU-PE						
Type I	5.5	500	545	–	560	Alberte et al. (1984)
	6.0	495	–	565	–	Hirose et al. (1969)
		497	543	–	563	Kursar et al. (1981)
	7.0	498	–	565	–	Rippka et al. (1974)
		501	–	564	573	Bryant et al. (1981)
		498	540	567	–	Stadnichuk et al. (1985)
Type II	6.5	500	547	565[a]	573	Fujita & Shimura (1974)
		496	536	558[a]	–	Cox et al. (1985)
Type III	7.0	492	543	–	565	Ong et al. (1984)
R-PE	6.0	–	–	–	573	Adams et al. (1979)
Type I	5.5	498	545	565	574	Honsell et al. (1984)
	6.3	490–498	532–544	560–568	573–581	Van der Velde (1973a,b)
	6.8	498	542	565	575	Gantt & Lipschultz (1980)
		498	540	565	576	Jiang et al. (1983)
		–	–	565	579	Zeng et al. (1986)
	7.0	493–499	534–544	564–568	572–577	Glazer et al. (1982)
		498	540	565	–	Stadnichuk et al. (1984)
Type II	6.3	490–498	532–544[a]	560–568	573–577	Van der Velde (1973a,b)
	6.8	485	545[a]	565	–	Honsell et al. (1984)
		498	540[a]	560	–	Jiang et al. (1983)
		–	–	560	576	Zeng et al. (1986)
	7.0	496–497	538–551	565–566	574–577	Glazer et al. (1982)

Table 5–1 (*Cont.*)

Biliprotein	pH	Absorbance Maxima (nm) 1st peak	2nd peak	3rd peak	Fluorescence maximum (nm)	Reference
Type III	7.0	497	539	565	–	Spencer et al. (1981)
		496	539–541	564–567	572	Glazer et al. (1982)
r-PE	6.8	498	542	560	575	Gantt & Lipschultz (1980)
PEC	5.5	515–530[a]	568–577	595–600	–	Bryant (1982)
	6.0	–	575	595[a]	625	MacColl et al. (1981)
	6.8	–	568	590[a]	607–610	Bryant et al. (1976)
	7.0	–	570	590[a]	630	Tyagi et al. (1980)
		–	576	–	–	Fuglistaller et al. (1981)
	8.0	530[a]	575	595[a]	625	Nies & Wehrmeyer (1981)

[a]Shoulders.

than at 488 nm will underestimate the abundance of cyanophytes containing CU-PE.

Parry (1984) reported what appeared to be CU-PE types I and II in cyanophytes epiphytic on ascidians, and Cox, Hiller, and Larkum (1985) found CU-PE type II and C-PE in a unicellular cyanophyte similar in structure to *Prochloron,* also symbiotic in colonial ascidians. Larkum et al. (1987) have also found both CU-PE and C-PE in cyanophytes in symbiosis with sponges and an ascidian; the CU-PE contained PUB on both α and β subunits, unlike other CU-PE (see Table 5–3), though the PUB/PEB ratios have not yet been measured. The nonmarine cyanophytes containing CU-PE include the abnormal terrestrial *Gloeobacter* (Rippka et al., 1974; Bryant, 1981) and *Oscillatoria princeps* from a thermal spring (Stadnichuk, Romanova, and Selyakh, 1985).

Phycoerythrocyanin, first described by Bryant, Glazer, and Eiserling (1976), is restricted to the Cyanophyceae and does not occur in the same species as PE (Bryant, 1982). Though it occurs in some marine species, the maximum absorption at about 570 nm is not as effective as some types of CU-PE with absorption at 500 nm (Table 5–1) in harvesting green light in the ocean. Bryant (1982) found that PEC tended to be limited to subgroups of the Cyanophyceae forming heterocysts, or to the genus *Chroococcoidiopsis* with possible affinities to heterocyst-forming groups.

The Cryptophyceae contain only PC or PE, differing from the equivalent pigments in the classes containing phycobilisomes in several respects. Scheer (1981, 1982) has used the prefixes Cr- or K- for cryptophyte biliproteins, and the former will be used here (Table 5–2). In an early report, O'hEocha and Raftery (1959) claimed that traces of Cr-PC accompanied the Cr-PE extracted from cultures grown in a south-facing window, but the slight absorbance that they observed at 620–630 nm might be similar to that found in Cr-PE_{545} extracted from *Chroomonas* sp. CS 24 by Martin and Hiller (1987); it is generally assumed now that a cryptophyte contains only one biliprotein. Symbioses between Dinophyceae and Cryptophyceae causing anomalous distribution of biliproteins in the latter class has been discussed by Wilcox and Wedemayer (1984, 1985) and Wilcox (1985). Hu, Yu, and Zhang (1980) have extracted Cr-PC_{645} from *Gymnodinium cyaneum,* and Zhang et al. (1983) have extracted Cr-PC_{630} from another species of *Gymnodinium* (Table 5–2). Barber, White, and Siegelman (1969) extracted Cr-PE_{545} from symbionts from the marine protozooan *Cyclotrichium meunieri*.

Small amounts of polypeptides containing covalently bound PCB occur in the core of PBSs, discussed in Section 5.9.1, below. These are known by their molecular weights and designated 75-K polypeptide, 18.3-K polypeptide, and so on. Gantt (1981) has listed those identified so far.

5.4. Absorption and fluorescence-emission spectra

Table 5–1 consolidates many of the more recent reports of absorption and fluorescence-emission spectra of the various types of the biliproteins in the Cyanophyceae and Rhodophyceae. Some workers have identified two or more variants within a type, shown in Fig. 5–2. Gantt and Lipschultz (1973, 1974, 1980) have used lowercase prefixes (r or b) to denote pigments with similar spectra to those with uppercase prefixes but of lower molecular weight. Variation in absorption maxima between different laboratories, shown in Tables 5–1 and 5–2, reflect the effects described in Sections 5.6.1–3, below or, perhaps, the accuracy of the spectrophotometers used. The absorption spectra of the biliproteins vary in complexity, showing one to three distinct peaks (Figs. 5–2 and 5–3), in particular, in the phycoerythrins (Glazer, 1984). Frequently, ill-defined peaks or shoulders occur, though these may be useful characteristics of certain pigments, such as Cr-PE_{545} and Cr-PE_{645} (Fig. 5–3) or B-PE I and II, and APC (Fig. 5–2). The number and types of phycobilin chromo-

Table 5–2. *Maxima of the absorption and fluorescence-emission spectra of the biliproteins of the Cryptophyceae*

Biliprotein	pH	Absorbance maxima (nm)			Fluorescence	Reference
		1st peak	2nd peak	3rd peak	maximum	
Cr-PC$_{570}$	6.8	*569*[a]	630	–	650–656	Hill & Rowan (1989)
Cr-PC$_{612}$	5.5	577	*612*	–	–	Glazer & Cohen-Bazire (1975)
	6.0	585	*612*	–	634	MacColl & Guard-Friar (1983a)
	6.4	585	*615*	–	–	O'hEocha & Raftery (1959)
	6.8	585	*615*	–	–	O'hEocha et al. (1964)
		578	*614*	–	641	Hill & Rowan (1989)
Cr-PC$_{630}$	6.7	584	*629*	–	649	Zhang et al. (1983)
	6.8	583	*625–630*	–	–	O'hEocha et al. (1964)
		580	*630*	–	648–657	Hill & Rowan (1989)
Cr-PC$_{645}$	H_2O	580	620.5	*645*	–	Allen et al. (1959)
	5.5	581	–	*641*	–	Glazer & Cohen-Bazire (1975)
	6.0	585	620[a]	*645*	655	MacColl et al. (1973)
		585	630[a]	*645*	661	MacColl et al. (1976)
		583	620[a]	*645*	662	Jung et al. (1980)
	6.7	580	–	*645*	–	Hu et al. (1980)
		583	620[a]	*645*	–	Meyer & Pienaar (1984)
	6.8	583	620[a]	*643*	–	O'hEocha et al. (1964)
		582	625[a]	*645–646*	661	Hill & Rowan (1989)
	7.0	584	620	*645*	660	Mörschel & Wehrmeyer (1975)
		584	620	*648*	662	Holzwarth et al. (1983)
Cr-PE$_{545}$	H_2O	*545*	–	–	–	Haxo & Fork (1959)
	5.5	*542*	562[a]	–	–	Brooks & Gantt (1973)
	6.0	*545*	560[a]	–	–	MacColl et al. (1976, 1983)
	6.4	*545*	–	–	–	O'hEocha & Raftery (1959)
	6.8	*542–544*	560–570[a]	–	–	O'hEocha et al. (1964)
		538–550	562–566[a]	–	583–588	Hill & Rowan (1989)
		550	–	644	584, 675	Hill & Rowan (1989)
	7.0	*544*	555–560[a]	–	584–586	Mörschel & Wehrmeyer (1975)
		545	–	–	580	Stewart & Farmer (1984)
		545	564[a]	644	586, 648	Martin & Hiller (1987)
Cr-PE$_{555}$	6.4	*554–556*	–	–	–	O'hEocha & Raftery (1959)
	6.8	*553–554*	–	–	–	O'hEocha et al. (1964)
		554	–	–	578	Hill & Rowan (1989)
Cr-PE$_{566}$	H_2O	*568*	612	–	–	Haxo & Fork (1959)
		565	600[a]	–	–	Allen et al. (1959)
	5.5	*567*	–	–	–	Glazer et al. (1971)
		566	–	–	–	Brooks & Gantt (1973)
	6.0	*560*	600[a]	–	617	MacColl et al. (1976)
	6.8	*565–566*	604[a]	–	617	Hill & Rowan (1989)
	7.4	*565*	–	–	–	Lichtlé et al. (1987)

Note: The wavelength of the major peak is italicized.
[a]Shoulder.

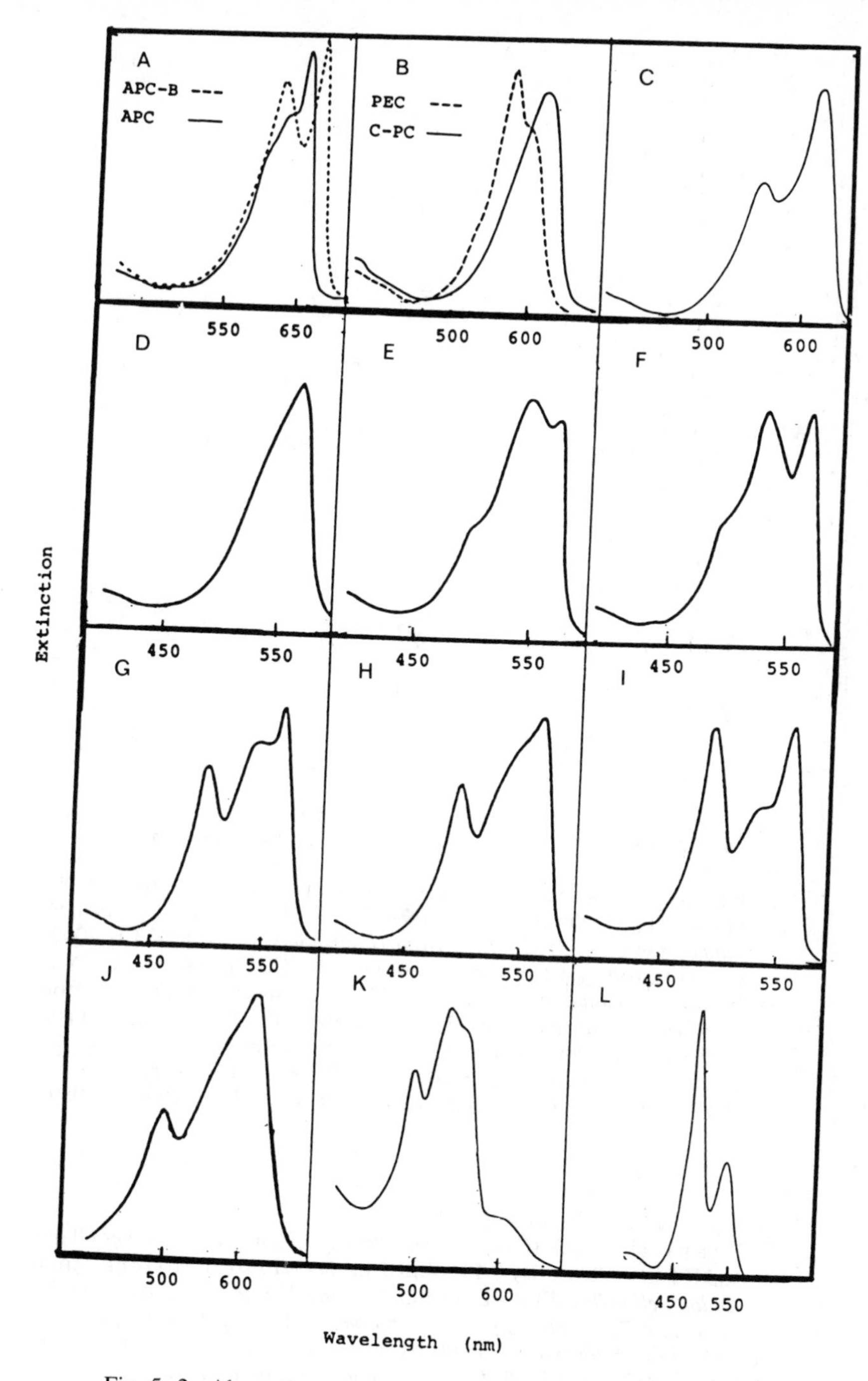

Fig. 5–2. Absorption spectra of the biliproteins from the Cyanophyceae and Rhodophyceae. (A) APC and APC-B from *Anacystis variabilis* (after Glazer and Bryant, 1975); (B) C-CP and PEC from *Anabaena* (after

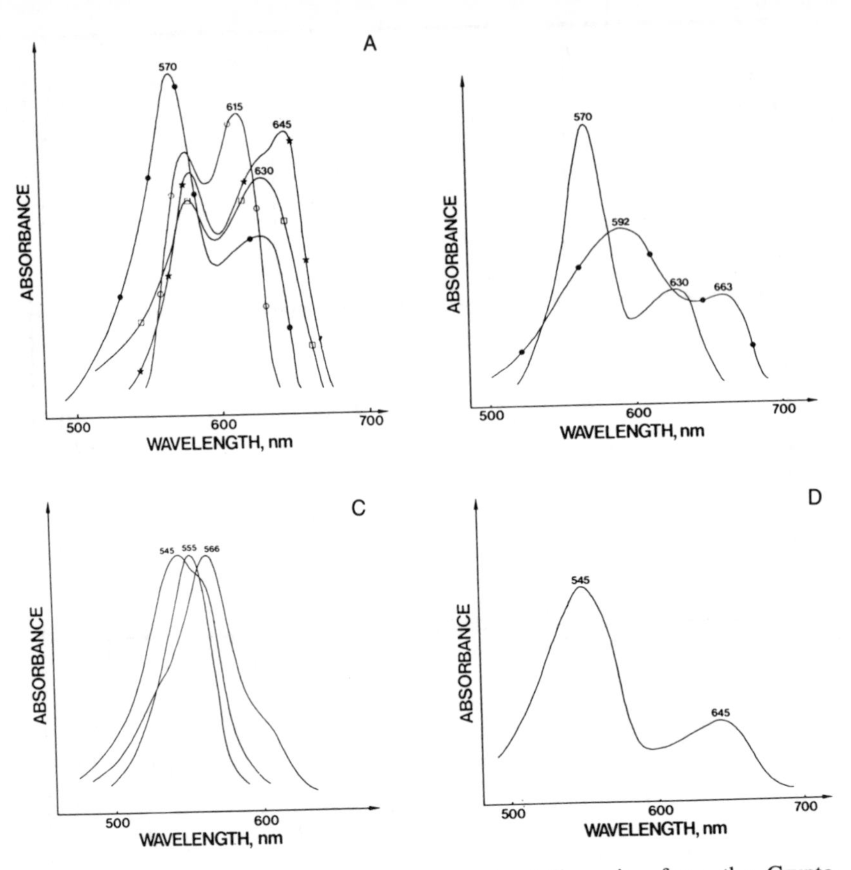

Fig. 5–3. Absorption spectra of the biliproteins from the Cryptophyceae. (A) Phycocyanins. (●—●) Cr-PC_{570} from *Chroomonas daucoides;* (○—○) Cr-PC_{615} from *Hemiselmis* sp. (UTEX LB 2002); (□—□) Cr-PC_{630} from *Chroomonas* sp. (NEPCC 51); (★—★) Cr-PC_{645} from *Chroomonas caerulea*. (B) Absorption spectra of Cr-PC_{570} from *Chroomonas daucoides* before (—) and after (●—●) treating with acid 8 M urea. (C) Phycoerythrins. Cr-PE_{545} from *Rhodomonas salina;* Cr-PE_{555} from *Hemiselmis brunnescens;* Cr-PE 566 from *Cryptomonas ovata*. (D) Cr-PE_{545} from *Cryptomonas acuta*. (Data of Hill and Rowan, 1989.)

Fig. 5–2. cont. from page 175
Bryant et al., 1976); (C) R-PC from *Porphyridium cruentum* (after Glazer and Hixson, 1975); (D) C-PE from *Synechocystis* 6701; (E) B-PE I from *Audouinella* 679; (F) B-PE II from *Goniotrichopsis sublittoralis;* (G) R-PE I from *Botryocladia pseudodichotoma;* (H) R-PE II from *Antithamnionella glandulifera;* (I) R-PE III from *Dasya* 2166; (D–I after Glazer et al., 1982); (J) CU-PE I from *Gloeobacter violaceus* (after Bryant et al., 1981); (K) CU-PE II from *Synechocystis trididemni* (after Cox et al., 1985); (L) CU-PE III from *Synechococcus* WH 8103 (after Ong et al., 1984).

phores covalently bound to the apoproteins largely determine the number of peaks in the absorption spectra, though, as we shall see in Sections 5.6.2–3, interaction between chromophores and protein and the state of polymerization of the units will affect the shape of the spectrum. The spectra of the phycoerythrins of the Cyanophyceae and the Rhodophyceae are more complex than those of the phycocyanins (Fig. 5–3), but cryptophycean phycocyanins have more complex spectra than the phycoerythrins, with up to three distinct chromophores in the molecule (Fig. 5–3). Glazer, West, and Chan (1982) examined the phycoerythrins in over 100 species of the Rhodophyceae; they distinguished three types of R-PE by the intensity rather than the wavelength of the three peaks of the spectra. These three types – I, II, and III – include those identified by Hirose et al. (1969) and Van der Velde (1973a,b). Similarly, B-PE has been divided into two types separated by the relative intensity of the two peaks of their spectra. APC also occurs in multiple types – I, II, III, and B; these are not all distinguished by their spectra, as only B has a distinct maximum and the others must be separated by column fractionation; however the fluorescence emission of I and B separate them from II and III. The cyanophyte phycoerythrins, designated CU-PE, now also fall into three types, based on the relative sizes of the peaks at about 495 and 565 nm (see Sec. 5.6.1).

The biliproteins of the Cryptophyceae (Table 5–2), reviewed by Gantt (1979), Glazer (1981), and Hill and Rowan (1989) have attracted less attention than those of the other two groups, probably because there are few of them and they do not occur in the phycobilisomes that are being investigated actively at present. Gantt (1979) tabulated virtually all of the spectral data available at that time. Whereas Cr-PE is usually similar to C-PE in that it has only one maximum in the spectrum, Cr-PC is more complex than C-PC and has two peaks, and a shoulder in one type. Earlier observations showed Cr-PE with a single peak of absorption at 545, 555, or 566 nm, falling cleanly into three types (Table 5–2). The spectra of Cr-PC were more complex, with two peaks, and a distinct shoulder in one type. They have been placed in up to three types, but the workers concerned have not agreed about the number (two or three) and the criteria for placing them in the types: One type has received three designations (I, II, or III) by various workers. In addition, we have recently found a Cr-PC in *Chroomonas daucoides* with the major peak at 569 nm, shown to contain more CV than PCB when treated with acid 8 M urea (Fig. 5–3). Confusion about the nomenclature of the biliproteins of the Cryptophyceae could be avoided by identifying them according to the value of the wavelength of the maximum absorption in the visible spectrum, as adopted by MacColl and Guard-Friar (1987) and Hill and Rowan (1989). Thus there would be Cr-PE$_{545}$, Cr-PE$_{555}$, Cr-PE$_{566}$, Cr-PC$_{570}$, Cr-PC$_{612}$, Cr-PC$_{630}$, and Cr-PC$_{645}$. Cr-PE$_{545}$ has a shoulder at the red side of the peak of absorption (Fig. 5–3), but Cr-PE$_{555}$ does not. Accordingly, pigments with this shoulder are placed in Cr-PE$_{545}$ when their maxima fall between 545 and 555 nm (Table 5–2).

Martin and Hiller (1987) and Hiller and Martin (1987) have reported a detailed study of a complex Cr-PE$_{545}$ in a *Chroomonas* sp. CS 24 with a definite shoulder at 644 nm. This pigment contains PEB on both the β subunits, but four types of α subunits were separated by ion-exchange chromatography, and eight or nine species of Cr-PE isoproteins were separated by isoelectric focusing, all with similar absorption spectra. They found that the chromophore on the α subunit changed reversibility on titration between a PCB-like form at pH 7 and a CV-like form at pH 1.9. The pigment was unusual in having two distinct maxima in the fluorescence-emission spectrum (586 and 648 nm), consistent with failure to transfer light energy from the PEB chromophores to the CV/PCB chromophores. Hill and Rowan (1989) have extracted a pigment from *Cryptomonas acuta* with the major absorption peak at 545 nm but a second substantial peak at 645 nm (Fig. 5–3) and with fluorescence emissions at 584 and 675 nm. The emission at 675 nm is at a much higher wavelength than that in *Chroomonas* CS24 (Martin and Hiller, 1987), suggesting that the chromophores on the α subunit differ in the two species. Fluorescence at such a high wavelength occurs only in APC I and B among other biliproteins (Table 5–1).

Many workers (see review by Scheer, 1981) have agreed that biliproteins of the same group (APC, PC, PEC, or PE) from the Cyanophyceae and the Rhodophyceae are closely related immunologically when tested by the Ouchterlony double-diffusion techniques, but those of different type do not cross-react. Unlike those of the other classes, the biliproteins of the Cryptophyceae showed cross-reactivity between Cr-PC and Cr-PE, and both reacted with B-PE from *Porphyridium cruentum* (MacColl, Berns, and Gibbons, 1976) but not with biliproteins from other sources (Berns, 1967; Glazer, Cohen-Bazire, and Stanier, 1971a,b). In contrast to these earlier results, Guard-Friar, Eisenberg, Edwards, and MacColl (1986) found cross-reactivity of cryptophyte biliproteins with C-PC. As similar amino-terminal sequences occur in Cr-PC and biliproteins from the Rhodophyceae and Cyanophyceae, Glazer and Apell (1977) concluded that the

Cryptophyceae were closer to both these classes than was previously believed.

5.5. Structure

The four established types of biliproteins are phycocyanin, phycoerythrin, allophycocyanin, and phycoerythrocyanin. As already mentioned, the biliproteins are now known to be conjugate proteins containing one or more phycobilins as prosthetic groups. The chemical structure, bonding to the apoprotein, and absorption spectra (Tables 5–1 and 5–2) of the phycobilins are known with varying degree of certainty. The experiments following the first demonstration, by Lemberg in 1928, that the prosthetic groups of the biliproteins were related to the bile pigments have been described by O'Carra and O'hEocha (1976). The chemical structures and bonding to the apoproteins proposed for phycocyanobilin and phycoerythrobilin by Killilea, O'Carra, and Murphy (1980) are not consistent with the linkages found by Glazer and co-workers (Glazer, 1985) between phycobilins and isolated peptides. They found that all were bound through thioether linkages to rings A, D, or both. The structure of phycoerythrobilin, phycocyanobilin, and phycourobilin, and the linkages found are shown in Fig. 5–1. O'hEocha, O'Carra, and Mitchell (1964) found evidence for a fourth phycobilin similar to the bilins of the class violin, in certain cryptophytes, coining the name "cryptoviolin" for it. However, when Bryant et al. (1976) first detected phycoerythrocyanin in Cyanophyceae, they found evidence for a chromophore in it similar to cryptoviolin in addition to PCB. Thus although now this phycobilin is not restricted to the Cryptophyceae, MacColl and Guard-Friar (1987) have used this name in their recent text and it will be used here. The abbreviation PXB has also been used for it in publications from several laboratories. Bishop et al. (1987) have examined the properties of CV isolated from PEC using NMR and MS, and have proposed a structure for it with a probable protein – thioether bond to ring A (see Fig. 5–1), though the linkage to ring D cannot yet be discounted. Investigations of the structure of the bilins and their linkages to the apoprotein have been discussed in detail by MacColl and Guard-Friar (1987).

One strain of the cryptophyte *Hemiselmis virescens* contains yet another unknown phycobilin with an absorption maximum at about 695 nm in acid urea (Glazer and Cohen-Bazire, 1975), and Jung et al. (1980) and MacColl and Guard-Friar (1983b) have also found a similar phycobilin in *Chroomonas* phycocyanin with apparent absorption maximum at 697 nm; the chemical structure is also unknown, and the latter authors have called it 697-nm bilin.

The biliproteins are readily depolymerized and early determinations of molecular weights varied with treatments used during extraction. O'Carra (1970) briefly reported splitting PE, PC, and APC into two subunits of slightly different molecular weights, but Glazer and Cohen-Bazire (1971) were the first to describe the two unequal chromopeptides formed by treating biliproteins with SDS as the α and β subunits, where α was the smaller – moving faster by PAGE. Subsequently, several workers found α and β subunits of equal size in some biliproteins, and Glazer et al. (1976) have suggested defining the α subunit as that with the amino-terminal sequence similar to the α subunit of *Synechococcus* sp. Although the α subunit in the Cyanophyceae and the Rhodophyceae is only slightly smaller than the β in the Cryptophyceae, α is only about half the molecular weight of β (Gantt, 1979). A third bilin-carrying subunit (γ) has been described in phycoerythrins by Glazer and Hixson (1977) and Ong, Glazer, and Waterbury (1984) and in allophycocyanin by Troxler, Greenwald, and Zilinskas (1980) and Zilinskas (1982). The techniques used for assigning chromophores to the subunits will be discussed in Section 5.6.1. Gantt (1981), Glazer (1981, 1982, 1985), Scheer (1981, 1982), Bryant (1982), and MacColl and Guard-Friar (1987) have summarized the phycobilins believed to occur in the subunits, shown in Table 5–3 with more recent references.

5.6. Factors affecting absorption spectra

Glazer (1982) and Glazer et al. (1982) have pointed out that some absorption spectra reported are of relatively crude extracts and that these must be considered suspect for one or more of the following reasons: (1) The extract can contain more than one biliprotein, even though the spectrum approximates that of the dominant pigment. (2) The method of preparation of the extract could lead to degradation through action of proteases in the tissue or from contaminating microorganisms, or low ionic strength of the medium. (3) Low ionic strength, lack of control of optimum pH (5–7), and dilute concentration could lead to dissociation of the polymer. Denatured biliproteins lose fluorescence (O'hEocha and O'Carra, 1961), and Glazer et al. (1982) have checked the E_{565}/E_{500} and E_{540}/E_{500} ratios of PE solutions prepared from over 100 species of Rhodophyceae, measured by both absorption spectroscopy and fluorescence excitation spec-

Table 5–3. *Distribution of the phycobilins among the subunits of the biliproteins*

Phycobilin	Aggregation state	Bilin content per subunit			References
		α	β	γ	
APC	$(\alpha\beta)_3$	PCB	PCB	–	Glazer & Fang (1973); A. S. Brown et al. (1975); Bryant et al. (1976); MacColl et al. (1978)
Type I	$(\alpha\beta)_3\gamma$	PCB	PCB	PCB	Zilinskas et al. (1978); Troxler et al. (1980);
Type II	$(\alpha\beta)_3$	PCB	PCB	–	
Type III	$(\alpha\beta)_3$	PCB	PCB	–	
Type B	$(\alpha\beta)_3$	PCB	PCB	–	Glazer & Bryant (1975); Ley et al. (1977)
C-PC					
pH 7	$(\alpha\beta)_{n=1-6}$	PCB	2PCB	–	Hattori et al. (1965); Neufeld & Riggs (1969); Bennett & Bogorad (1971); Glazer & Fang (1973); Bryant et al. (1976, 1978); Gardner et al. (1980)
pH 5–6	$(\alpha\beta)_6$	PCB	2PCB	–	Hattori et al. (1965); Neufeld & Riggs (1969); Saito et al. (1974); Glazer (1976)
Red light	$(\alpha_1\beta_2)_3$	PCB	2PCB	–	Bryant & Cohen-Bazire (1981); Bryant (1981)
Green light	$(\alpha\beta)$				
Cr-PC_{612}	$(\alpha_2\beta_2)$	PCB	2PCB CV	–	MacColl & Guard-Friar (1983a); Guard-Friar & MacColl (1986)
		PCB	PCB	–	Glazer & Cohen-Bazire (1975)

Table 5–3 (*Cont.*)

Phycobilin	Aggregation state	Bilin content per subunit α	β	γ	References
645	$(\alpha\alpha'\beta_2)$	CV PCB 697	CV PCB CV	–	Jung et al. (1980)
		697	2PCB	–	MacColl & Guard-Friar (1983b); Guard-Friar & MacColl (1986)
R-PC					
Type I	$(\alpha\beta)_3$	PCB	PCB PEB	–	Glazer & Hixson (1975); Bryant et al. (1978)
Type II	$(\alpha\beta)_2$	PCB	PCB PEB	–	Ong & Glazer (1987)
B-PE	$(\alpha\beta)_6\gamma$	2PEB	3PEB	2PEB 2PUB	Lundell et al. (1984)
b-PE	$(\alpha\beta)$	2PEB	3PEB	–	Lundell et al. (1984)
C-PE	$(\alpha\beta)_{n=1-6}$	2PEB	3/4 PEB	–	Muckle & Rüdiger (1977)
		2PEB	3PEB	–	Glazer (1985)
Cr-PE					
545	$(\alpha_2\beta_2)$	PEB	PEB	–	Mörschel & Wehrmeyer (1977)

	$(\alpha\alpha'\beta_2)$	2CV PEB	PEB	–	MacColl et al. (1983); Guard-Friar & MacColl (1986)
566	$(\alpha\beta)$	PEB	PEB	–	Glazer et al. (1971b)
CU-PE					
Type I	$(\alpha\beta)$	3PEB	3PEB PUB	–	Bryant et al. (1981)
		2PEB	2PEB PUB	–	Kursar et al. (1981)
		2PEB	3PEB PUB	–	Alberte et al. (1984)
		2PEB	3PEB PUB	–	Stadnichuk et al. (1985)
R-PE	$(\alpha\beta)_6\gamma$	2PEB	2PEB PUB	PEB 3PUB	Klotz & Glazer (1985), Glazer (1985)
		2PEB	3PEB PUB	3PEB 2PUB	Stadnichuk et al. (1984)
	$(\alpha\beta)_6\gamma'$	2PEB	3PEB PUB	1/2 PEB PUB	
Type I	$(\alpha\beta)_6$ $(\alpha\beta)$?	PUB/PEB: PUB/PEB: PUB/PEB:	0.23–0.28 0.23–0.25 0.36		Glazer et al. (1982)
PEC	$(\alpha\beta)_3$	CV	2PCB	–	Bryant et al. (1976)

tra, thus confirming that their samples of PE were not denatured or contaminated by PC (Fig. 5–2). Glazer (1982) has pointed out that three factors govern the spectroscopic properties of native biliproteins.

5.6.1. Type and number of the chromophores

We now know that binding to the apoprotein of the biliprotein distorts the absorption spectrum of the phycobilin to some extent. Phycoerythrobilin and phycocyanobilin are removed from their conjugate protein in their least altered state only by brief treatment of the biliproteins with cold, concentrated hydrochloric acid (O'hEocha, 1958, 1963; O'Carra et al., 1964; O'Carra, Murphy, and Killilea, 1980). The spectra of these released pigments closely corresponded with the spectra of the biliproteins denatured by low pH or by 8 M urea (Table 5–4); the unfolding of the polypeptide chain of the apoprotein exposes the enclosed chromophore, and the effect is readily reversed when low pH or urea is removed (O'Carra et al., 1980).

O'Carra et al. (1980) discussed work in other laboratories on the "purple pigment" and "blue pigment" released from biliproteins (phycoerythrin and phycocyanin) by treating with hot methanol: The spectra of these two pigments do not correspond with those of the denatured biliproteins, and the structure and spectra of the chromophores released by cold acid are now accepted (Glazer, 1982). Phycourobilin is not released by treating phycoerythrin with cold acid, but the products of hydrolysis with trypsin can be resolved into two chromopeptides containing PEB or PUB (O'Carra and O'hEocha, 1976; O'Carra et al., 1980). The absorption spectrum of denatured R-phycoerythrin (i.e., with no absorption maximum around 500 nm) leaves a spectrum equivalent to i-urobilin (O'Carra et al., 1980; Bryant et al., 1981). The recently detected phycobilins, CV, and the 697-nm bilin (see Sec. 5.4) were also identified by difference spectra, that is, by subtracting the spectrum equivalent to the amount of PCB known to be present from the total absorption spectrum of the PC (Bryant et al., 1976; O'Carra et al., 1980; MacColl and Guard-Friar, 1983a,b). MacColl and Guard-Friar (1983a,b) preferred using chromatic equations for calculating molar ratios of phycobilins on α and β subunits in Cr-PC from *Hemiselmis virescens* (Cr-PC_{612}) and *Chroomonas* (Cr-PC_{645}). They preferred their model for *Chroomonas* to that proposed by Jung et al. (1980), who made direct measurements of each bilin with numerous technical difficulties, inevitable in handling small samples.

Assigning chromophores to the subunits of the biliproteins became possible when Glazer and co-workers (Glazer, 1981) measured the extinction coefficients of PCB and PEB by denaturing appropriate biliproteins with acid 8 M urea. Although the covalent bonds between the apoprotein and the phycobilins still slightly distorted absorption maxima, the spectrum of the denatured biliprotein was essentially similar to the free phycobilin (Table 5–4), thus permitting identification of the phycobilins present. They then measured the molecular weight of the isolated subunits, and, using the extinction coefficients of the subunits denatured with urea, were able to determine the number of chromophores present on each (Table 5–3). However, when Lundell, Glazer, DeLange, and Brown (1984) determined the number of bilins linked to peptides in the β subunit from B-PE, they found only three PEB groups, rather than the four previously determined by spectroscopic analysis, and concluded that the extinction coefficient of the PEB determined by Glazer and Hixson (1975) was low and that the higher value determined by Muckle and Rüdiger (1977) was correct. A more recent estimate made on small bilin peptides containing PEB gave an even higher value, and a direct estimate for peptide-bound PUB was also made by Klotz and Glazer (1985) (see Table 5–6), who stated that C-, B-, and R-PE all have five bilins per αβ monomer (Table 5–3), and that the previous estimate of four PEB on the β subunit of C-PE by Glazer and Hixson (1975) was incorrect. Similar analyses of peptides from subunits from the other biliproteins – C-PC, R-PC, and PEC – have shown that the numbers of phycobilins assigned to them were correct. Multiple forms of γ subunits have been reported for B-PE (Redlinger and Gantt, 1981) and R-PE (Yu, Glazer, Spencer, and West, 1981; Stadnichuk, Odintsova, and Strongin, 1984; Klotz and Glazer, 1985) and thus these PEs may be a mixture of two or more molecules, as may be Cr-PE_{545} from *Chroomonas* (Martin and Hiller, 1987; Hiller and Martin, 1987) mentioned above. Muckle and Rüdiger (1977) have shown that unfolding C-PE by treating with urea, and by digesting with trypsin, caused equal decreases in the extinction coefficients of the biliproteins, thus showing that urea does not bleach the chromophore. MacColl and Guard-Friar (1987) have warned that bilins can be bleached in acid urea under certain conditions; high bilin concentration and pH values near 2 retain absorbance, but reducing agents slowly induce loss of absorbance, and mercaptoethanol modifies PCB so that its absorbance overlaps that of CV.

Although most workers propose fixed ratios of chromophores on the subunits, Yu et al. (1981) found differences in PUB/PEB ratios in R-PE extracted from *Callithamnion roseum* cultured in light intensities between 6 and 37 $\mu E \cdot m^{-2} \cdot s^{-1}$. The PUB/PEB ratio in R-PE I and R-PE II varied

Table 5–4. *Absorption maxima of bilins and denatured phycobiliproteins*

Compound	Maxima of absorption spectra (nm)					Solvent	Reference
Bilins							
i-Urobilin	495	–	–	–	–	0.1 M HCl	O'Carra et al. (1980)
Phycoerythrobilin	–	550	–	–	–	Acetic acid (pH 3)	Muckle & Rüdiger (1977)
	–	556	–	–	–	0.1 M HCl	O'Carra et al. (1980)
Mesobiliviolin	–	–	590–600	–	–	0.1 M HCl	O'Carra et al. (1980)
Cryptoviolin	–	–	591	–	–	5% HCl/methanol	Chapman et al. (1967)
	–	–	590(α)	–	–	Acid urea	Bryant et al. (1976)
Phycocyanobilin	–	–	–	658(β)	–	Acid urea	Bryant et al. (1976)
	–	–	–	655	–	HCl/methanol	O'Carra et al. (1980)
Denatured biliproteins: Cyanophyceae and Rhodophyceae							
APC	–	–	–	662.5	–	Acid urea	Glazer & Fang (1973)
APC-B	–	–	–	662.5	–	Acid urea	Glazer & Bryant (1975)
C-PC	–	–	–	662.5	–	Acid urea	Glazer & Fang (1973)
	–	–	–	665–660	–	HCl/methanol	O'Carra et al. (1980)
R-PC	–	560	–	662	–	Acid urea	Glazer & Hixson (1975)
B-PE	500(s)	550	–	–	–	Acid urea	Glazer & Hixson (1977)
b-PE	–	550	–	–	–	Acid urea	Glazer & Hixson (1977)
C-PE	–	555	–	–	–	Acid urea	Glazer & Hixson (1975)
	–	550	–	–	–	Acetic acid (pH 3)	Muckle & Rüdiger (1977)
	–	550	–	–	–	0.1 M HCl	O'Carra et al. (1980)
	–	558	–	–	–	Acid urea	Alberte et al. (1984)

CU-PE	497	555	–	–	–	Acid urea	Bryant et al. (1981)
	500	555	–	–	–	Acid urea	Alberte et al. (1984)
	496	550	–	–	–	0.1 M acetic acid	Larkum et al. (1987)
R-PE	498	556	–	–	–	0.1 M HCl	O'Carra et al. (1980)
	498	556	–	–	–	Acid urea	Jiang et al. (1983)
	500	550	–	–	–	Acid urea	Kursar et al. (1983a)
	498	555	–	–	–	Acid urea	Stadnichuk et al. (1984)
PEC	–	–	600	660		Acid urea	Bryant et al. (1976)
Denatured biliproteins: Cryptophyceae							
Cr-PC_{570}	–	–	592	663	–	Acid urea	Hill & Rowan (1989)
Cr-PC_{612}	–	–	590	660	–	0.1 M HCl	O'Carra et al. (1980)
	–	–	590(β)	662(β)	697(α)	Acid urea	MacColl & Guard-Friar (1983a)
	–	–	605	663	–	Acid urea	Hill & Rowan (unpublished)
Cr-PC_{645}	–	–	600(β)	660(β)	694(α)	Acid urea	Glazer & Cohen-Bazire (1975)
	–	–	597	660	694	Acid urea	Jung et al. (1980)
	–	–	590(β)	662(β)	697(α)	Acid urea	MacColl & Guard-Friar (1983b)
	–	–	605	664	–	Acid urea	Hill & Rowan (unpublished)
Cr-PE_{545}	–	565(α)	–	–	–	Acid urea	Mörschel & Wehrmeyer (1977)
	–	555(β)	–	–	–		
	–	550(αβ)	590(s)(α)	–	–	Acid urea	MacColl et al. (1983)
Cr-PE_{555}	–	552	–	–	–	Acid urea	Hill & Rowan (unpublished)
Cr-PE_{566}	–	565	–	–	–	Acid urea	Hill & Rowan (unpublished)

Note: α and β indicate measurements made on separated subunits of the biliproteins; (s) indicates a shoulder on the absorption spectrum.

between 0.36 and 0.46 and between 0.27 and 0.38, respectively, but the total number of phycobilins calculated from the spectra in acid urea remained at 34 in both treatments. They postulated that an isomerase modulated by light could interconvert PUB and PEB. The range of PUB/PEB ratios of R-PE I from *C. roseum* (0.36–0.46) was considerably higher than those for R-PE I in the survey carried out by Glazer et al. (1982), where the pigment was located in R-PE III.

Comparing some types of phycoerythrins shows how the chromophore content can influence the absorption spectrum. R-PE I, with a theoretical PUB/PEB ratio of 9:25 (Table 5–3), has a distinct peak at ca. 497 nm, and this is even more distinct in R-PE III where the PUB/PEB ratio is higher than in R-PE I (Glazer et al., 1982); in B-PE with a PUB/PEB ratio of 2/32, only a shoulder appears at ca. 497 nm, and the shoulder at 540 nm in the R-PEs is now a distinct peak at 543 nm. C-PE, with no PUB, has a single peak only at 565 nm. These comparisons show that PUB is responsible for the ca. 497-nm and PEB for the 565-nm peaks (Fig. 5–2).

The pigments designated CU-PE by MacColl (1982) are those phycoerythrins extracted from Cyanophyceae having an absorption maximum at about 500 nm, indicating a urobilin chromophore in the pigment; the intensity of the other peaks at about 545 and 565 nm varies considerably in relation to the intensity of the peak at 500 nm (Fig. 5–2). These pigments have also been described as R-PE (Glazer, 1984). Based on the relative heights of the peaks, three types of CU-PE may be distinguished, separated by the ratio of the peak height at 500 nm to the other major peak at 545 or 565 nm. The ratio for type I would be about 0.5; for type II, about 1.0; and for type III, greater than 1.0. The type III pigment recently described by Ong et al. (1984) differs from the other types in having a γ subunit, and in having an uncommonly high proportion of PUB to PEB (20.5/13) per molecule; the distribution of the phycobilins on the subunits of the pigment (α, β, γ) remains to be determined.

5.6.2. Conformation and environment of the chromophore within the subunit

The absorption spectra of the phycobilins may be altered both through the covalent bonds between them and the apoprotein (Glazer, 1981, 1984, 1985), and through the noncovalent protein–chromophore interactions (Scheer, 1981, 1982) when the conformation of the linear tetrapyrrole varies from the cyclical to the elongate form. Free bilins may exist as mixtures of configurational, conformational, and tautomeric forms in solution (Glazer, 1985). Thus the absorption spectrum of a bilin may change markedly, depending on the configuration it is in (extended or helical). Transition from the extended to the helical configuration increases the ratio of the absorption maxima in the visible to the UV region of the absorption spectrum. Binding of the bilin to the apoprotein by covalent bonding alters the absorption spectrum of the free bilin, and the effect can be reversibly removed by treating with acid 8 M urea or acid (O'hEocha, 1963; O'Carra et al., 1964).

The importance of the interactions between the protein and the chromophore is indicated in those biliproteins containing only PCB (APC or C-PC) or PEB (C-PE) at a given state of polymerization (Table 5–3). Whereas the absorption maximum of C-PC occurs at about 620 nm, APC I, II, and III have peaks at about 655 nm, though shoulders sometimes occur at 620 nm; APC-B has a major peak at 670 nm in addition to a peak at 620 nm and a shoulder at about 645 nm, but has the same chromophores as APC II and III. Thus the number of PCBs per biliprotein does not determine these variations, because APC II, III, and B all have a single PCB on the α and β subunits, and APC I, having PCB on the α, β, and γ subunits, is little different from APC II and III with only α and β subunits. This not to say that the state of polymerization does not also affect the conformation of the chromophores, as shown in Section 5.6.3.

The spectra of the dimeric phycoerythrins of the Cryptophyceae fall into three distinct types 10 nm apart (Table 5–2), but usually the only phycobilin in them is PEB (Table 5–3); these differences are likely to be due to different interaction between protein and PEB. The variation in the type of linkage between the phycobilins and the apoproteins and the difference in amino acids about the site of attachment recently determined by Glazer and associates (Glazer, 1985) may well account for some of the above variation, because at least two types of linkage can occur with the phycobilin of a given type of biliprotein.

Zilinskas, Greenwald, Bailey, and Kahn (1980) have examined the spectra of APC I, II, III, and B from *Nostoc* using Gaussian analysis, resolving the spectra into six to eight components. Although APC II, III, and B contain trimers of α and β subunits with only one chromophore (PCB), the spectrum and Gaussian curve analysis of APC B differs substantially from II and III, whereas II and III differ slightly. They believe that these variations arise from the flexible structure of the open chain of the tetrapyrroles. Bryant (1982) attributes spectroscopic differences in the

C-PE of some Cyanophyceae to chromophore – chromophore or chromophore–protein interactions.

5.6.3. *Changes in the environment of the chromophore produced by aggregation of the monomers*

The state of aggregation of the biliproteins varies with the conditions in which they are isolated – in particular, the pH, ionic strength, temperature, and protein concentration (Glazer, 1981, 1982). In general, aggregation shifts the maximum of the absorption spectrum (E_{max}) to a longer wavelength through different interactions between the subunits. MacColl, Csatorday, Berns, and Traegar (1980) suggested that when the trimer of APC forms, the α and β subunits of the adjacent monomers interact to shift E_{max} from 615 to 648 nm. Neufeld and Riggs (1969) converted hexameric C-PC from *Anacystis nidulans* by dilution to the monomer at low pH and the dimer at high pH; E_{max} shifted from 613 nm at pH 3.5 to 622 nm at pH 8.5.

Glazer, Fang, and Brown (1973) converted the hexamer of C-PC from *Synechococcus* 6301 to the monomer by raising the pH from 5.5 to 7 and lowering the protein concentration 20-fold; E_{max} changed from 621 to 615 nm. The value for the trimer was intermediate. Saxena (1988) found that C-PC from *Synechocystis* 6201 was in a stable hexameric state at pH 5.0 to 6.0 at a concentration of 1–10 mg/ml, and was primarily in a trimeric state at pH 8 at about 5 mg/ml. Gray, Cosner, and Gantt (1976) separated two forms of C-PC from *Agmanellum quadruplicatum* with values of E_{max} of 623 nm (PC_{623}) and 637 nm (PC_{637}); on fourfold dilution, the excitation and emission spectra of the fluorescence of PC_{637} slowly changed to values similar to PC_{623}. MacColl and Guard-Friar (1987) have discussed the effects of physical factors on the aggregation state of C-PC. R-PE I (hexamer) and R-PE II (monomer) are spectrally distinct, but both have the same PUB/PEB ratio when the E_{495}/E_{565} is measured (Glazer et al., 1982). Type I spectra show a distinct peak at 543 nm, but type II show a shoulder at only 551 nm (Fig. 5–2); this difference apparently reflects the state of polymerization of the R-PE rather than the ratio of the chromophores PUB and PEB.

Murakami, Mimuro, Okki, and Fujita (1981) found that the ratio of the extinctions of the main absorption maxima (E_{650}/E_{620}) of APC of *Anabaena* increased with high protein concentration and ionic strength through polymerization, and decreased when pH was lowered from 6 to 5 through depolymerization; glucose, sucrose, and glycerol (1–5 M) induced an increase in the ratio without changing the polymerization, though propylene and ethylene glycol caused dissociation into the monomer. They postulated that a protein field favorable to the 650-nm band forms when protein takes a "tight" state and that the monomer represents a "relaxed" (linear) state of the protein, resembling the electronic state in PC.

Glazer (1984) has discussed the different aggregation states possible in extracts, depending on pH, ionic strength, concentration, and source of the biliprotein. The structure of the phycobilisome places constraints on the biliproteins, and they must exist in aggregates of three or six (αβ) units imposed by the linker polypeptides. He pointed out that when the absorption maxima of the separated α and β units of APC, C-PC, and C-PE are compared with the same proteins assembled within the PBS, they are blue-shifted, the extinction coefficient is lowered, and no effect of strong interaction is found. It is only when aggregation of subunits goes beyond the monomer (αβ) that the environment of the subunit is altered sufficiently to affect the absorption spectrum.

Preparing pure biliproteins free from linker polypeptides is not always easy, and much work appears to have been done with pigments retaining part of their linker polypeptides (Glazer, 1984). Fuglistaller, Suter, and Zuber (1986) prepared linker-free biliproteins from the PC and PEC of *Mastigocladus laminosus* by gel filtration in 50% formic acid. Glazer and associates (Glazer, 1983, 1984, 1985; Lundell, Yamanaka, and Williams, 1983) have determined the role of the linker polypeptides in the structure of the phycobilisomes and their effect on the spectral properties of the pigments at the different positions in their rods and cores (Sec. 5.9.1).

The α and β subunits do not interact, as Cohen-Bazire et al. (1977) have shown that the absorption spectrum of the monomer of APC is the sum of those of the two subunits, but the maximum of the trimer of APC shows a pronounced shift from 620 to 650 nm, due to exciton interaction between subunits of adjacent monomers or to changes in monomer conformation following assembly of the trimer. The spectroscopic effects of denaturation (D) and aggregation of C-PE on E_{550}/E_{305} and E_{max} are shown below (Glazer, 1985):

	α_D	β_D	α	β	$(\alpha\beta)$	$(\alpha\beta)_3$	$(\alpha\beta)_6$	$(\alpha\beta)_{12}$
$\frac{E_{550}}{E_{305}}$	1.1	1.1	6.3	6.7	6.4	8.9	8.7	7.9
E_{max}	–	–	575	574	575	577	578	581

Watson, Waaland, and Waaland (1986) have isolated R-PC with and without a linker polypeptide of 30 K, having E_{max} values of 624–625 and 617 nm, respectively. Both had the same PCB/PEB ratio, as judged by the E_{665}/E_{570} ratio (Glazer and Fang, 1973).

5.7. Extraction and purification

The techniques used to extract and purify proteins are also applied to the biliproteins. The pigments are usually extracted into phosphate or acetate buffer at pH from 5.5 to 8, often containing mercaptoethanol to prevent oxidation of thiol groups (Bryant et al., 1976; Bryant, Hixson, and Glazer, 1978; Nies and Wehrmeyer, 1980; Fugistaller et al., 1981), and azide to inhibit bacterial growth during manipulations (Glazer and Hixson, 1975; Bryant et al., 1976; Bryant, 1982). Recently, Guard-Friar and MacColl (1984) have shown that mercaptoethanol can alter the absorption spectra of biliproteins and used a method without it; as the α and β subunits still separated on ion-exchange columns, they concluded that the subunits were not joined by disulfide bonds when mercaptoethanol was not present.

Freezing and thawing usually precedes extraction of pigments from Cryptophyceae (Glazer et al., 1971b; Brooks and Gantt, 1973; Glazer and Cohen-Bazire, 1975; Mörschel and Wehrmeyer, 1975), sometimes followed by homogenization; Cyanophyceae are usually ruptured with a French press, sometimes preceded by homogenization (Glazer and Bryant, 1975; Bryant et al., 1976, 1978; Nies and Wehrmeyer, 1980; Bryant, 1982) or by ultrasonication (Hattori and Fujita, 1959; Neufeld and Riggs, 1969; Brown et al., 1975; Murakami et al., 1981; Padgett and Krogmann, 1987). Fujita and Shimura (1974) and Adams, Kao, and Berns (1979) ground Cyanophyceae with glass beads. Pigments from Rhodophyceae have usually been extracted by homogenization (Van der Velde, 1973b; Glazer et al., 1982), by sonication (Koller and Wehrmeyer, 1975), or by freezing and extraction with a French press (Glazer and Hixson, 1975, 1977). Stewart and Farmer (1984) and Berns (1979) ground cyanophytes with glass beads. Padgett and Krogmann (1987) have described a large-scale preparation of phycobiliproteins from a bloom of *Microcystis aeruginosa;* for this species, multiple freezing and thawing gave good extraction of pigments but was not satisfactory for other species. They found that manipulating ionic strength during extraction could remove the PC from the rods of the phycobilisomes but would leave intact the cores of APC. Stewart and Farmer (1984) collected phytoplankton by filtration on glass-fiber filters and homogenized them in a tissue grinder in buffer containing lysozyme, followed by incubation in darkness at 37°C for 2 h, then at 1.5°C for 20 h. Isolating phycobilisomes sometimes precedes preparation of extracts from Cyanophyceae and Rhodophyceae (Gantt and Lipschultz, 1974, 1980; Grabowski and Gantt, 1978; Bryant and Cohen-Bazire, 1981).

The pigments extracted are usually salted out with ammonium sulfate; although this treatment precipitates PC at a lower concentration than the other pigments (Rippka et al., 1974; Bryant et al., 1981), the fractionation is not clean, and for effective purification one or more column separations are usually employed after dialysis to remove the ammonium sulfate.

Gantt and associates have relied on preliminary separations on brushite columns (neutral, crystalline calcium phosphate) using gradients of phosphate buffer between 10 and 250 mM, containing 100 mM NaCl; this high ionic strength may stabilize APC (Zilinskas, Zimmerman, and Gantt, 1978). In particular, this method separates the PCs from APCs as the latter are eluted from the column only with higher concentrations of buffer; the exact concentration required has varied with different experiments (Brown et al., 1975; Gray et al., 1976; Zilinskas et al., 1978; Troxler et al., 1980), probably as a result of differences in batches of brushite (Siegelman and Firer, 1964). They have then further purified the pigments by PAGE and/or sucrose density gradient centrifugation (Gantt and Lipschultz, 1974; Gray and Gantt, 1975; Zilinskas et al., 1978) or by gel filtration or further separation on hydroxyapatite (Gray et al., 1976). Gel filtration using Sephadex G-200 did not clearly separate B-PE, b-PE, R-PC, and APC extracted from *Porphyridium* (Gantt and Lipschultz, 1974) or APC I, II, and III from *Nostoc* sp. (Zilinskas et al., 1978).

Glazer and Bryant (1975) and Bryant et al. (1976) purified small amounts of APC-B and PEC from a mixed biliprotein fraction from *Anacystis nidulans* (syn. *Synechococcus*) containing considerably larger amounts of APC and C-PC. These papers give a valuable example of the approach to the problem of separating a complex mixture of the pigments, particularly in the use of appropriate absorbance ratios in monitoring fractions eluted from columns, and in the use of gradient elution with various columns (hydroxyapatite, Sephadex, DEAE-Sephadex, or DEAE-cellulose). Glazer et al. (1982) developed sucrose-gradient centrifugation as a micromethod for purifying R-PE from a large number of species of the Rhodophyceae. As a final step, PAGE has frequently been used, particularly when subunits of the phycobilisomes are being studied. Larkum et al. (1987) used chromatofocusing by adsorbing partially purified solutions of cyanophyte PE onto polybuffer exchanger (PBE 94) and developing with Polybuffer 74 (Pharmacia) brought to an initial pH of 3.95.

5.8. Estimation of biliproteins and phycobilins

5.8.1 Biliproteins

As with the lipid-soluble pigments, the concentration of biliproteins in solution can be calculated spectroscopically by using known extinction coefficients (Table 5–5). However, the properties of the biliproteins in solution change with variation of pH, concentration, ionic strength of the medium, and thus the conditions for assay should be similar to those for which the extinction coefficient used was measured and the pigment must obey Beer's law at the concentration range examined. When the pigments have been purified after extraction, by chromatography or electrophoresis, the concentration in solution can be simply calculated from the value of the known extinction coefficient (molecular or specific). However, extracts from the Cyanophyceae and Rhodophyceae contain two or more biliproteins, and several workers have presented chromatic equations based on known extinction coefficients (assuming the pigments obey Beer's law and that their absorptions are additive) and derived by solving simultaneous equations for the absorption at the wavelengths of maximum absorption for each of the components of the mixture. Bennett and Bogorad (1973) gave the following equations for C-PC, APC, and C-PE based on specific extinction coefficients for pigments from the cyanophyte, *Fremyella:*

$$\text{C-PC} = \frac{E_{615} - 0.474\,(E_{652})}{5.34} \quad \text{mg}\cdot\text{l}^{-1}$$

$$\text{APC} = \frac{E_{652} - 0.208\,(E_{615})}{5.09} \quad \text{mg}\cdot\text{l}^{-1}$$

$$\text{C-PE} = \frac{E_{562} - 2.41\,(\text{PC}) - 0.849\,(\text{APC})}{9.62} \quad \text{mg}\cdot\text{l}^{-1}$$

Siegelman and Kycia (1978) reproduced these formulas, and Tandeau de Marsac (1977) used them when examining chromatic adaptation in *Gloeobacter*.

Table 5–5. *Extinction coefficients of biliproteins*

Biliprotein	pH	E_{max}	Extinction coefficient: Molecular ($M^{-1} \cdot cm^{-1} \times 10^{-4}$)	Extinction coefficient: Specific ($E^{1\%}_{1cm}$)	Reference
APC	5.5	652	21	–	Bryant & Cohen-Bazire (1981)
	6.8	650	–	50.0	Gantt & Lipschultz (1974)
		648	16.3–17.7	–	Bryant et al. (1976)
		648	20.9–23.5	–	Cohen-Bazire et al. (1977)
		650	23.0	–	Cohen Bazire et al. (1977)
		651	13.1	–	Grabowski & Gantt (1978)
	7.0	652	–	56.5	Bennett & Bogorad (1973)
			–	73.0	Brown et al. (1975)
		650	23.2	–	Cohen-Bazire et al. (1977)
		650	24.0	–	Bryant et al. (1981)
		650	–	60.3	Kursar & Alberte (1983)
Type I	7.0	654	–	39.0	Canaani & Gantt (1980)
Type II	7.0	650	–	54.0	
Type III	7.0	650	–	52.0	
Type B	6.8	671	17.2	–	Glazer & Bryant (1975)
		670	20.0	–	Bryant (1977)
	7.0	673	–	33.0	Canaani & Gantt (1980)
C-PC	5.5	621	33.3	–	Glazer et al. (1973)
	6.5	615	–	65.0	O'Carra (1965)
	6.8	614	27.9–29.0	–	Bryant et al. (1976)
	7.0	615	–	59.2	Bennett & Bogorad (1973)
		615	23.0	–	Glazer et al. (1973)
		625	26.3	–	Troxler et al. (1975)
		622	9.44	–	Gardner et al. (1980)
		617	36.4	–	Bryant et al. (1981)
		618	–	72.4	Kursar & Alberte (1983)
	7.2	615	9.34	–	Hattori & Fujita (1959)
Cr-PC$_{645}$	6.0	645	–	114	MacColl & Berns (1978)
R-PC	6.5	615	–	66.0	O'Carra (1965)
		618	–	64.5	
	6.8	620	–	41.6	Gantt & Lipschultz (1974)
		620	13.9	–	Grabowski & Gantt (1978)
	7.0	618	25.5	70.0	Glazer & Hixson (1975)
B-PE	5.5	545	241	100.4	Glazer & Hixson (1975)
		563	233	97.0	
	6.5	546	–	82.3	O'Carra (1965)
Type I	7.0	542	–	55.4	Koller & Wehrmeyer (1975)
Type II	7.0	542	–	56.3	Koller & Wehrmeyer (1975)
B + b-PE	6.8	54	–	52.6	Gantt & Lipschultz (1974)
b-PE	5.5	543	34.1	97.4	Glazer & Hixson (1977)
C-PE	5.5	559	45.5	–	Bryant & Cohen-Bazire (1981)
	6.8	563	–	125	O'Carra (1965)
		562	–	96.2	Bennett & Bogorad (1973)
	7.0	562	46.8	127	Glazer & Hixson (1975)

Table 5–5 (*Cont.*)

Biliprotein	pH	E_{max}	Extinction coefficient: Molecular ($M^{-1} \cdot cm^{-1} \times 10^{-4}$)	Specific ($E^{1\%}_{1cm}$)	Reference
		562	–	127	Tandeau de Marsac (1977)
		560	42.2–52.1	–	Muckle & Rüdiger (1977)
	7.2	565	28.5	–	Hattori & Fujita (1959)
Type I	6.8	560	11.0	–	Grabowski & Gantt (1978)
Type II		563	6.9	–	
Cr-PE$_{545}$	6.0	545	–	126	MacColl et al. (1976)
Cr-PE$_{566}$	6.0	566	–	99	MacColl & Berns (1978)
CU-PE	7.0	564	45.6	–	Bryant et al. (1981)
Type II	7.0	492	278	–	Ong et al. (1984)
Type III		543	114	–	
PEC	6.8	568	30.7	–	Bryant et al. (1976)
R-PE	6.5	563	–	80.2	O'Carra (1965)
Type I	6.5	564	–	81.5	
Type II	6.8	565	–	196	Yu et al. (1981)
	7.0	564	220	–	Hixson (1976)

Gantt and Lipschultz (1974) used the following equations for estimating pigments in *Porphyridium cruentum* containing both B- and b-PE, which they were unable to estimate separately:

$$\text{R-PC} = \frac{E_{620} - 0.666\,E_{650}}{5.26} \quad \text{g}\cdot\text{l}^{-1}$$

$$\text{APC} = \frac{E_{650} - 0.105\,E_{620}}{4.65} \quad \text{g}\cdot\text{l}^{-1}$$

$$\text{B- and b-PE} = \frac{E_{545} - 0.572\,E_{620} - 0.246\,E_{650}}{5.26} \quad \text{g}\cdot\text{l}^{-1}$$

Kursar, Van der Meer, and Alberte (1983a) used similar types of equations for the red alga, *Gracillaria:*

$$\text{R-PC} = 151.1\,E_{614} - 99.1\,E_{651} \quad \text{mg}\cdot\text{l}^{-1}$$

$$\text{APC} = 181.3\,E_{651} - 22.3\,E_{614} \quad \text{mg}\cdot\text{l}^{-1}$$

$$\text{R-PE} = 155.8\,E_{498.5} - 40.0\,E_{614} - 10.5\,E_{651} \quad \text{mg}\cdot\text{l}^{-1}$$

For measuring concentrations of biliproteins in *Anacystis nidulans* and the red alga *Neoagardiella bailei,* Kursar and Alberte (1983) used the following equations:

$$\text{C-PC} = 166\,E_{618} - 108\,E_{650} \qquad \text{mg}\cdot\text{l}^{-1}$$
$$\text{APC} = 200\,E_{650} - 52.3\,E_{618} \qquad \text{mg}\cdot\text{l}^{-1}$$
$$\text{R-PE} = 169\,E_{493} - 8.64\,E_{618} - 1.76\,E_{650} \qquad \text{mg}\cdot\text{l}^{-1}$$

They substituted E_{615} for E_{618} in the equations for C-PC and APC with the red alga, using the equation for C-PC, not R-PC, since they thought that *Neoagardiella* contained C-PC.

Using molecular extinction coefficients based on protomers ($\alpha\beta$), Bryant et al. (1979) presented the following general equations:

$$\text{PC} = \frac{E_{620} - \left(\dfrac{\epsilon_{620}^{\text{APC}}}{\epsilon_{650}^{\text{APC}}} \cdot E_{650}\right)}{\epsilon_{620}^{\text{PC}} - \left(\dfrac{\epsilon_{620}^{\text{APC}} \cdot \epsilon_{650}^{\text{PC}}}{\epsilon_{650}^{\text{APC}}}\right)}$$

$$\text{APC} = \frac{E_{650} - \left(\dfrac{\epsilon_{650}^{\text{PC}}}{\epsilon_{620}^{\text{PC}}} \cdot E_{620}\right)}{{}_{650}^{\text{APC}} - \left(\dfrac{\epsilon_{650}^{\text{PC}} \cdot \epsilon_{620}^{\text{APC}}}{\epsilon_{620}^{\text{PC}}}\right)}$$

$$\text{PE} = \frac{E_{560} - (\text{PC} \cdot \epsilon_{560}^{\text{PC}}) - (\text{APC} \cdot \epsilon_{560}^{\text{APC}})}{\epsilon_{560}^{\text{PE}}}$$

With these equations, wavelengths and extinction coefficients can be altered to suit the spectral properties of any biliprotein. MacColl and Guard-Friar (1987) have published a table of specific absorptivities of biliproteins at six wavelengths, and have used them to calculate equations for estimating the concentrations of the PC, PE, and APC of the various types.

Beer and Eshel (1985) used the formulas of Bennett and Bogorad (1973) for calculating concentrations of R-PC and R-PE in crude extracts by altering the wavelengths of the maxima and the extinction coefficients to those for red algae. They modified the equations by assuming either an exponential or linear baseline of nonspecific absorption to the absorption spectra, preferring correction using the linear baseline. Stewart and Farmer (1984), utilizing fluorometry, estimated biliproteins extracted from phytoplankton, preparing calibration graphs by using solutions standardized by spectroscopy.

Table 5–6. *Molecular extinction coefficients (ϵ) of phycobilins in 8 M urea (pH 3)*

Phycobilin	Wavelength (nm)	$\epsilon \times 10^4$ mol$^{-1}\cdot$cm^{-1}	Reference
Phycocyanobilin	662.5	3.32–3.475	Glazer & Fang (1973)
		3.55 (PC)	
		3.52 (APC)	
Phycoerythrobilin	550	4.28	Glazer & Fang (1973)
	490	13.7	Glazer & Hixson (1975)
	555	4.33	
	550	4.28	Glazer & Hixson (1977)
	550	4.55–4.74	Muckle & Rüdiger (1977)
	550	5.37	Klotz & Glazer (1985)
Phycourobilin	495	9.4	Klotz & Glazer (1985)
Cryptoviolin (derived)	590	4.3	Bryant et al. (1976)
Bilin 697	697	3.55	Jung et al. (1980)

5.8.2. Phycobilins

Glazer and Fang (1973) showed that 8 M urea at pH 3 eliminated non-covalent perturbations of the spectrum of the chromophore of C-PC (PCB), allowing them to measure the molecular extinction coefficient (ϵ) of PCB, using APC, C-PC, and the α and β subunits of C-PC. Subsequently, Glazer and Hixson (1975) measured ϵ for PEB in C-PE, using this value to calculate a value of ϵ for the polypeptide-bound PUB. There are apparently two other phycobilins of unknown structure in PEC and Cr-PC – that is, CV and the 697-nm bilin; values of ϵ have also been calculated by subtracting the absorption of the known component of each biliprotein to give the spectrum of the unknown. Values found, with references, are shown in Table 5–6. With these values, several sets of chromatic equations have been used in assaying the number of chromophores on the α and β subunits of the biliproteins; these are shown below:

1. Cr-PE_{545} (MacColl, Guard-Friar, and Csatorday, 1983)

$$[\mathrm{PEB}] = \frac{43{,}000\,E_{555} - 37{,}840\,E_{590}}{1{,}244 \times 10^9}$$

$$[\mathrm{CV}] = \frac{E_{555} - 43{,}000\,[\mathrm{PEB}]}{37{,}480}$$

2. Cr-PC_{645} (MacColl and Guard-Friar, 1983b)

$$[697\text{-nm bilin}] = (3.982 \times 10^{-5} E_{697}) - (1.468 \times 10^{-5} E_{662}) - (1.138 \times 10^{-6} E_{590})$$

$$[PCB] = (3.037 \times 10^{-5} E_{697}) + (4.067 \times 10^{-5} E_{662}) - (3.153 \times 10^{-6} E_{590})$$

$$[CV] = (3.177 \times 10^{-6} E_{697}) - (1.174 \times 10^{-5} E_{662}) + (2.417 \times 10^{-5} E_{590})$$

3. Cr-PC_{612} (MacColl and Guard-Friar, 1983a)

$$[PCB] = \frac{(4.30 \times 10^4 E_{662}) - (3.3. \times 10^3 E_{590})}{1.47 \times 10^9}$$

$$[CV] = \frac{E_{662} - (3.55 \times 10^4 [PCB])}{3.3 \times 10^4}$$

5.9. Phycobilisomes

Electronmicrographs of virtually all Cyanophyceae and Rhodophyceae show bodies up to 60 nm wide appressed to the thylakoid membranes. Elizabeth Gantt and co-workers (Gantt and Conti, 1966a,b) identified these when they first used the new fixative glutaraldehyde, thus preventing dissociation of the particles. Subsequently, extraction with Triton X-100 and high concentration of phosphate buffer (0.5 M) using a French press or sonication (Gantt and Lipschultz, 1972; Gantt, 1980) gave undissociated particles that could be examined biochemically; this was not possible in fixed samples. The extraction was carried out at room temperature because below 10°C, dissociation occurs. After removal of cell debris by centrifugation, pure preparations of phycobilisomes were usually made by centrifugation on sucrose gradients, thus separating them from chlorophyll, free biliproteins, and other cell contaminants (Gantt, 1980, 1986). Hiller, Post, and Stewart (1983) have isolated phycobilisomes by treating thylakoids from *Griffithsia monilis* with trypsin, producing particles indistinguishable from those prepared with Triton X-100. The criteria used for testing the purity of samples of PBS are absorption and fluorescence spectroscopy, position on a sucrose gradient, and the appearance of particles under the electron microscope (Gantt, 1980). Gantt (1980, 1981, 1986), Cohen-Bazire and Bryant (1982), Glazer (1982, 1984, 1985), MacColl (1982), and Zilinskas and Greenwald (1986) have written reviews devoted to PBS in particular, and these are dealt with also in the reviews listed at the start of this chapter.

5.9.1. **Structure**

Two structural types of phycobilisomes have been elucidated by examining electronmicrographs of sections cut at three dimensions relative to the thylakoid membranes; these are now known as hemiellipsoidal and hemidiscoidal (MacColl and Guard-Friar, 1987). Gantt, Lipschultz, and Zilinskas (1976) proposed the first model of a PBS from their work with *Porphyridium cruentum,* consisting of a core of APC surrounded by successive hemispherical layers of PC and PE. This model was deduced from the rate of release of the biliproteins from the PBS when the ionic strength of the medium was lowered and from the rate of reaction with antisera specific for PE, PC, and APC. Forming ferritin conjugates with the reacted antisera provided an electron-dense marker showing PE on the outer surface of undissociated PBS (Gantt et al., 1976). In the red alga, *Rhodella violacea,* Mörschel, Koller, Wehrmeyer, and Schneider (1977) prepared the first electronmicrographs showing a central core carrying radiating rods. The central core appeared as a triangle of ring-shaped aggregates, and the radiating rods as stacks of ring-shaped aggregates consisting of tripartite units. Nies and Wehrmeyer (1981) proposed a model for the PBS of the cyanophyte *Mastigocladus laminosus,* with a core of APC, as in previous models, but with PEC as the terminal hexamers of the rods in place of the usual PE units in most other PBS.

Although early reports stated that all the protein of the PBS was phycobilin, Tandeau de Marsac and Cohen-Bazire (1977) found a number of colorless polypeptides in dissociated PBS, accounting for about 15% of the total protein. Many reports have confirmed that they occur in the PBS, and they are now accepted as links between the components of the rods and core. They were identified by visualizing on PAGE gels using Coomassie blue as stain. Zilinskas and Greenwald (1986) have discussed the function of three classes of linker polypeptides, not all of which were colorless. Group I (70–120K) attach the PBS to the thylakoid membranes and contain PCB. Group II (30–70K) maintain rod structure by linking hexamers (Zilinskas and Howell, 1983). Group III (25–30K) attach rods to the APC core of the PBS, and appear to terminate elongation of rods. Riethman, Mawhinney, and Sherman (1987) have detected glycoproteins in the linker polypeptides from *Anacystis nidulans*.

Dissociation of PBS has shown that a single disk (Fig. 5–4) of the core or rod of the PBS is a phycobilin trimer (Bryant et al. 1976, 1979) and that the hexamers in solution must represent a pair of disks (Table 5–3). Bryant et al. (1979) have proposed a model for the PBS from the *Lyng-*

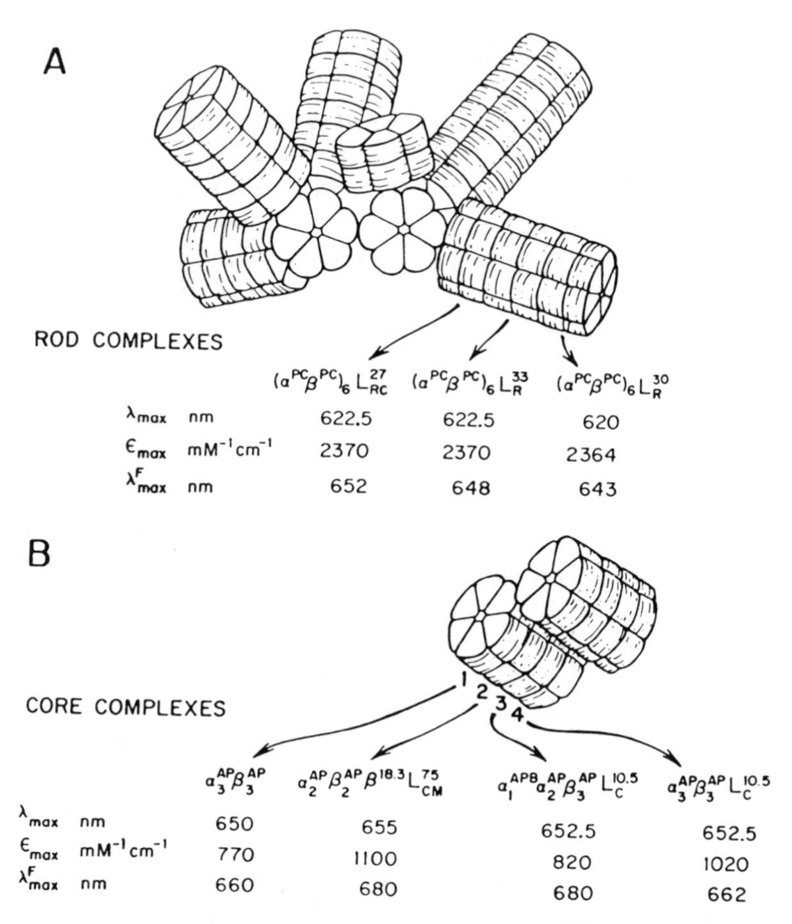

Fig. 5–4. A model of the rod and core complexes of phycobilisomes of *Synechococcus* 6301, showing the arrangement of the subunits of the biliproteins and linker polypeptides. For abbreviations, see Table 5–8. (Reproduced by permission from the *Annual Review of Biophysics and Biophysical Chemistry*, Vol. 14, ©1985 by Annual Reviews, Inc.)

bya–Plectonema–Phormidium group of the Cyanophyceae based primarily on electronmicroscopy and the pigment content of cells in which length of rods were altered by culture in red or green light. Rosinski, Hainfeld, Rigbi, and Siegelman (1981) have proposed a similar model. Honsell, Ghirardelli, and Avanzini (1985) have reported that when *Nitophyllum punctatum* grows below 8°C, biliproteins do not aggregate into phycobilisomes, and that on subsequent removal to a higher temperature the biliproteins still do not form phycobilisomes.

In a brilliant series of investigations, Glazer and associates have determined the structure of the PBS of *Synechococcus* 6301, reviewed by Glazer et al. (1983) and Glazer (1984, 1985). This species was chosen for its relatively simple structure and composition (only two core units, not the usual three, and only C-PC in the rods, not C-PE as well). Components of the PBS were resolved by SDS–PAGE and isoelectric focusing, and the quantities of each component were determined by supplying the growing cells with $^{14}CO_2$, thus labeling the proteins of the PBS. The components of the wild-type and the mutant AN 112 are shown in Table 5–7. Note that the 18.3K and 75K polypeptides contain one PCB per molecule.

Rod structures were investigated by the following means:

1. *Nitrogen starvation*. This treatment induced degradation of the PBS and inhibited synthesis of the biliproteins. The sedimentation coefficients of the PBS decreased, the PC/APC ratio fell, and the 30K colorless polypeptide decreased. These results implied that the 30K polypeptide was associated with PC disks.
2. *Mutation*. The AN 112 mutant produced small PBS low in PC, and lacked 30K and 33K polypeptides. Electronmicrographs of the PBS showed rods of only one disk in length. Another mutant, AN 135, contained some 33K polypeptide but only the 30K when grown in unfavorable conditions. This showed that the 33K polypeptide was associated with the second disk, absent in AN 112.
3. *Culture conditions*. In a favorable environment, PC/APC increased in parallel with 30K and 33K polypeptides, showing that one of either polypeptide occurred per disk, since the PC concentration would be equivalent to disk number. Although PC per cell increased in AN 112, PC per PBS did not, showing that the absent 30K and 33K polypeptides were required for extension of the rods.
4. *Assembly of rod structures*. With the use of purified polypeptides and PC in acid urea, rods could be reconstituted after serial dialysis. The 27K, 30K, and 33K polypeptides gave ordered aggregates, whereas the 75K polypeptide was inactive. As the 27K polypeptide alone gave hexameric disks only, and when included with the 30K and 33K polypeptides, produced shorter rods than when it was not included, they concluded that the 27K polypeptide terminated assembly of the rods. The colorless polypeptides were therefore described as "linker polypeptides."

The core substructure was determined by using the AN 112 mutant with reduced rods. The PBSs from this mutant were partially dissolved to give

Table 5–7. *Composition of the phycobilisomes of Synechococcus 6301*

Component[a]	Number of copies: Wild type	Number of copies: Mutant AN 112	Phycocyanobilins per polypeptide
Rod substructure			
α^{PC}	100	36	1
β^{PC}	100	36	2
L_R^{33}	6	0	0
L_R^{30}	3–6	0	0
L_R^{27}	6	6	0
Core substructure			
α^{AP}	20	20	1
β^{AP}	22	22	1
α^{APB}	2	2	1
$\beta^{18.3}$	2	2	1
L_{CM}^{75}	2	2	1
$L_C^{10.5}$	4	4	0

[a]Abbreviations of Glazer (1985), as shown in Table 5–8.
Source: Glazer et al. (1983).

three fractions by density-gradient centrifugation: 18*S*, 11*S*, and 6*S*. The 18*S* fraction contained the 18.3K and 75K polypeptides and some of the 27K polypeptide, the 11*S* fraction contained mainly the 27K polypeptide, and the 6*S* fraction contained most of the APC-B α subunit but no 27K polypeptide. The 18*S* particle appeared homogeneous, with fluorescence-emission peaks at 680 nm (APC-B) and 660 nm (APC), but contained only half the total APC. This, combined with quantitative analysis of the components, showed that the 18*S* assembly was half a PBS. As the ratio of the polypeptide-terminating rods (27K; see above) to PC was 1:3, not 1:6 as in the intact wild-type PBS, they concluded that the 18*S* particle contained two trimeric units of PC, not the hexameric unit in the intact PBS. Analysis of three subcomplexes of the 18*S* particle produced by tryptic digestion, to which the biliproteins are resistant, gave rise to the composition assigned to the trimeric complexes 1 and 2, shown in the diagram for wild-type PBS in Fig. 5–4 (Glazer, 1984, 1985). The 18.3K polypeptide contained in complex 2 is homologous with the β subunit of APC but is a unique polypeptide containing one PCB molecule (Lundell and Glazer, 1983a,b). The 40K, 18.3K, and 11K polypeptides were derived from the 75K polypeptide. The 6*S* fraction gave rise to two similar but not identical complexes, with complex 3 containing a unique α sub-

Table 5–8. *Abbreviations for biliprotein subunits and for linker polypeptides*

Type of polypeptide	Abbreviation
Phycoerythrin subunits	α^{PE}, β^{PE}, γ^{PE}
Phycoerythrocyanin subunits	α^{PEC}, β^{PEC}
R-Phycocyanin subunits	α^{RPC}, β^{RPC}
C-Phycocyanin subunits	α^{PC}, β^{PC}
Allophycocyanin subunits	α^{AP}, β^{AP}
Allophycocyanin B α subunit	α^{APB}
β-type core biliprotein subunit	β^{MW}
Rod linker polypeptides	L_R^{MW}
Linker attaching rod elements to core	L_{RC}^{MW}
Core linker polypeptides	L_C^{MW}
Linker attaching core to membrane	L_{CM}^{MW}

Abbreviations: PE, phycoerythrin; PEC, phycoerythrocyanin; RPC, R-phycocyanin; PC, C-phycocyanin; AP, allophycocyanin: APB, allophycocyanin-B; L, linker polypeptide; MW, apparent molecular weight, R, rod; C, core; M, thylakoid membrane.
Source: Glazer (1985).

unit, which, when isolated from the cell, was associated with a β subunit of APC; the resulting trimer has been described as APC-B, but Glazer et al. (1983) consider that it is an artifact of isolation (see below). The 10.5K polypeptide is considered to stabilize the complexes 3 and 4, preventing subunit exchange between APC α- and APC-B α-containing complexes (Glazer, 1984). The structure proposed in Fig. 5–4 has not been rigorously defined, as the two core cylinders might not be antiparallel, and the order of the complexes might not be exactly as shown. However, a larger particle containing complexes 1, 2, and 3 has been isolated by Lundell and Glazer (1983c), using partial tryptic digestion. This has shown that four 27K polypeptides, those contained in 18*S* particles, are not released, whereas two attached to units 3 and 4 are released. As the APC 10.5K trimer is released, this must be a terminal unit, No. 4. Glazer (1985) has proposed the following abbreviations for the biliprotein subunits and linker polypeptides of the phycobilisomes (Table 5–8).

Glazer (1984) has suggested that one function of the 75K polypeptide (Lundell, Yamanaka, and Glazer, 1981a,b) is in attaching the PBS to the thylakoid membrane ("anchor" polypeptide), probably with the 18.3K fraction initiating the assembly. Tandeau de Marsac and Cohen-Bazire (1977) originally proposed this role for a large polypeptide found both in PBS and in a membrane fraction in Cyanophyceae. Subsequently, Redlinger and Gantt (1982), Ruskowski and Zilinskas (1982), and Hiller et al. (1983) have made similar suggestions, and Zilinskas and Howell (1986) have found similar polypeptides in 18 species. The relationship of the pluglike structure, seen in electronmicrographs of PBS by Wanner and Kost (1980), to the large polypeptides remains undetermined.

Glazer and associates have also investigated the structure of the *Synechocystis* 6701 PBS (Gingrich, Lundell, and Glazer, 1983; Glazer 1984, 1985); this is more complicated than the PBS from *Synechococcus* 6301 in that it contains three core units and contains C-PE in the rods. Again a mutant, CM 25, without C-PE at the end of the rods, was used. They exchanged the PBS from the mutant into phosphate–NaCl–glycerol medium and separated the APC from the core from the PC from the rods by differential elution on columns of hydroxyapatite; electrophoresis and isoelectric focusing were used as before. The APC fraction was resolved into four fractions by DEAE cellulose chromatography. Analysis of these four fractions and analogy with the corresponding complexes from *Synechococcus* 6301 (e.g., a similar 18*S* complex was found), showed that the main differences were in the third core containing subunits equivalent to the subnits 1 and 4 in *Synechococcus* 6301, and that a 99K polypeptide replaced the 75K "anchor" polypeptide.

The structure of the PBS of a *Nostoc* sp. investigated by Gantt, Zilinskas, and associates (Troxler et al., 1980, Canaani and Gantt, 1982, 1983; Glick and Zilinskas, 1982; Ruskowski and Zilinskas, 1982; Zilinskas, 1982; Zilinskas and Howell, 1983) and *Pseudoanabaena* 7409 (Bryant and Cohen-Bazire, 1981) are similar to the PBS of *Synechocystis* 6701. A study of the rods from PBS of *Anabaena variabilis* has shown the importance of the colorless polypeptides in determining the orientation and sequence of the disks of trimers within the rods and in modulating (red shifting) the spectroscopic properties of the PC within the rods, thus forming a gradient of increasing absorption maxima in the rods toward the core (Yu, Glazer, and Williams, 1981; Yu and Glazer, 1982; Glazer, 1984, 1985). The "anchor" polypeptide of *A. variabilis* is larger than usual (115K), and the core cylinders contain five, rather than the usual four disks (Isono and Katoh, 1987).

The PC and PE extracted from the Cyanophyceae and Rhodophyceae are derived from disks of the rods of PBS (Fig. 5–4), but APC forms complete disks only at the ends of the core. Gingrich et al. (1983) believe that APC types I, II, and III reported by Zilinskas et al. (1978) from *Nostoc* are different components of the core; thus APC I would be analogous to $(\alpha_2^{AP}\beta_2^{AP})L_C^{18.3} \cdot L_{CM}^{75}$ trimer No. 2 of *Synechococcus* 6301; APC II, to $(\alpha_3^{AP}\beta_3^{AP})$, trimer No. 1; and APC III, to $(\alpha_3^{AP}\beta_3^{AP})L_C$, trimer No. 4.

APC-B is equivalent to ($\alpha_3^{APB}\beta_3^{AP}$); this structure does not occur in the PBS, and they believe that it is derived from ($\alpha^{APB}\alpha^{AP}\beta^{AP}$)$L_C^{10.5}$ (trimer No. 3) by loss of the $L_C^{10.5}$ and subsequent monomer exchanges, thus an artifact of extraction, as already mentioned. (See Table 5–8 for the terminology used above.) Zuber and associates (Fuglistaller, Mimuro, Suter, and Zuber, 1987; Rumbeli, Wirth, Suter, and Zuber, 1987) have examined assemblies of biliproteins and linker polypeptides in the PBS of the cyanophyte *Mastigocladus laminosus*. This PBS contains a core of three cylinders of APC with six rods composed of PC and PEC. This differs from the PBS of *Synechocystis* 6701 (Glazer, 1985) in having linker polypeptides of smaller size in both rods and core and in having PEC in place of PE in the rods. The core cylinders are more stable than in other species, and APC-B, believed to be a degradation product, is not found. Elmorjani, Thomas, and Sebban (1986) have assigned component polypeptides to the substructure of the PBS by isolating mutants of *Synechocystis* PCC 6803 lacking PC, thus providing PBS cores uncontaminated by rods.

Incubating mixtures of biliproteins can lead to the reassociation of the pigments into their original arrangement in the PBS. Canaani, Lipschultz, and Gantt (1980) separated an APC and a PC-PE fraction from *Nostoc* PBS, showing that on mixing again, the fluorescence was that of an intact PBS, not the fluorescence of the two fractions before mixing. Lipschultz and Gantt (1981) separated PC and PE from *Porphyridium sordidum* by gradient centrifugation, finding a 30K polypeptide associated with the PC; remixing the two components gave rise to rodlike complexes, with the PE fluorescence at 575 nm decreasing and the PC fluorescence at 655 nm increasing as the pigments combined. The 30K polypeptide was essential for this change in fluorescence to occur. As already mentioned, Lundell, Williams, and Glazer (1981a) assembled a range of mixtures containing the highly purified PC isolated from *Synechococcus* 6301, *Anabaena variabilis,* and *A. quadruplicatum* and one or more of the colorless polypeptides. Mixtures containing the 27K, 30K, and 33K polypeptides gave reasonable yields of aggregates, but the 75K polypeptide did not react, as might be expected from its known location in the core of the PBS. Reassembling PC with the 30K and 33K polypeptides gave hexameric disks and rods of stacked disks similar to the rods of intact PBS: Electronmicrographs of them are shown by Glazer et al. (1983). With the 27K polypeptide only, disks formed, not rods, and when the three polypeptides were combined, the rods were shorter than without the 27K polypeptide. The authors believe that this result indicates that the 27K polypeptide terminates rod assembly, an action consistent with its known position in the 18*S* particle from the PBS, where it has been assigned to the basal hexamer unit of the rod (Glazer et al., 1983; Glazer, 1984). The outer unit is believed to contain the 30K polypeptide, and the middle unit the 33K polypeptide, by virtue of the gradient of absorption and fluorescence-emission maxima this arrangement provides (Lundell et al., 1981a). Glick and Zilinskas (1982) reconstituted functional PBS by combining APC with mixtures of PE and PC from *Nostoc,* as judged by spectroscopic criteria and structure in electronmicrographs. No reconstitution occurred without a colorless 29K polypeptide. Kirilovsky, Kessel, and Ohad (1983) dissociated PBS attached to thylakoid membranes from *Fremyella diplosiphon* by lowering the salt concentration in Tris-HCl at pH 6. The components (PBS and membranes) reassociated when the salt concentration was raised again, giving membrane-bound particles identical to the original complex both in structure on electronmicrographs and in spectral properties, and showing evidence of energy transfer from PE to chlorophyll *a*.

Kirilovsky and Ohad (1986) reassociated phycobiliproteins (prepared from extracted PBS) with intact thylakoids or particles containing PS I and PS II. They demonstrated energy transfer from PE to chlorophyll *a* with particles containing PS II extracted from Cyanophyceae and Chlorophyceae, or thylakoids extracted from Cyanophyceae, but LHC II antennae in eukaryotic thylakoids hindered transfer. Transfer did not occur to chlorophyll *a* in PS I with thylakoids or isolated reaction centers. Lu and Yu (1986) demonstrated reconstitution of PBSs following their strong dissociation by dialysis, detected by the strong fluorescence identical with that of intact PBS at 686 nm at liquid nitrogen temperature. When the dissociated particles were first separated by sucrose-gradient centrifugation, some essential factor was lost and PBS were not reconstituted.

5.9.2. Energy transfer

The general principles of energy transfer between pigments have been discussed previously (Chapter 1, Sec. 1.4); the PBS provides the most complex system of this type, acting as the light-harvesting mechanism for PS II in Rhodophyceae and Cyanophyceae. Dale and Teale (1970; Teale and Dale, 1970) presented evidence that the α and β components of each biliprotein acted as sensitizing (s) and fluorescing (f) chromophores, passing energy from the former to the latter by radiationless Förster resonance transfer. This is now widely accepted and has been confirmed by many workers. Larkum and Barrett (1983) have discussed possible mechanisms of energy transfer in the PBS (Förster resonance transfer or exciton interaction), and Glazer (1984, 1985) and Zilinskas and Greenwald (1986) have discussed experiments examining picosecond time-resolved energy transfer in intact cells and isolated PBS, concluding that homotransfer within a biliprotein domain is more rapid than between different pigments. Although most of the fluorescence emitted by radiated phycobilisomes is at about 675 nm, because of the terminal acceptors L_{CM} and APC-B (see Table 5–8), the fluorescence from other members of the array amounts to about 10% of the total and the path of energy along the array can be determined by using laser pulses and sensitive detectors to follow the risetimes and decay of fluorescence when the PBS is irradiated with light at the absorption maximum of the initial light-acceptor, PC or PE (Searle, Barber, Porter, and Tredwell, 1978; Yamazaki et al., 1984; Glazer, Yeh, Webb, and Clark, 1985a, Glazer et al., 1985b; Glazer and Clark, 1986). These experiments have confirmed that light energy flows along arrays of the components of the phycobilisomes in the sequences deduced from the absorption and fluorescence spectra of the components (Fig. 5–4). Glazer et al. (1985a,b) have confirmed the following sequence for *Synechocystis* 6701 (for terminology, see Table 5–8):

$$(\alpha^{PE}\beta^{PE})_6{\cdot}L_R^{30.5} \rightarrow (\alpha^{PE}\beta^{PE})_6{\cdot}L_R^{31.5} \rightarrow (\alpha^{PC}\beta^{PC})_6{\cdot}L_R^{33.5} \rightarrow (\alpha^{PC}\beta^{PC})_6{\cdot}L_{RC} \rightarrow$$

$$(\alpha^{AP}\beta^{AP})_3 \rightarrow (\alpha^{AP}\beta^{AP})_2{\cdot}\beta^{18.5}{\cdot}L_{CM}^{99}$$

$$\upharpoonleft\downharpoonright$$

$$\rightarrow (\alpha_1^{APB}\ \alpha_2^{AP}\ \beta_3^{AP}){\cdot}L_C^{10}$$

Figure 5–4 shows the structure of the phycobilisome of *Synechococcus* 6301 proposed by Glazer (1985); the absorption and fluorescence maxima show gradients from E_{max}^{620} and F_{max}^{643} at the end of the rods, rising to E_{max}^{665} and F_{max}^{643} at the disk containing the linker polypeptide, L_{CM}^{75}. These spectral properties show the modulating effect of the linker polypeptides, because each hexamer of the rods has six PC molecules, but the spectra vary. Similar schemes can be drawn up for other algae, such as *Porphyridium cruentum* (Ley et al., 1977; Redlinger and Gantt, 1981).

	B-PE → b-PE	R-PC →	APC →	APC-B
E_{max}	563	550,620	620,650	640–670
F_{max}	575	637	660	680

The light-harvesting mechanism in the Cryptophyceae differs from that in the Cyanophyceae and Rhodophyceae; they do not carry phycobilisomes at the thylakoid surface and contain only a single biliprotein, Cr-PC or Cr-PE; many contain the unusual phycobilin, cryptoviolin or, occasionally, the 697-nm bilin (Table 5–3). The biliproteins are in solution at the thylakoid surface, thus passing light energy directly to PS II, not along the complex pathway found in the phycobilisomes. Gantt, Edwards, and Provasoli (1971) demonstrated loss of electron-dense matrix within the thylakoid lumina when the tissue was treated with pronase for 2 h; the cells were then free of biliprotein which had diffused into the medium. Alternatively, chlorophylls and carotenoids were extracted with methanol–acetone, removing the thylakoid membranes but leaving the matrix. Mörschel and Wehrmeyer (1979) believed that their electronmicrographs of thin sections of chloroplasts from *Hemiselmis* showed biliproteins dispersed within the thylakoid lumens or orientated in fine columns at right angles to the inner surface of the thylakoid membranes.

As APC does not occur in the Cryptophyceae, chlorophyll c_2 has been suggested as acting in the same way as the immediate energy accepter before the RC of PS II. However, the emission maximum of the fluorescence of the Cr-PC of some species of Cryptophyceae is as high as 660 nm, and MacColl and Berns (1978) have calculated spectral overlap between three cryptophyte biliproteins (Table 5–2): Cr-PC$_{645}$ from a *Chroomonas* species (F_{max} = 661 nm), Cr-PE$_{545}$ from *Rhodomonas lens* (F_{max} = 585 nm), and Cr-PE$_{566}$ from *Chroomonas ovata* (F_{max} = 617 nm). Calculating the overlap of the fluorescence spectra of the donor pigments with the absorption spectra of chlorophyll *a* (663 nm) and chlorophyll c_2 (630 and 580 nm) in acetone showed that the overlap between Cr-PC$_{645}$ and chlorophyll *a* was much greater than with chlorophyll c_2, and this implies that energy transfer direct from Cr-PC$_{645}$ to chlorophyll *a* is more efficient than to chlorophyll c_2. However, overlap between Cr-PE$_{545}$ and chlorophyll c_2 was of the same order as with chlorophyll *a*, with Cr-PE$_{566}$ intermediate between Cr-PE$_{545}$ and Cr-PC$_{645}$. Thus the role of chlorophyll c_2 in the transfer of energy from the biliproteins to chlorophyll *a* in the Cryptophyceae will vary, depending on the value of the fluorescence emission maximum of the biliprotein, as these range from about 585 nm (Cr-PE$_{545}$) to 660 nm (Cr-PC$_{645}$) (Table 5–2).

Lichtlé and colleagues (Lichtlé, Jupin, and Duval, 1980; Lichtlé, Duval, and Lemoine, 1987) have presented a scheme for the arrangement of three pigment–protein complexes in the thylakoid membranes of *Cryp-*

tomonas rufescens. Fraction 1 contained CC I and A_1 (LHC I: chlorophylls *a* and c_2 and α-carotene). A photochemically active fraction (2) appeared to contain fraction 3 (CC II, Cr-PE_{566}, chlorophyll c_2, and xanthophylls), A_0 (a chlorophyll antenna containing chlorophyll c_2 and xanthophylls), and A_2 (LHC II containing chlorophyll c_2 and xanthophylls). Membranes from each fraction separated by density-gradient centrifugation were examined by negative staining under electronmicroscopy. Fractions 1 and 2 consisted of closed vesicles, 70–130 nm and 25–60 nm, respectively. Fraction 3, containing the Cr-PE_{566}, consisted of vesicles that were slightly smaller than those in fraction 2, but were covered in polyhedric units similar to those in preparations of membrane-free PE (fraction 4). These units were similar to a single disk from PE rods of the groups containing PBS. These units were occasionally seen as stacks of three or four attached to the thylakoid vesicles of fraction 3. The authors believe that high molecular weight polypeptides (97K and 87K) found in fraction 3 may link stacks of PE disks to the thylakoid membranes, pointing out that this arrangement of these stacks is consistent with energy transfer occurring between different forms of PE (Hiller and Martin, 1987).

5.10. Floridorubin

In 1843, Kutzing described a pigment isolated from *Rytiphlaea* as "phycohaematin," in addition to describing "phykoerythrin" and "phycokyan" (phycocyanin). Subsequently, Feldmann and Tixier (1947a,b) extracted the same pigment from *Rytiphlaea tinctoria* and *Vidalia obtusiloba*, naming it floridorubin, and claimed that, as they could not extract phycoerythrin from the algae, this pigment acted in its place in harvesting light energy for photosynthesis. The pigment was also isolated from *Rhodomela confervoides* (Augier, 1953), *Polysiphonia nigrescens* (O'hEocha, 1961), and *Lenormandia prolifera* (Saenger, Rowan, and Ducker, 1969). Saenger (1970) identified floridorubin in all the 12 members of the Amansia group that he examined, and its absence from two species contributed to the evidence for transferring them out of the group (Saenger, Ducker, and Rowan, 1971). The pigment was localized in the sheath of *L. prolifera* (Pedersen, Saenger, Rowan, and Hofsten, 1979) and has been identified as a mixture of bromophenols by Saenger, Pedersen, and Rowan (1976) and Chevelot-Magueur et al. (1976). Chevelot-Magueur, Lavorel, and Potier (1974) have continued to support the earlier claims that floridorubin was a light-harvesting pigment in *Rytiphlaea tinctoria*, but the transmission spectrum of the thallus that they have published shows three peaks at 495, 535, and 565 nm, precisely those found in R-phycoerythrin (Van der Velde, 1973a, b; Glazer et al., 1982). The peak in the fluorescence-emission spectrum from the intact thallus at 585 nm is also close to that of R-PE, whereas at the wavelength of the maximum emission of floridorubin at 550 nm (Saenger et al., 1969), the spectrum shows a trough. As Peyriere (1968) published electronmicrographs clearly showing an arrangement of thylakoids and PBS typical of the Rhodophyta, there seems no further reason to believe that light harvesting in *R. tinctoria* is unusual.

6

Pigments of the chlorophyll–protein complexes

Thornber (1986) has written a valuable chapter summarizing the pigment–protein complexes of all photosynthetic organisms; the field is confused by the different results found by workers using different methods for extraction and separation. However, after Markwell, Thornber, and Boggs (1979) ran extracts from *Zea mays* without generating free chlorophyll, the evidence is strong that all chlorophyll exists bound to protein in the thylakoids and that free chlorophyll on gels arises from degradation (Satoh, 1986). The nomenclature used in the different laboratories has also varied, and the table provided by Machold, Simpson, and Møller (1979) has compared that used in seven laboratories. At a meeting held at Zurich in 1983 (Thornber, 1986), research workers in the field felt that the four pigmented components shown in Table 6–1 occur in the photosynthetic apparatus. The core complexes (CC I and II) consist of a reaction center (RC I or II) containing the primary electron donor and associated antenna chlorophyll (Larkum and Barrett, 1983). LHC I has only recently been recognized in algae (Anderson, 1985; Chu and Anderson, 1985; Anderson and Barrett, 1986; Herrin et al., 1987), and in earlier papers the designation LHCP must still be used (see Table 6–2). In spite of the Zurich conference, the designations CC I and CC II have rarely been used yet.

6.1. Properties of the pigment–protein complexes

6.1.1. Photosystem I

A. *Core complex I (CC I)*

The primary electron donor of the reaction center, RC I, is P_{700}, a form of chlorophyll detected by difference absorption spectra (Setif and Mathis, 1986). Although it was originally believed to contain a dimer of P_{700}, Setif and Mathis (1986) discuss evidence for only one molecule of chlorophyll. Larkum and Barrett (1983) have discussed the properties of this

Table 6–1. *Synonyms that have been used for the pigmented components of the photosynthetic apparatus, the core complexes (CC), and the light-harvesting pigment–protein complexes (LHC)* .

Component	Synonyms
CC I	chl–P_{700} *a*–protein; CP 1a,b,c,d,e; Pa,b
LHC I	Antenna chl-protein of PS I; IIa,b; LHCP; CPO
CC II	CPa; CP III, IV; CP_{47}, CP_{43}; CP_2
LHC II	LHCP; light-harvesting chl *a*/*b*–protein; CP II, LHC_1, LHC_3

complex in algae. The new chlorophyll RC I (Chapter 3) is stoichiometric with P_{700}, and Katoh et al. (1985) have suggested that P_{700} and chlorophyll RC I are identical, though Dörnemann and Senger (1986) are still not convinced. It is now well established that CC I contains β-carotene (or α-carotene in siphonous green algae) though Table 6–2 shows that β-carotene has not always been reported as present in the complexes extracted from algae. Anderson and Barrett (1986) give the approximate molar ratios for the pigments in the complex as chlorophyll a/P_{700} = 40–65:1 and chlorophyll a/β-carotene = 4:10, though Table 6–2 shows that some chlorophyll a/P_{700} ratios found are as high as 100 or over. CC I is not always separated without LHC I (Anderson and Barrett, 1986); the fraction from *Chlamydomonas* designated CP 1a by Ish-Shalom and Ohad (1983) appeared to break down into two complexes, one containing RC I and core antenna (CC I) and the other, connecting antenna and LHC I; the latter complex is a fraction, CPO, isolated by Wollman and Bennoun (1982). In Table 6–2, complexes apparently containing RC I, as judged by the presence of P_{700} or electrophoretic behavior, have been included under CC I; thus the xanthophylls and chlorophylls *b* and *c* are probably from fractions equivalent to the chlorophyll–protein complex (CPO) fraction from *Chlamydomonas*. The number and molecular weights of the polypeptide subunits associated with CC I varies according to the method of PAGE used (Wollman, 1986).

B. Light-harvesting complex (LHC I)

A specific light-harvesting complex associated with CC I has not been identified in algae until recently (Anderson and Barrett, 1986). The first of these complexes in algae appears to have been fraction CPO isolated by Wollman and Bennoun (1982) and Ish-Shalom and Ohad (1983). This

Table 6–2. *Pigment–protein complexes extracted from algae*

Algal source	Designation	Measurements	Polypeptides (K)	Reference
		Photosystem I reaction center complexes (CC I)		
Cyanophyceae				
Phormidium luridum	CP1	Contains P_{700}	118	Reinman & Thornber (1979)
	H	$a/P_{700} = 43$	229, 210, 194, 172	Huang et al. (1984)
P. laminosum	A,A-1	$a/P_{700} = 60$	–	Stewart (1980)
	Chlorophyll–protein complexes	$a/P_{700} = 54$	105	Ford (1987)
		$a/P_{700} = 57$	150	
		$a/P_{700} = 129$	165	
		$a/P_{700} = 85$	205	
		$a/P_{700} = 54$	450	
Plectonema boryanum	II	$a/P_{700} = 73$ $\beta/a = 0.21$	112	Hladik et al. (1982)
Synechococcus sp.	CP1a, b, c, d, e	All contain P_{700}	62, 60, 14, 13, 10	Takahashi et al. (1982)
Prochlorophyceae				
Prochloron sp.	CP1	$a + b/P_{700} = 100$	400 (27, 22, 20, 16 on rerunning)	Hiller & Larkum (1985)
		$a/b = 3.8$		
	CP1, 1a	–	40 to 10	
	Pa	$a + b/P_{700} = 123$	70, 16, 10, 8	Schuster et al. (1985)
	Pb	$a + b/P_{700} = 43$	70, 16, 10, 8	
Prochlorothrix hollandica	CP1	–	58, 57	Bullerjahn et al. (1987)
Rhodophyceae				
Porphyridium cruentum	CP1	$a{:}\beta{::}1{:}.07$	–	Redlinger & Gantt (1983)

Cryptophyceae				
Chroomonas sp.	CP1	Contains P_{700} & c_2	100	Ingram & Hiller (1983)
Cryptomomas maculata	PS I particles	$a/P_{700} = 100$	140	Rhiel et al. (1987)
C. rufescens	PS I complex	$a/P_{700} = 150$	69, 65, 47–43, 24–15	Lichtlé et al. (1987)
Dinophyceae				
Gonyaulax polyedra	I	$a/P_{700} = 50$	19–13 (5 complexes)	Prézelin & Alberte (1978)
Glenodinium sp.				
Glenodinium sp.	I	–	160	Boczar et al. (1980)
	I	Contains c_2 & β	115	Boczar & Prézelin (1986)
Synurophyceae				
Synura petersenii	CPI	$a/P_{700} = 108$	–	Wiedemann et al. (1983)
Tribophyceae				
Tribonema aequale	CPI	$a/P_{700} = 60$	108	Wiedemann et al. (1983)
Pleurochloris meiringensis	CPI	–	116	Wilhelm et al. (1986)
	CPIa	–	127	
	CPI	–	110	Wilhelm et al. (1988)
Eustigmatophyceae				
Nannochloropsis salina	PS I complex	$a/P_{700} = 160$ $\beta/a = 0.29$	–	J. S. Brown (1987)
Bacillariophyceae				
Phaeodactylum tricornutum	RC1	$a/P_{700} = 55$	–	Owens & Wold (1986)
	1	$a/P_{700} = 220$ $a:c:\beta:D:Fx::$ $1:0.07:0.1:0.17$	61, 54–52, 18, 16	Caron & Brown (1987)
	2	$a/P_{700} = 105$	61, 54–52, 34, 18, 16	

Table 6–2 (*Cont.*)

Algal source	Designation	Measurements	Polypeptides (K)	Reference
		$a:c:\beta:D:Fx::1:0.07:0.02:0.14$		
Cylindrotheca fusiformis	PS I complex	$a/P_{700} = 180$	93, 81, 61, 31, 22.5, 18, 16	Hsu & Lee (1987)
		$a/c = 9.1$		
Phaeophyceae				
Acrocarpia paniculata	CP1	$a/P_{700} = 38$	–	Barrett & Anderson (1980)
Fucus serratus	PS I complex	$a/P_{700} = 180$	–	Caron et al. (1985)
		$a/c = 31$		
	PS I complex	$a/P_{700} = 160$	–	
		$a/c = 42$		
Euglenophyceae				
Euglena gracilis	CP1	–	73, 58, 47, 28, 15	Ortiz & Stutz (1980)
	CP1	–	110	Schuler et al. (1982)
	CP1	–	100	Lee et al. (1985)
	CP1	Contains *b*	68	Pineau et al. (1985)
	CP1a	Contains *b*	68, 28, 20, 17, 5	
	CP1	$a/\beta = 0.06–0.12$	66	Cunningham & Schiff (1986a,b)
Prasinophyceae				
Micromonas sp.	CP1	–	107	Wilhelm et al. (1986)
	CP1a	–	138	
Chlorophyceae				

Chlamydomonas reinhardii	RCI + core antenna	–	68–66	Ish-Shalom & Ohad (1983)
Chlamydobotrys stellata	CP1a	–	155	Brandt et al. (1982)
	CP1b	–	140	
	CP1c	–	135	
Chlorella	CP1	$a/b = 4.9–18.5$	–	Wild & Urschel (1980)
	CP1a	$a/b = 2.8$		
	CP1	–	101	Wilhelm et al. (1986)
Ulva mutabilis	CP1	–	71	Hushovd et al. (1982)
Bryopsis maxima	CP1	$a{:}b{:}\alpha{::}1{:}0.09{:}0.05$	100	Nakayama et al. (1986)
	CP1	$a{:}b{:}\alpha{::}1{:}0.14{:}0.10$	67	Itagaki et al. (1986)
Caulerpa cactoides	CP1	$a/P_{700} = 177$	–	Anderson et al. (1980)
	CP1a	$a/P_{700} = 189$	–	
Codium fragile	CP1	$a/b = 1.4$	100	Benson & Cobb (1983)
	CP1a	$a{:}b{:}\alpha{::}1{:}1{:}0.1$	100	
	CP1	–	94	Zhou et al. (1983)
	CP1a	–	94	
Codium sp.	CP1	$a/b = 20$	–	Anderson (1983)
	CP1a	$a/b = 5–7$	–	
	CP1	$a/b = 2.34$	66	Anderson (1985)
		Photosystem II reaction center complex (CC II)		
Cyanophyceae				
Synechococcus sp.	CPa_i	–	85	Oquist et al. (1981)
	CPa_{ii}	–	43.5	
	CP2a	–	2 × 47	Yamagishi & Katoh (1983, 1984), Breton & Katoh (1987)
	CP2b	–	47	
	CP2c	–	40	
	CP2d	–	47	

Table 6–2 (*Cont.*)

Algal source	Designation	Measurements	Polypeptides (K)	Reference
Prochlorophyceae				
Prochloron sp.	CPa	–	–	Hiller & Larkum (1985)
Prochlorothrix hollandica	CP4	–	48, 45	Bullerjahn et al. (1987)
Rhodophyceae				
Porphyridium cruentum	III	a/carotenoid = 20	52	Redlinger & Gantt (1983)
	IV	a/carotenoid = 20	52, 48, 40	
Cryptophyceae				
Chroomonas sp.	CPIV		42	Ingram & Hiller (1983)
Cryptomonas rufescens	PS II complex	$a:c_2:\alpha:A:Cr::1:0.13:0.02:0.39:0.10$	100–97, 87–85, 79–77, 77–75, 47–39, 24, 19	Lichtlé et al. (1987)
Tribophyceae				
Tribonema aquale	CPa	–	50	Wiedemann et al. (1986)
Pleurochloris meiringensis	CPa	–	49	Wilhelm et al. (1986)
	CPa	–	45	Wilhelm et al. (1988)
Eustigmatophyceae				
Nannochloropsis salina	PS II complex	$\beta/a = 0.15$	–	J. S. Brown (1987)
Bacillariophyceae				
Phaeodactylum tricornutum	3	$a:c:\beta:D:Fx::1:0.07:0.17:0.07:0.16$	61, 54, 52, 43, 34, 18, 16	Caron & Brown (1987)
Phaeophyceae				
Fucus serratus	PS II complex	$a/c = 17$	–	Caron et al. (1985)

Euglenophyceae				
Euglena gracilis	CPa	–	57, 48, 41, 28, 10	Ortiz & Stutz (1980)
	CPa	–	56, 48	Schuler et al. (1982)
	CPa	$a/\beta = 0.18–0.25$	54	Cunningham & Schiff (1986a, b)
Prasinophyceae				
Micromonas sp.	CPa	–	44	Wilhelm et al. (1986)
Chlorophyceae				
Acetabularia cliftonii	CPa-1	$a/b = 8.5$	41	Green et al. (1982)
	CPa-2	$a/b = 10$	37	
Chlorella fusca	CPa	$a/b = 2.4–2.8$	–	Wild & Urschel (1980)
Bryopsis maxima	CPa	$a:b:\alpha::1:0.21:0.14$	48	Nakayama et al. (1986)
Caulerpa cactoides	CPa	$a/b > 40$	–	Anderson et al. (1980)
Codium fragile	CPa	–	40	Zhou et al. (1983)
Codium sp.	CPa	$a/b = 6.1$	–	Anderson (1983)
	CPa	–	50–40	Anderson (1985)
		Light-harvesting pigment–protein complexes (LHCP)		
Prochlorophyceae				
Prochloron sp.	Chl *a–b*	$a/b = 2.2–2.4$	35–30	Hiller & Larkum (1985)
	Chl *a/b*–protein	–	34	Schuster et al. (1985)
Prochlorothrix hollandica	CP2, 3, 5	$a/b = 4$	33–30	Bullerjahn et al. (1987)
Cryptophyceae				
Cryptomonas maculata	LHC	$a/c_2 = 1.7$	22, 19, 18	Rhiel et al. (1987)
C. rufescens	LHCP	$a:c_2:A:Cr::1:0.2:0.36:0.07$	69, 47–39, 24–19	Lichtlé et al. (1987)
Dinophyceae				

Table 6–2 (*Cont.*)

Algal source	Designation	Measurements	Polypeptides (K)	Reference
Glenodinium sp.	PCP	Per/a = 4	35	Prézelin & Haxo (1976)
Gonyaulux polyedra				
Glenodinium sp.	II	c_2/a = 4.8	20	Boczar et al. (1980)
	II (LHC II)	c_2/a = 4.2	71	Boczar & Prézelin (1986)
	III	Per/a = 3.4	61	
	IV	Per/a = 3.2	47	
Gonyaulux polyedra	I/I′	c_2/a = 0.20/0.22	133/125	Boczar & Prézelin (1987)
	II/II′	c_2/a = 3	92/95	
	III/III′	c_2/a = 1/0.4	71/78	
	IV/IV′	c_2/a = 0.25	51	
	V	c_2/a = 0.25	46	
Tribophyceae				
Tribonema aequale	LHCP	–	32	Wiedemann et al. (1983)
Pleurochloris meiringensis	LHC_3	–	27	Wilhelm et al. (1986)
	LHC	a:c:H:D:Vau::1:0.22:0.15:0.25:0.13	30	Wilhelm et al. (1988)
Eustigmatophyceae				
Nannochloropsis salina	LHC II	a:Vx:Vau::4:1:1	–	J. S. Brown (1987)
Prymnesiophyceae				
Pavlova gyrans	LHCP	–	20.5, 18, 17, 16.5	Fawley et al. (1987)
P. lutherii	LHCP	a/c = 4.7	21	Hiller, Larkum, & Wrench (1988)
Bacillariophyceae				

Skelotenema costatum	II	$a/c = 0.5$	35–40	Alberte et al. (1981)
Phaeodactylum tricornutum	LHCP	$a{:}c{:}\mathrm{Fx}{::}2{:}1{:}5$	17.5–18.0	Friedman & Alberte (1984)
	LHCP	$a{:}c{:}\mathrm{Fx}{::}2{:}1{:}4$ or $a{:}c{:}\mathrm{Fx}{::}5{:}1{:}4$	15	Gugliemelli (1984)
	2	$a{:}c_1{:}c_2{::}1{:}0.23{:}0.26$	–	Owens & Wold (1986)
	3	$a{:}c_1{:}c_2{:}\mathrm{Fx}{::}1{:}0.09{:}0.28{:}2.22$	–	
	4	$a{:}c{:}\mathrm{D}{:}\mathrm{Fx}{::}1{:}0.3{:}0.22{:}1.25$	41–37, 20–16, 15	Caron & Brown (1987)
	5	$a{:}c{:}\mathrm{D}{:}\mathrm{Fx}{::}1{:}0.5{:}0.17{:}1.25$	56, 20–16, 15	
	6	$a{:}c{:}\mathrm{D}{:}\mathrm{Fx}{::}1{:}0.5{:}0.21{:}0.18$	41–37, 20–16, 15	
	LHCP	–	17.5, 17, 16.4, 16	Fawley et al. (1987)
Cylindrotheca fusiformis	LHCP (a/c)	$a/c = 4.6$	52, 42, 22.5, 18, 16	Hsu & Lee (1987)
	LHCP (a/c/Fx)	$a/c = 2.8$	52, 42, 22.5, 18, 16	
Phaeophyceae				
Acrocarpia paniculata	LHCP	$a{:}c_2{:}\mathrm{Fx}{::}1{:}0.5{:}1$	–	Barrett & Anderson (1980)
	LHCP	$a{:}c_1{:}c_2{:}\mathrm{Vx}{::}8{:}1{:}1{:}1$	–	
Several sp.	II	$a/c = 0.5$	40–35	Alberte et al. (1981)
Fucus serratus	LHCP	$a/c = 6$	–	Caron et al. (1985)
Euglenophyceae				
Euglena gracilis	LHCP	–	26, 23	Ortiz & Stutz (1980)
	LHCP	–	26, 24	Schuler et al. (1982)
	LHCP	–	29, 26	Lee et al. (1985)
	LHCP	$a/b = 3$	29, 26	Pineau et al. (1985)
	LHCP	$a{:}b{:}\mathrm{D}{:}\mathrm{Nx}{::}12{:}6{:}4{:}1$	28, 26.5, 26	Cunningham & Schiff (1986a,b)

Table 6–2 (*Cont.*)

Algal source	Designation	Measurements	Polypeptides (K)	Reference
	$LHCP_2$	a:b:D:Nx::12:4:3:1	28, 26.5, 26	
Prasinophyceae				Fawley et al. (1986)
Asteromonas gracilis	LHCP	–	35–25	
Mantoniella squamata	LHCP	–	55	
Micromonas pusilla	LHCP	–	54	
Microthamnion sp.	LHCP	–	35–25	
Nephoselmis rotunda	LHCP	–	35–25	
Tetraselmis suecica	LHCP	–	35–25	
Micromonas sp.	LHC_1	a/b = 0.8; a/c = 7	65	Wilhelm et al. (1986)
	LHC_3	–	28	
Mantoniella squamata	LHC	a:b:c:β:Nx:Px:Vx::1 0.07:0.12:0.08:0.09: 0.28:0.07	–	Wilhelm & Lenarz-Weiler (1987)
Chlorophyceae				
Chlamydomonas reinhardii	LHCP	a/b = 1	28	Kan & Thornber (1976)
		a:β:L:Nx:Vx:: 1:0.05:0.12:0.06:0.10		
	LHCP	–	29, 25, 24	Michel et al. (1981)
	LHC	a/b = 1.75	30.5, 26.5, 25	Wollman & Bennoun (1982)
	CPO	a/b = 6.3	27.5, 27, 25, 23, 19	
	Connecting antenna	–	27.5, 27, 25, 19	Ish-Shalom & Ohad (1983)
	CPO LHC I }	–	15	

	LHCP	–	35–25	Fawley et al. (1986)
	LHC I	$a/b = 4$	27, 26, 24, 20	Herrin et al. (1987)
Chlamydobotrys stellata	LHCPa (LHC I)	a	36, 32	Brandt et al. (1982, 1983)
	LHCPb (LHC II)	$a + b$		
Chlorella fusca	$LHCP_1$	$a/b = 1.8$	–	Wild & Urschel (1980)
	$LHCP_2$	$a/b = 1.9$	–	
	$LHCP_3$	$a/b = 1.3$	–	
	LHC_1	–	65	Wilhelm et al. (1986)
	LHC_3	–	28	
	LHC	$a:b:\beta:L:Lx:Nx:Vx::$ 1:0.33:0.02:0.06:0.08 0.02:0.03	–	Wilhelm & Lenarz-Weiler (1987)
Ulva mutabilis	CP II	–	21	Hushovd et al. (1982)
Acetabularia cliftonii	CP II_1	–	27	Green & Camm (1982)
	$CPII_{11}$	–	26	
Bryopsis maxima	CP1a	$a:b:\alpha:Sp:Sx::$ 1:0.41:0.05:0.15:0.06	130	Nakayama et al. (1986)
	$LHCP_1$	$a:b:Nx:Sp:Sx::$ 1:1.2:0.12:0.15:0.28	72	
	$LHCP_2$	$a:b:\alpha:Nx:Sp:Sx::$ 1:0.91:0.08:0.19: 0.19:0.25	55	
	$LHCP_3$	$a:b:\alpha:Nx:Sp:Sx::$ 1:0.66:0.06: 0.15:0.19:0.10	31	
	$CP1a^1$	$a:b:\alpha:Nx::1:$ 0.45:0.01:0.17	26, 25, 24, 22	Itagaki et al. (1986)
	$CP1a^2$	$a:b:\alpha:Nx:Sx::$ 1:0.28:0.07:0.02:0.03 0.03	26, 25, 24, 22	
	$CP1a^3$	$a:b:\alpha:Nx:Sx$ ::1:0.5:0.07:0.02:0.04	26, 25, 24, 22	

Table 6–2 (*Cont.*)

Algal source	Designation	Measurements	Polypeptides (K)	Reference
	CP1a[4]	a:b:α:Nx:Sx ::1:0.5:0.07:0.03:0.04	26, 25, 24, 22	
	CP1a[5]	a:b:α:Nx:Sx::1: 0.36:0.06:0.03:0.04	26, 25, 24, 22	
	CP1a[6]	a:b:α:Nx:Sx ::1:0.4:0.07:0.03:0.04	26, 25, 24, 22	
Caulerpa cactoides	$LHCP_1$	$a/b = 0.62$	–	Anderson et al. (1980)
	$LHCP_2$	$a/b = 0.72$	–	
	$LHCP_3$	$a/b = 0.74$		
Codium fragile	$LHCP_1$	a:b:Sx::1:1.4:0.02	–	Benson & Cobb (1983)
	$LHCP_2$	$a/b = 0.85$ Sx/Sp = 5	–	
	$LHCP_1$	$a/b = 0.67$	81	Zhou et al. (1983)
	$LHCP_2$	$a/b = 0.83$	52	
	$LHCP_3$	$a/b = 0.81$	25	
Codium sp.	LHC I	$a/b = 1.7$	24.5, 23, 22.5, 20, 19	Chu & Anderson (1985), Anderson (1985)
	LHC II	–	66, 63, 35.5 34, 27	
Charophyceae				
Chara sp.	LHCP	–	35–25	Fawley et al. (1986)
Klebsormidium flaccidum	LHCP	–	35–25	

Note: Concentrations of carotenoids are not included unless the ratio to chlorophyll a is significant.

Abbreviations: a, b, c, c_1, c_2: chlorophyll a, b, c, c_1, c_2; α, α-carotene; β, β-carotene; A, alloxanthin; Cr, crocoxanthin; D, diadinoxanthin; Fx, fucoxanthin; H, heteroxanthin; L, lutein; Lx, loroxanthin = trihydroxycarotene; Nx, neoxanthin; Per, peridinin; Px, prasinoxanthin + epoxide; Sp, siphonein; Sx, siphonaxanthin; Vau, vaucheriaxanthin; Vx, violaxanthin.

fraction always includes the appropriate accessory pigments – chlorophyll *b*, siphonaxanthin, peridinin, or fucoxanthin.

6.1.2. Photosystem II

A. Core complex II (CC II)

In higher plants and Cyanophyceae, this core complex contains polypeptides of five sizes ranging from 7K to 50K (Satoh, 1986; Namba and Satoh, 1987; Satoh, 1988). Some workers proposed that the 47K subunit (Table 6–2) was the site of charge separation in PS II (Camm and Green, 1983; Yamagishi and Katoh, 1983, 1984; Breton and Katoh, 1987) and that the spectra of the core particle could be constructed from the 40K and 47K subunits, the CP2b and CP2c of Breton and Katoh (1987). However, Namba and Satoh (1987) present evidence that this core complex from spinach contains chlorophyll *a*, pheophytin *a*, and β-carotene in the ratio 5:2:1 and that the chlorophyll *a* and pheophytin are associated with the two subunits of about 30K (D_1 and D_2). The difference spectrum induced in this complex by light in the presence of viologen and dithionite shows a peak of reduced pheophytin at 682 nm, consistent with this complex of D_1 and D_2 being the site of charge separation in PS II (Satoh, 1986), and Van Dorssen et al. (1987) believe that there is convincing evidence that the complex contains the reaction center of PS II. Yamagishi and Katoh (1983) have proposed a scheme for the stepwise degradation of CC II in SDS. The nature of CC II in eukaryotic algae remains uncertain.

B. Light-harvesting complex (LHC II)

This was one of the first complexes to be isolated and comprises about 50% of total pigment in most Chlorophyta (Anderson and Barrett, 1986). This fraction varies in the different classes of algae through the wide range of accessory pigments in them. The Dinophyceae differ from other classes in containing a hydrophilic complex, whereas all others require extraction from the thylakoid membranes with detergents.

6.2. Separation and identification of the components of the pigment–protein complexes

Algal cells vary considerably in the ease with which cell walls are broken, and enzymic as well as physical techniques have been used to isolate the thylakoids from which the complexes are extracted. The pressure cell (French or Yeda) in which the cell suspensions are forced through a narrow pathway has been widely used, for instance, by Reinman and Thornber (1979), Alberte et al. (1981), Wollman and Bennoun (1982), Ingram and Hiller (1983), Redlinger and Gantt (1983), Friedman and Alberte (1984), and Brown (1987). Sonication, sometimes after freezing and thawing and sometimes after homogenization, has also proved effective (Wiedeman et al., 1983), particularly with dinophytes (Prézelin and Haxo, 1976; Boczar, Prézelin, Markwell, and Thornber, 1980; Boczar and Prézelin, 1986, 1987), with *Chlamydomonas* (Kan and Thornber, 1976) and Cyanophytes (Reinman and Thornber, 1979). Isolating sphaeroplasts by dissolving the cell wall followed by osmotic shock (Stewart, 1980; Hladik, Pancoska, and Sofrova, 1981), or passage through a pressure cell (Oquist, Fork, Schoch, and Malmberg, 1981; Takahashi, Koike, and Katoh, 1982) has been used. High-performance tissue disintegraters (Reinman and Thornber, 1979; Alberte et al., 1981; Gugliemelli, 1984) have been used with cyanophytes, brown algae, and diatoms. Barrett and Anderson (1980) utilized several different homogenizers followed by osmotic shock in extracting complexes from *Acrocarpia* and other brown algae. Extracting thylakoid membranes from siphonous green algae requires less drastic treatment, and homogenization or cutting (Nakayama, Itagaki, and Okada, 1986), followed by extraction with appropriate detergents, is usually employed (Anderson, 1980, 1983, 1985; Anderson, Waldron, and Thorne, 1980; Green and Camm, 1982; Chu and Anderson, 1985). Recently, detergents less drastic than SDS have been used to reduce degradation during extraction from thylakoids – in particular, LDS, LDAO, β-octyl glucoside, digitonin, and deriphat. Satoh (1986) has pointed out that using the older detergents during PAGE can also damage the fractions, in particular CC II (Larkum and Barrett, 1983), and that sucrose-gradient centrifugation and isoelectric focusing preserve structures more successfully. Ion-exchange chromatography (Satoh, 1986) has also been used with algal extracts (Larkum and Barrett, 1983; Gugliemelli, 1984; Siefermann-Harms, 1985).

6.3. Pigment–protein complexes found in the algal classes

Figure 6–1 shows a typical separation by PAGE of the pigment–protein complexes using the old nomenclature (Table 6–1). Table 6–2 shows the complexes reported in algae in more recent papers; some earlier reports

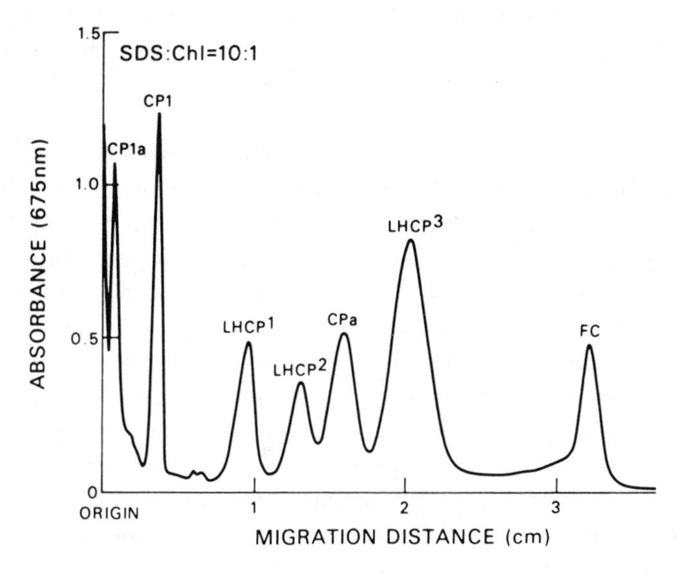

Fig. 6–1. Chlorophyll–protein complexes extracted from spinach thylakoid membranes resolved by discontinuous SDS–PAGE after solubilizing in SDS (SDS:chlorophyll = 10:1). (From Anderson et al., 1978, by permission.)

will be found in Larkum and Barrett (1983) and Siefermann-Harms (1985). This table shows the variation in results found in different laboratories. Green and Camm (1982) have summarized results obtained by examining LHCP by three experimental procedures:

1. SDS–PAGE yields one, two, or more polypeptides.
2. Triton X-100 and sucrose-gradient centrifugation followed by cation precipitation yields two major and one minor polypeptide, whereas fractionation on controlled-pore glass yields one major polypeptide.
3. Digitonin extraction, isoelectric focusing, and column chromatography yield a single polypeptide.

Inspecting Table 6–2 shows the variations found in the complexes, bearing in mind that a few of the values for CC I are for preparations run before denaturation. Larkum and Barrett (1983), Siefermann-Harms (1985), and Anderson and Barrett (1986) have discussed the composition of the complexes, though Siefermann-Harms deals only with those showing evidence of containing carotenoids. The table shows that there are few quantitative analyses for xanthophylls in the complexes; qualitative analyses will be considered below. Siefermann-Harms (1985) has used the size of the ratio of absorbance of pigments in the complexes at 495 nm (chlorophyll *a*) and 435 nm (β-carotene), corrected by subtracting absorbance at 550 nm, as an indication of the amount of carotenoids in the complexes.

The major pigments of the four types of complex have been discussed previously in Section 6.1. Most workers have been interested in the chlorophylls and, to a lesser extent, in carotenes, but the concentrations of xanthophylls have been measured in more recent papers, particularly in the siphonous Chlorophyceae and the Dinophyceae. Siefermann-Harms (1985) has reviewed evidence for carotenoids in the various complexes. Table 6–2 contains results of quantitative analyses made on the complexes where the carotenoids are about 3% or more of the concentration of chlorophyll *a*. Minor identifications will be discussed below. Low levels of xanthophylls in some fractions or failure to identify them as present may well be due to instability during PAGE, as free pigment fractions on the gels may contain substantial amounts of xanthophylls.

6.3.1. Cyanophyceae

Although this class contains many xanthophylls, echinenone is the only one positively identified (Siefermann-Harms, 1985); canthaxanthin, caloxanthin, nostoxanthin, and zeaxanthin are other xanthophylls that are likely to occur in the complexes, though some occur in the extrathylakoid space (Chapter 2). Breton and Katoh (1987) have detected β-carotene in fractions CP2b and c (CC II) from *Synechococcus*.

6.3.2. Prochlorophyceae

No carotenoids have been identified in fractions from this class.

6.3.3. Rhodophyceae

Few workers have examined complexes in this class, and only β-carotene has been identified in CC I and CC II (Redlinger and Gantt, 1983). Zeaxanthin is almost universal in this class and might be expected in low concentration in an antenna or light-harvesting complex.

6.3.4. *Cryptophyceae*

The LHCP separated by sucrose-gradient centrifugation by Ingram and Hiller (1983) contained a xanthophyll with spectral characteristics of diadinoxanthin and seems distinct from alloxanthin or monadoxanthin, other possible xanthophylls that might occur in this class. Lichtlé et al. (1987) suggest that a peak on their HPLC scan is of chlorophyll RC I; they have also identified *cis*-alloxanthin in their PS II complex, shown pooled together with the trans form in Table 6–2. They also measured crocoxanthin in this complex, though they did not find monadoxanthin or diadinoxanthin.

6.3.5. *Dinophyceae*

The Dinophyceae are unusual among the Chromophyta in containing peridinin as the major light-harvesting xanthophyll in place of fucoxanthin. The bulk of the peridinin is contained in a pigment–protein complex that is unusual in that it is water-soluble, and has been investigated in detail by Prézelin and associates (Prézelin and Haxo, 1976; Prézelin and Alberte, 1978). Now up to five other hydrophobic LHCPs have been isolated by Boczar and Prézelin (1986, 1987) from *Gonyaulux polyedra* grown in low light intensity. All contain chlorophyll c_2, and, in addition, fractions I and II contain xanthophylls whereas fractions III, IV, and V contain peridinin.

6.3.6. *Synurophyceae*

Wiedemann et al. (1983) identified chlorophyll *c* and fucoxanthin in a LHCP fraction from *Synura* but did not measure pigment ratios.

6.3.7. *Tribophyceae*

Wiedemann et al. (1983) found half the chlorophyll *a* and some xanthophylls in the band of free pigment when they examined the pigment–protein complexes from *Tribonema aequale* by PAGE. They believed that these free pigments were derived from LHCP by decomposition during extraction and separation. Subsequently, Wilhelm, Büchel, and Rousseau (1988) have separated CC I, CC II, and LHCP in extracts from *Pleu-*

rochloris meiringensis by using PAGE. The LHCP contained the xanthophylls common to the class – diadinoxanthin, heteroxanthin, and vaucheriaxanthin – and chlorophyll *c* (Table 6–2). They considered that chlorophyll *c* mediated energy transfer from the xanthophylls to chlorophyll *a*.

6.3.8. *Eustigmatophyceae*

Brown (1987) found small amounts of neoxanthin, violaxanthin, and vaucheriaxanthin bound to CC I and II extracted from *Nannochloropsis salina;* the latter two were about one-third of the carotenoids in LHC II.

6.3.9. *Prymnesiophyceae*

Romeo (1981) isolated fractions from *Pavlova lutheri* with characteristics of CC I, CC II, and LHCP; the latter contained chlorophyll *c* and fucoxanthin. In *P. gyrans,* Fawley et al. (1987) identified fucoxanthin in LHCP, and diadinoxanthin as the furanoid rearrangement, diadinochrome.

6.3.10. *Bacillariophyceae*

Gugliemelli, Dutton, Jursinic, and Siegelman (1981), Gugliemelli (1984), and Owens and Wold (1986) identified β-carotene, diadinoxanthin, and diatoxanthin in LHCP from *Phaeodactylum tricornutum,* and Owens and Wold (1986) identified β-carotene in CC I.

6.3.11. *Phaeophyceae*

Caron, Dubacq, Berkaloff, and Jupin (1985) summed the concentrations of fucoxanthin and violaxanthin in complexes from *Fucus serratus,* finding chlorophyll *a*/xanthophyll ratios of 5 in CC I and II and of 2 in LHCP.

6.3.12. *Prasinophyceae*

The chlorophyll *c*-like pigment first separated by Ricketts (1966a) as MgDVP, and subsequently considered to be chlorophyll c_1 by Wilhelm et al. (1986) and Wilhelm (1987), occurs in a LHCP fraction (LHC I) in *Mantoniella* (Wilhelm et al., 1986), but carotenoids such as α-carotene, neoxanthin, siphonaxanthin, prasinoxanthin, and lutein have not been found.

6.3.13. Chlorophyceae

In *Chlamydomonas,* Michel, Schneider, Tellenbach, and Boschetti (1981) reported lutein in LHCP, and Herrin et al. (1987) reported β-carotene in LHC I. In this case, the siphonous algae have received the most attention, and, in addition to the quantitative results shown in Table 6 – 2, Benson and Cobb (1983) reported neoxanthin and violaxanthin in the two LHCP fractions isolated from *Codium fragile* but could find no violaxanthin in CP1 and CP1a. Nakayama, Itagaki, and Okada (1986) identified neoxanthin, siphonaxanthin, and siphonein in CC I and II, and Itagaki, Nakayama, and Okada (1986) found the same pigments in CC I but only siphonein in CP1a, in extracts from *Bryopsis maxima*.

6.4. Chlorophyll–protein oligomers

Varying numbers of bands containing chlorophyll are separated by PAGE; The number varies with different species and, in particular, with different methods of extraction, as shown in Table 6–2. Figure 6–1 shows a typical scan of a gel of pigment–proteins extracted from spinach thylakoids with mild SDS (SDS/chlorophyll ratio = 10) by Anderson, Waldron, and Thorne (1978); CP1a is an oligomer of CP1 (CC I), and peaks $LHCP_1$ and $LHCP_2$ are oligomers of $LHCP_3$ (LHC II). When the initial bands are eluted and run again by PAGE after mild heating, oligomeric forms break down to a number of smaller chlorophyll–proteins. Guikema and Sherman (1983) found six bands (I–VI) in varying amounts after mild extraction from *Anacystis nidulans* R_2, using lysozyme to release sphaeroplasts, followed by osmotic shock to release thylakoid membranes. A mild detergent was used to dissociate the membranes, and the complexes were then collected by sucrose-gradient centrifugation and run on LDS–PAGE. The apparent molecular weights of bands I, II, and III were 360K, 250K, and 140K. Band IV (95K), band V (75K), and band VI (45K) occurred in smaller amounts or were unstable. When bands I, II, III, IV, and VI were eluted, heated for 10 min at 70°C, and rerun, two to seven proteins were found in the different bands between 64K and 10K (Table 6–3). By separating proteins from detergent-treated membranes using dif-

Table 6–3. *Polypeptide composition of chlorophyll-containing bands isolated from* Anacystis nidulans *by Guikema and Sherman (1983)*

		Chlorophyll–protein complex				
Band No.	K	I	II	III	IV	VI
PS I						
25	64	+	+	+	+	–
46	36	–	–	+	+	–
47	35	–	–	+	–	–
74	16.5	+	+	+	–	–
75	16	+	+	+	–	–
89	≦10	+	+	+	–	–
90	≦10	+	+	–	–	–
91	≦10	+	–	–	–	–
PS II						
36	48	–	–	–	–	+
38	45	–	–	–	–	+

ferences in density (PS I > PS II) and by detecting surface-exposed particles by iodination, they proposed the model for chlorophyll–protein complexes in the thylakoid membrane shown in Fig. 6–2.

Takahashi, Koike, and Katoh (1982) separated seven complexes containing chlorophyll from a *Scenedesmus* species following extraction essentially similar to that used by Guikema and Sherman (1983). Five contained P_{700} and were designated CP1a, b, c, d, and e; after incubation of the complexes in 10% SDS, 8 M urea, and 10% mercaptoethanol for 1 h, they were run on PAGE–2 M urea and up to five polypeptides were found in each. They deduced that the large complex CP1e was common to all and that CP1a was a dimer of CP1b because the polypeptides were identical, and that a complex with similar mobility to CP1a appeared when CP1b was rerun on PAGE. They proposed the following scheme for the changes induced by SDS (molecular masses [K] in parentheses):

$$\begin{array}{ccccccc} \text{CP1a} & \rightarrow & \text{CP1b} & \rightarrow & \text{CP1c,d} & \rightarrow & \text{CP1e} \\ 2 \times (62,60,14,13,10) & \rightleftarrows & (62,60,14,13,10) & & (62,60,14,10) & & (62,60) \end{array}$$

CP1c and d differ only in SDS-binding or conformation of protein. As CP1a is the major product after solubilization by SDS, Takahashi et al. believe that it reflects the state of chlorophyll *a*–P_{700}–protein in the reaction center of CC I.

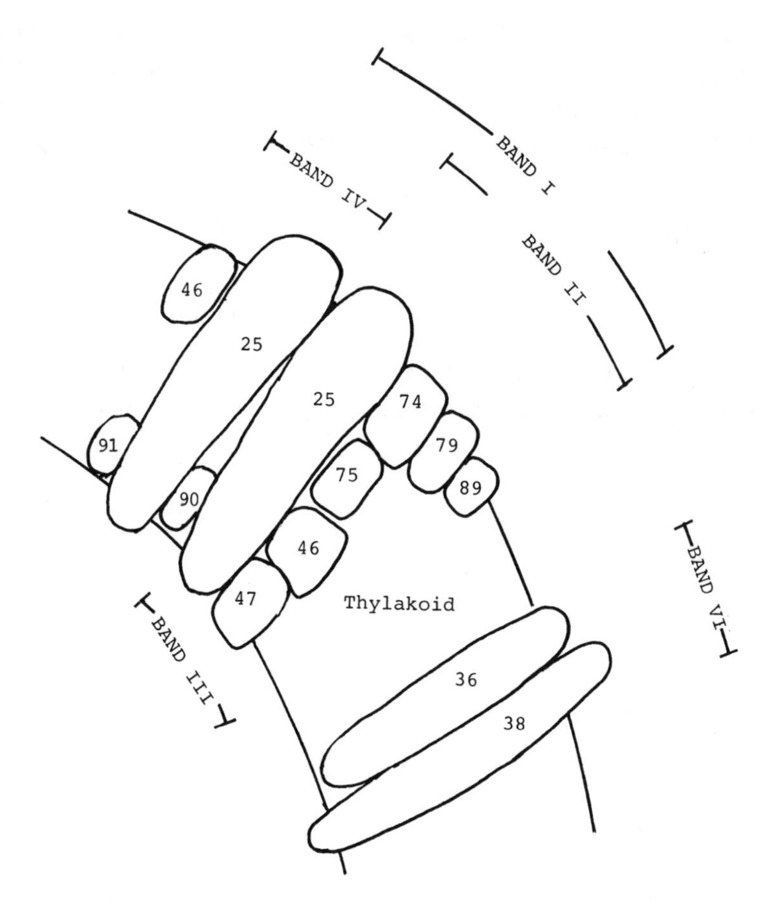

Fig. 6–2. A model proposed for the relationship of polypeptides from chlorophyll-containing bands isolated from *Anacystis nidulans*. Numbers represent the polypeptides separated by PAGE after labeling of cells with ^{35}S; their distribution in the bands containing chlorophyll–protein complexes is shown in Table 6–3. (Redrawn from Guikema and Sherman, 1983.)

Yamagishi and Katoh (1983) purified the reaction center complex of CC II from particles prepared in digitonin from *Synechococcus* by sucrose-gradient centrifugation, followed by digitonin–PAGE, giving a preparation free of P_{700}. Rerunning on SDS–PAGE produced CP2a, b, and c, as previously described by Takahashi et al. (1982). By repeating extractions in varying ratios of SDS to chlorophyll, Yamagishi and Katoh decreased the proportion of CP2b and c to a ratio of 20, whereas CPa increased. Rerunning the CP2 fractions showed that a and b gave rise to a new form, d, whereas b was partly converted to a; CP2c was not altered. Major polypeptides were 47K and 40K, with minor polypeptides of 66K and 31K. They deduced the following scheme for the disintegration of the PS II reaction center complex by SDS:

PS II	→	CP2b	→	CP2a	→	CP2d
(66,47,40,31)		(66,47,31)		(2 × 47)		(47)
		+		+		
		CP2c				
		(40)		(31,66)		

As mentioned in Section 6.1.1A, Yamagishi and Katoh (1983, 1984) believed the 47K subunit to be the site of charge separation in PS II.

7

Photosynthetic pigments in relation to phylogeny and taxonomy of algae

7.1. Phylogeny

The path of evolutionary history of the algae comes from diverse evidence, and the photosynthetic pigments contribute to a small but important part of it. The so-called accessory pigments determine the location of a class of algae in the simple evolutionary tree such as that of Christensen (1962, 1964), where the prokaryotic Cyanophyceae are shown preceding the Rhodophyceae, which in turn precede the Chromophyta and the Chlorophyta. Location in the tree would be most readily determined by knowing the accessory pigments, the biliproteins, or chlorophylls *b* or *c*, and schemes more complicated than that of Christensen also rely on knowing the accessory pigments. The following scheme proposed by Christensen does not account for classes with only chlorophyll c_2, or with both chlorophyll c_2 and biliproteins, and for the Chlorophyta with xanthophylls typical of the Chromophyta.

Chlorophyta (chlorophyll *b*)

Cyanophyceae (biliprotein) → Rhodophyta (biliprotein) ↗ ↘

Chromophyta (chlorophyll *c*)

Ragan and Chapman (1978) have shown five other phylogenetic systems, and Stransky and Hager (1970c), Lee (1972), Cavalier-Smith (1975, 1982, 1986, 1987a), Taylor (1976, 1978), Liaaen-Jensen (1978), Whatley and Whatley (1981), and Larkum and Barrett (1983) have published evolutionary schemes based on photosynthetic pigments. The three types of photosynthetic pigments – chlorophylls, carotenoids, and biliproteins – all contribute to these schemes.

Measuring homology of the base sequence of comparable nucleic acids or the amino acid sequences of comparable proteins is the most objective method for determining phylogeny (Pigott and Carr, 1972; Dayhoff and

Schwartz, 1980; Fox et al., 1980; Doolittle and Bonen, 1981; Cavalier-Smith, 1982; Gray and Doolittle, 1982; Hori, Lim, and Osawa, 1985; Lim, Kawai, Hori, and Osawa, 1986), though this approach assumes that the rate of mutation has been constant.

The 16*S* rRNA, the major component of rRNA, provides the most satisfactory comparison (Woese, 1987) and because the molecule is large, comparisons are usually made by sequencing the oligonucleotides released from it by ribonuclease T1 rather than by sequencing the whole molecule (Doolittle and Bonen, 1981). Many workers have used 5*S* rRNA for comparisons. However, this rRNA occurs in both organelles and cytoplasm, and furthermore, extracts are often made from intact tissue, not isolated chloroplasts; therefore, it is not satisfactory for comparing the sequences of the presumed symbionts thought to contribute to the chloroplasts by the xenogenous hypothesis. This type of analysis of intact cells has widely separated *Chlamydomonas* from other Chlorophyta (see Sec. 7.4.5). In addition, as the molecule is one-tenth the size of 16*S* rRNA, it must be completely sequenced; it is also highly conserved (Doolittle and Bonen, 1981). Schwartz and Dayhoff (1981) have used amino acid sequences from ferridoxin and cytochrome *c* to construct phylogenetic trees, but again comparisons with these molecules are not satisfactory for comparing chloroplasts, and both are relatively small molecules (Doolittle and Bonen, 1981).

7.2. Algal phylogeny and the photosynthetic pigments

As already mentioned, the pigments are just one of a number of properties used in building phylogenetic trees: Structure of flagella, chloroplast, cell walls, and mitochondria; methods of reproduction; type of nuclear and cell division; type and location of food reserves; sequence data of nucleic acids and proteins; and extent of immunological cross-reactivity of biliproteins and pigment–protein complexes can also be used for this purpose.

7.2.1. Value of the three types of pigment in determining phylogeny

A. Chlorophylls

Chlorophyll *a* is ubiquitous in algae and is of no value in phylogeny. Chlorophyll *b* delimits the small division, Prochlorophyta, in the Prokaryota (Lewin, 1976, 1977), and the Chlorophyta in the Eukaryota (Christensen, 1962, 1964). After the two forms of chlorophyll *c* were discovered, the original distribution was thought to be chlorophyll c_2 only in the Cryptophyceae and Dinophyceae, and both c_1 and c_2 in the remainder of the Chromophyta. However, we now know that there is no c_1 in some pennate marine diatoms (Jeffrey and Stauber, 1985) and no c_2 in the new class, Synurophyceae (Andersen, 1987) (see Table 3–1). If we regard chlorophyll c_2 as primitive and c_1 as derived, we must decide whether the pennate diatoms without c_1 are primitive, or, as seems more likely, the absence of c_1 is a secondary derived state. In the Synurophyceae, absence of c_2 is certainly a derived state. The value of the distribution of chlorophyll c_3 (Vesk and Jeffrey, 1987) in phylogeny is uncertain.

B. Carotenoids

Carotenes. With the exception of the Cryptophyta, β-carotene is ubiquitous (Table 4–1) and thus of little phylogenetic value. The concentration of α-carotene is relatively high in some Chlorophyceae, and its occasional dominance over β-carotene in some taxa could be of value as a derived state. If we regard absence of α-carotene as primitive, its uneven distribution in the Chromophyta is interesting (Table 4–1).

Xanthophylls. Liaaen-Jensen (1977, 1978, 1979) and Ragan and Chapman (1978) have stressed that the biosynthetic pathways of carotenoid synthesis rather than the individual pigments are of phylogenetic value. The important reactions are the synthesis of the triple bond between C-7 and C-8 or C-7′ and C-8′, allene bond formation at C-7 or C-7′, and synthesis of the C-8 carbonyl and C-19 hydroxyl groups. Table 7–1 shows the important pathways of synthesis in the classes of the eukaryotic algae.

C. Biliproteins

The major biliproteins of the Cyanophyceae and Rhodophyceae are essentially similar, and Ouchterlony tests have shown cross-reaction between the similar pigments (PC, PE, and APC) in the two classes. The biliproteins of the Cryptophyceae, originally believed not to react with the equivalent pigments from the other two classes (Chapter 5), are now known to show some reactivity with them, for Guard-Friar et al. (1986) have found strong cross-reactivity between B-PE and C-PE and both Cr-PC and Cr-PE. However, the cross-reactivity between C-PC and the cryp-

Table 7–1. *Important pathways of synthesis of carotenoid structures in the eukaryotic algae*

	Structure						
Class	ϵ-Ring	Triple bond	Allene	Epoxide	Acetyl	8-Keto	19-Hydroxy
Rhodophyceae	+	–	–	–	–	–	–
Cryptophyceae	+	+	–	–	–	–	–
Chrysophyceae[a]	–	–	+	+	+	+	–
Phaeophyceae	–	–	+	+	+	+	–
Prymnesiophyceae	+	+	+	+	+	+	–
Bacillariophyceae	+	+	+	+	+	+	–
Dinophyceae	–	+	+	+	+	+	–
Eustigmatophyceae	–	–	+	+	–	–	+
Tribophyceae	–	+	+	+	–	–	+
Euglenophyceae	–	+	+	+	–	–	–
Prasinophyceae	+	–	+	+	–	+	+
Chlorophyceae	+	–	+	+	–	+	+
Charophyceae	+	–	+	+	–	–	–

[a]Includes Raphidophyceae and Synurophyceae.

tophyte biliproteins was marginal. This result is consistent with the finding of similar amino-terminal sequences of the β-subunit of Cr-PC from *Hemiselmis* and C-PC and R-PC (Glazer and Apell, 1977); the α subunit of Cr-PC is much smaller than that from C-PC and R-PC and no homology was found, though Glazer and Apell (1977) have pointed out that there might be homology between the α subunit of Cr-PC and an internal sequence of the other pigments. Thus the relationship between the chloroplasts of the Cryptophyceae and the Rhodophyta is closer than originally believed.

Larkum and Barrett (1983) have suggested that the original oxygenic prokaryote contained chlorophyll *a* and carotenoids only, and that the biliproteins, evolved when green light was absorbed, gave survival value when the suspensions of the primitive phytoplankton became dense. Thus *Prochloron* could have evolved independently of the Cyanophyceae, from a primitive oxygenic prokaryote (urkaryote). Biliproteins discovered in marine organisms (PEC and CU-PE: see Table 5–1) also fill a part of the green window not covered by PC and PE, and could represent survivors of the early cyanophytes or their precursors. Organisms from the Cyanophyceae having CU-PE, a biliprotein containing urobilin, could be close to the symbiont envisaged by Cavalier-Smith (1982, 1986) as giving rise to the chloroplast of the Rhodophyceae (Sec. 7.3).

7.3. Hypotheses for the evolution of Eukaryota

The essential difference between the Prokaryota and the Eukaryota lies in the enclosure of the main part of the genome of the Eukaryota in a nucleus bounded by a double membrane (Stanier and Van Neil, 1962). The other important differences are that the Eukaryota contain membrane-bound organelles, primarily the mitochondria and plastids, both containing a small part of the total genome of the cell. The partitioning of the genome could have arisen in two ways (Gray and Doolittle, 1982):

1. The genomes of the nucleus, plastids, and mitochondria originally inhabited different sorts of cells (the endosymbiont or xenogenous hypothesis).
2. Nuclear, plastid, and mitochondrial genomes became separated into compartments within a single sort of cell (the direct or autogenous hypothesis).

Either hypothesis requires a common ancestor for nuclear and plastid genomes. If the genomes were known to arise from different classes of prokaryote, the xenogenous origin of plastids could be considered proven. Although the plastid genome of the Rhodophyceae probably derives from the Cyanophyceae (Bonen and Doolittle, 1975), proof of the xenogenous hypothesis also requires that the nuclear genome was derived from a lineage other than the Cyanophyceae. Evidence for polyphyletic origin of plastids would also provide strong evidence for the xenogenous hypothesis. The 18*S* ribosome appears to represent a lineage other than that from the Prokaryota from which the nuclear genome could be derived (Fox et al., 1980). Sequence data have shown that the plastids of the Rhodophyceae and Cryptophyceae are probably derived from the Cyanophyceae, whereas there is little homology between rRNA of Eubacteria and Eukaryota (Gray and Doolittle, 1982).

The xenogenous hypothesis was suggested as long ago as 1883 by Schimper and also more recently by other workers (Taylor, 1974), although it has received, and still receives, criticism (Cavalier-Smith, 1975; Conquist, 1981; Uzzell and Spolsky, 1981). Detecting DNA and 16*S* (as opposed to 18*S*) RNA in the mitochondria and plastids stimulated interest, and the hypothesis has been stated in detail by Sagan (1967), also under her later name, Margulis (1970, 1975, 1981), and has been supported by Raven (1970), Lee (1972), Loeblich (1976), Round (1980), Stewart and

Mattox (1980), Cavalier-Smith (1982, 1986, 1987a), Gray and Doolittle (1982), Taylor (1983), Whatley and Whatley (1984), Wilcox and Wedemayer (1984, 1985), Cavalier-Smith and Lee (1985), and Smith and Douglas (1987). The hypothesis states that the plastids, mitochondria, and flagella of the Eukaryota arise from entry into a primitive urkaryote by another type of cell, not necessarily a prokaryote. The evidence for the symbiont origin of the plastid is strongest and is grudgingly accepted as possible by those criticizing this origin of the mitochondria and flagella. Although the marked chemical and structural similarity between the Cyanophyceae and the chloroplast of the Rhodophyceae is generally accepted as evidence that the latter is descended closely from the former, this conclusion does not prove the endosymbiont origin of the plastid; in fact, this evidence equally supports both hypotheses, as primitive character states must be distinguished from derived states in determining phylogeny (Uzzell and Spolsky, 1981). Primitive character states persist through the phylogenetic tree, whereas derived states arise only in taxa descended from the common ancestor. Several RNA genes from *Anacystis nidulans* show higher homologies with those from chloroplasts than do those from other prokaryotes (Kumano, Tomioka, and Sugiura, 1983; Tomioka and Sugiura, 1983; Douglas and Doolittle, 1984).

Woese and Fox (1977a,b) have stressed that the terms "prokaryote" and "eukaryote" are not phylogenetic distinctions in the sense that the latter is descended from the former, but that the two are domains or kingdoms, each with its own phylogenetic classification. They have proposed that the prokaryotic eubacteria, arachaebacteria, and urkaryotes descended from a primitive entity they called a "progenote." Thus the line of descent leading to the eukaryotic cytoplasm diverged from the prokaryote line at the progenote, not the prokaryote stage. The nearest living organism to the hypothetical urkaryote appears to be the anaerobic amoeba *Pelomyxa palustris,* without mitochondria or Golgi membranes (Woese and Fox, 1977b). Gray and Doolittle (1982) have also stressed that prokaryotes are not necessarily more primitive than eukaryotes, but they also believe that plastids are descended from oxygenic prokaryotes, citing the similarities in association coefficients calculated from base sequences of 16*S* rRNA (Bonen and Doolittle, 1975; Fox et al., 1980; Doolittle and Bonen, 1981) and DNA/rRNA hybridization (Pigott and Carr, 1972). Cavalier-Smith (1987a) disagreed with the concept of a hypothetical "progenote" as the ancestor of both prokaryotes and eukaryotes, and proposed that the latter evolved from the former. He cited the fossil record as indicating that eukaryotes appeared only about 2000 million years after the prokaryotes, though this in itself is not a conclusive argument. He believed that a primitive archaebacterium was the precursor of the subkingdom Archezoa, containing primitive eukaryotes without mitochondria. Of this subkingdom, a member of the Archamoeba, such as *Pelomyxa* or one of the Mastigamoeba, would be the host for the endosymbiosis proposed between a negibacterium, giving rise to the mitochondria, and between a cyanophyte, giving rise to the chloroplasts. He believed that these two symbioses occurred at the same time, not serially at distantly separated times (Cavalier-Smith, 1987b).

Although many phycologists interested in phylogeny seized on the discovery of the prokaryote *Prochloron,* containing chlorophyll *b*, as an immediate precursor of the chloroplasts of the Chlorophyta, either by symbiosis or direct descent, the carotenoid pigments of this genus are entirely those of the Cyanophyceae and the pathways of carotenoid synthesis developed in the Chlorophyta are not present (Table 7–1). Also, the base sequence homology of the 16*S* rRNA between *Prochloron* and cyanophytes is high, but low between it and the 16*S* rRNA of the chlorophyte chloroplasts (Seewaldt and Stackebrant, 1982) and homology between 5*S* rRNA from *Chlorella* and *Prochloron* is only 58% (Yamada and Shimaji, 1986) but 75% between *Prochloron* and *Anacystis* (MacKay et al., 1982; Douglas and Doolittle, 1984). Bremer (1985) believes that the similarities between *Prochloron* and cyanophytes (Lewin, 1984) are shared primitive states, whereas those with the green algal chloroplast are shared derived states. The cladistic analysis by Wolters and Erdmann (1988) widely separated *Prochloron* from *Cyanophora* and several green algae. The fine structure is similar to that of the Cyanophyceae (Whatley, 1977).

Lewin (1983) has suggested infection with a plasmid from another organism as a possible origin of ability to form chlorophyll *b* in *Prochloron*. Hiller and Larkum (1985) have found that the LHCP-3 of other organisms and the equivalent chlorophyll *a*/*b*-binding protein of *Prochloron* are substantially different and are consistent with a polyphyletic origin of chlorophyll *b*. Cavalier-Smith (1982, 1986) has disavowed the view that *Prochloron* is the endosymbiotic origin of the chloroplast of the Chlorophyta and chloroplasts of all classes, as opposed to a multiple series of endosymbioses; certainly the 20 symbioses proposed by Sagan (1967) seem excessive (Cavalier-Smith, 1982). He has proposed a scheme, including the origin of the plastids of algae, based on the number of membranes around the plastid (Whatley and Whatley, 1981), the type of cristae of the mitochondria, and the type of cilia (Fig. 7–1). This scheme envisages only two symbioses, the first, a symbiosis between a Gram-

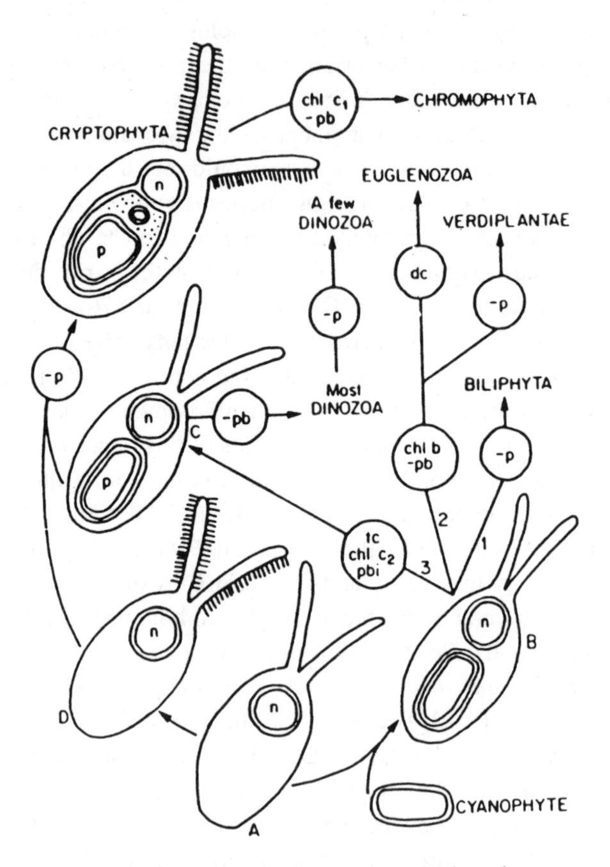

Fig. 7–1. A phylogeny for photosynthetic eukaryotes, assuming only two endosymbioses. (A) A nonphotosynthetic eukaryote with a single nucleus (n) and two anterior flagella with nontubular mastiogonemes. (B) A cyanophyte engulfed by A to form an endosymbiotic cyanelle (first symbiosis) bounded by three membranes. After many generations, the cyanelle evolved into the plastids typical of the red algae (Biliphyta), Chlorophyta, and the primitive dinozoan (C) by the mutations shown in the pathways 1, 2, and 3. p, phagosome membrane; tc, tubular cristae of mitochondria; dc, disk-shaped cristae; pb, phycobilins; pbi, phycobilins inside thylakoid membranes (intrathylakoid phycobilins). The second symbiosis between C and a nonphotosynthetic eukaryote with tubular mastigonemes gave rise to the Cryptophyceae. These in turn gave rise to the Chromophyta by loss of bp and gaining chlorophyl c_1. (From Cavalier-Smith, 1982, by permission.)

negative prokaryote and a colorless protozoan, giving three membranes around the plastid, due to the double membrane of the Gram-negative prokaryote in which the two membranes are similar to those of the chloroplast membranes (Keegstra, Werner-Washburne, Cline, and Andrews,

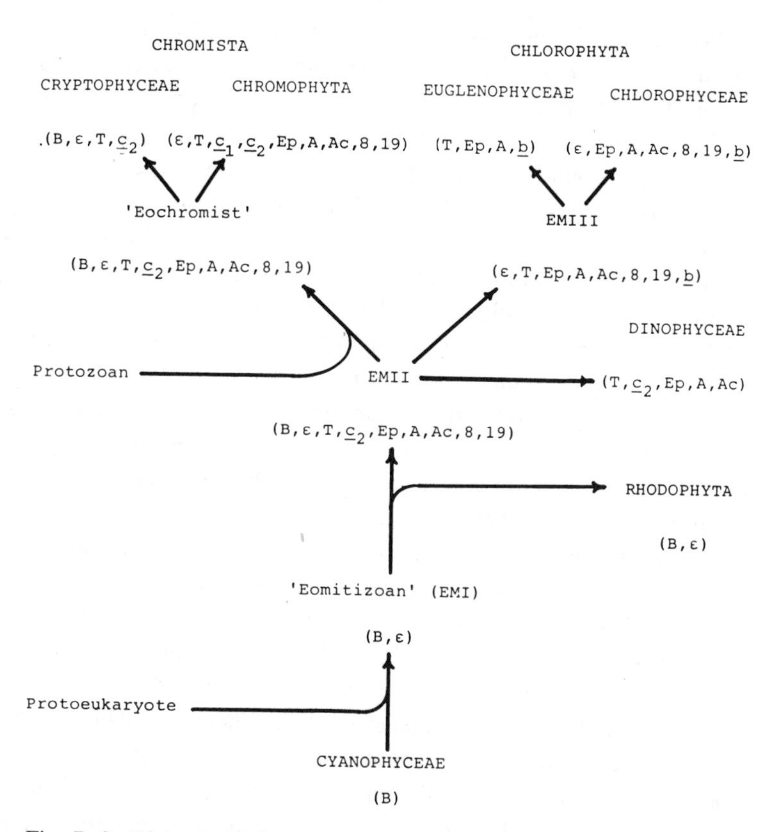

Fig. 7–2. The scheme for the phylogeny of the algae proposed by Cavalier-Smith (1986), modified to take account of the evolution of pathways of synthesis of the chlorophylls, biliproteins, and carotenoids. B, biliproteins; ϵ, ϵ-ring; T, triple bond (C-7 to C-8); c_1, chlorophyll c_1; c_2, chlorophyll c_2; Ep, 5,6-epoxide; A, allene bond; Ac, acetylation; 8, C-8 keto formation; 19, C-19 hydroxylation; *b*, chlorophyll *b*.

1984), and the plasma membrane of the host. This he named an eomitozoan, which was to develop into the Dinophyceae, Euglenophyceae, and Chlorophyceae (Fig. 7–2). In addition, it would give rise to the kingdom Chromista containing the Chromophyta (sensu Cavalier-Smith: this excludes the Dinophyceae and the Cryptophyceae) and the Cryptophyceae, by a second symbiosis with a second protozoan. He envisaged the eomitozoan as containing biliproteins and chlorophyll *c*, losing the biliproteins before the symbiosis, to give rise to his Chromophyta. Many changes in photosynthetic pigments would have to have occurred during these stages, with chlorophylls *b* and *c* and the numerous pathways of carotenoid synthesis arising at different stages. The simplest scheme without proposing polyphyletic origins is shown in Fig. 7–2.

Wilcox and Wedemayer (1985) have examined a freshwater dinoflagellate, *Amphidinium wigrense,* containing phycocyanin, as judged by its color. The thylakoids were similar to those of the Cryptophyceae, but the cells contained no cryptophyte nucleus or nucleomorph and only three membranes around the chloroplast. Cavalier-Smith (personal communication) has suggested that this could be closely related to his eomitozoan. Neither *A. wigrense* nor *Gymnodinium acidotum* (Wilcox and Wedemayer, 1984), another dinophyte containing an apparent cryptophyte symbiont, has been cultured, and evidence that they are stable symbionts rests only on their structure. Larsen (1988) has described a phagotrophic dinophyte, *Amphidinium poecilochroum,* that multiplies when feeding on small cryptophytes (*Hemiselmis* or *Chroomonas*) but cannot be isolated into unialgal culture. Only cryptophytes were accepted as food. Larsen (1988) has pointed out that *A. poecilochroum* and *G. acidotum* are structurally similar, and that the latter may equally well be a phagotrophic, not an autotrophic organism, and that both might represent an early stage in establishing a stable symbiosis. He has also pointed out that Hu et al. (1980) and Zhang et al. (1983) have not examined the fine structure of the *Gymnodinium* species they have shown to contain Cr-PC, and the two species have not been grown in culture; presumably there is no good evidence that they too are not phagotrophic. Wilcox and Wedemayer (1985) believe that the structure of their organism arose by degradation of a cryptophyte symbiont in which the nucleomorph degenerated, but the lack of the nucleomorph and three rather than five membranes around the chloroplast are consistent with the structure of Cavalier-Smith's eomitozoan (Fig. 7–1). Unfortunately, only a few cells of *A. wigrense* were collected, and pigments and DNA sequences could not be analyzed.

Critics of the xenogenous hypothesis point out that the various classes of algae appeared early in geological time (Taylor, 1978) and wonder why no more recent classes have arisen by symbiosis. In supporting the xenogenous hypothesis, Round (1980) argued that the primitive forms were much less specialized than modern forms and that the modern eukaryotic symbionts are always enclosed in a vacuole. One might also argue that a few symbioses, such as that between chrysophytes and heterotrophic Dinophyceae, might have arisen relatively recently without being detected in the fossil record. Cavalier-Smith's earlier argument against the hypothesis (Cavalier-Smith, 1975) was that it did not account for the formation of a nucleus; however, he now stresses that the endosymbiont hypothesis accounts not for the origin of the eukaryotes, but only for the symbiotic origin of chloroplasts and mitochondria (Cavalier-Smith, 1987b).

7.4. Evidence for the phylogeny of the algal classes

7.4.1 Rhodophyceae

The fine structure of the chloroplast of the Rhodophyceae is essentially similar to the whole cell of the Cyanophyceae, and the cross-reactivity between homologous biliproteins in the two groups is high (Chapter 5, Sec. 5.3). The sequence homology between 16*S* rRNA from the chloroplasts of *Porphyridium cruentum* and *Anacystis nidulans* was also high (Bonen and Doolittle, 1975) but low between cytoplasmic 18*S* rRNA from *P. cruentum* and 16*S* rRNA from chloroplasts and *A. nidulans*. Doolittle and Bonen (1981) subsequently found 16*S* rRNA from *P. cruentum* more closely related to that from a number of Cyanophyceae than to that from *Euglena gracilis,* though Takaiwa, Kusuda, Saga, and Sugiura (1982) found that homology between 5*S* rRNA from *Porphyra yezoensis* (Bangiales) and 5*S* rRNA from *E. viridis* and *Chlorella* was higher than that between *P. yezoensis* and *A. nidulans*. Lim et al. (1986) found sequence homology of 5*S* rRNA between Rhodophyceae and Chromophyta quite low (59–61%). *Porphyra tenera* emerged significantly earlier than most of the Florideophycidae and the one species of chrysophyte tested. Although the Bangiophycidae appear to be more ancient than the Florideophycidae, the phycoerythrins, B-PE and R-PE do not fall neatly into these two groups (Chapter 5, Sec. 5.3). The difference in the two pigments lies in part in the ratios of the phycobilins, PEB and PUB; this ratio is not under strict genetic control, as it varies with the environment in which the species grows (Yu et al., 1981). Kikuchi, Ashida, and Hirao (1979) have found B-PE only in mutants of *Porphyra yezoensis* (green and yellow), whereas the wild-type and deep red mutant contained R-PE. Larkum and Barrett (1983) believe that PUB evolved later than PEB, as PUB is found in only a few Cyanophyceae, those containing CU-PE (Table 5–1). Unless the endosymbiont giving rise to the chloroplasts of the Rhodophyceae came from this small group, we must assume a polyphyletic origin for the algae containing PUB.

Cavalier-Smith (1982) believed that the Glaucophyceae, a group of cyanelle-containing organisms including *Cyanophora* and *Glaucocystis,* were a primitive branch of the Rhodophyceae and thus represented a primitive stage of the evolution of the phylum, though the Glaucophyceae are usually flagellate and have affinities with the Chlorophyta (Moestrup, 1982). Critics of this hypothesis state that (1) the cyanelles have retained the typical cell wall of a cyanophyte (though *Glaucosphaera vacuolata* does not), (2) the division of cells and cyanelles is not synchronous (Trench, 1982), and (3) the pattern of food reserves differs from those of Rhodophyceae (Kremer, Kies, and Rostami-Rabet, 1979; Moestrup, 1982). Another difficulty is that the cyanelles in these Glaucophyceae never contain biliproteins with the phycobilins PEB and PUB (Chapman, 1966b; Trench and Ronzio, 1978; Kremer et al., 1979). If these organisms are regarded as giving rise to the Rhodophyceae, we must assume a polyphyletic origin for PEB and PUB, a fact neglected by the proponents of the serial endosymbiont hypothesis. Taylor (1978) objected to Glaucophyceae as descendants of a stock in a transitional state for more than 600 million years, but thought that these were rather an artificial assemblage of recent symbioses. The Glaucophyceae resemble *Prochloron* in that they have few carotenoids and none of the xanthophylls specific to the Cyanophyceae, such as echinenone and myxoxanthophyll.

7.4.2. Cryptophyceae

Although the biliproteins of this class were originally believed to be distantly removed from those of the Cyanophyceae and the Rhodophyceae, some immunological cross-reactivity between certain biliproteins of the Cryptophyceae and the other classes has now been found (Chapter 5, Sec. 5.3.). The hypothesis that the chloroplasts of the Cryptophyceae are from a eukaryotic symbiont has rested mainly on the nucleomorph found in the periplastic space and the four membranes surrounding the chloroplast (Greenwood, Griffiths, and Santore, 1977; Gillot and Gibbs, 1980; Santore, 1982); this recent immunological evidence suggests that the chloroplast is derived from a eukaryotic symbiont with the nucleomorph as the vestigial nucleus, a finding recently confirmed by Hansman, Falk, and Sitte (1985) and Ludwig and Gibbs (1985), who detected DNA in the nucleomorph of *Cryptomonas* by fluorescence occurring after treating thin sections with 4′,6-diamino-2-phenylindole. Although most authors have believed that the symbiont was a red alga, the hypothesis proposed by Cavalier-Smith (1982, 1986), that the symbiont was a primitive eomitozoan containing both biliproteins and chlorophyll c_2, is attractive (Figs. 7–1, 7–2). With either precursor, synthesis of phycobilisomes must have been lost after symbiosis; this amounts to losing the capacity for synthesis of the linker polypeptides (Chapter 5, Sec. 5.9.1.), as occurs in some Rhodophyceae in certain environments (Honsell et al., 1985) or certain mutants in the Cyanophyceae (Glazer et al., 1983).

7.4.3. Dinophyceae

Like the Cryptophyceae, the Dinophyceae usually lack chlorophyll c_2, but in the few species containing fucoxanthin or fucoxanthin-like pigments (Table 4–2) in place of peridinin, chlorophyll c_1 occurs also; in one species only (*Prorocentrum cassubicum* = *Exuviaella cassubica*) that contains peridinin, chlorophylls c_1 and c_2 occur together (Jeffrey et al., 1975). Of the species shown in Table 4–2, some contain a eukaryotic nucleus in addition to the usual primitive mesokaryotic nucleus found in all other members of this class, the exceptions being in the *Gymnodiniales* (Loeblich, 1976), and the symbiotic origin of the fucoxanthin is now widely accepted. Apart from the synthesis of peridinin, the Dinophyceae have not evolved the number of pathways of synthesis of xanthophylls found in many of the chromophyte classes. Peridinin and related xanthophylls are almost entirely specific to the class, involving butenolide formation and C_3 expulsion, reactions not found in the synthesis of other pigments (Liaaen-Jensen, 1977, 1978). This fact, combined with the lack of synthetic pathways found in many other classes, is consistent with the ancient origin of the Dinophyceae based on other features, particularly the prokaryotic affinities of the chromosomes (Loeblich, 1976). Steidinger and Cox (1980) have proposed a phylogenetic system with three branches based on the major xanthophyll present – peridinin – fucoxanthin, or what we now know to be fucoxanthin-like pigments (Liaaen-Jensen, 1985; see Table 4–2), with the fucoxanthin branch divided again into those species with either one or two nuclei. This proposal envisages a different symbiosis for each branch, thus giving rise to the three membranes surrounding the nucleus in most species, with four in the binucleate species. According to Cavalier-Smith (1982), only two membranes around the chloroplast in some species represent secondary loss of the phagosome membrane (Fig. 7–1).

7.4.4. Chromophyta

Discussed in this section are classes other than Cryptophyceae and Dinophyceae (Chromophyta sensu Cavalier-Smith, 1986; see Fig. 7–2). The remaining classes have developed pathways of synthesis of carotenoids not found in the previous two classes, and have chloroplasts surrounded by four membranes (Whatley and Whatley, 1981), implying a symbiosis with a primitive, extinct eukaryote containing chlorophyll c_1, such as the eomitozoan of Cavalier-Smith (1986). The new pathways for xanthophyll

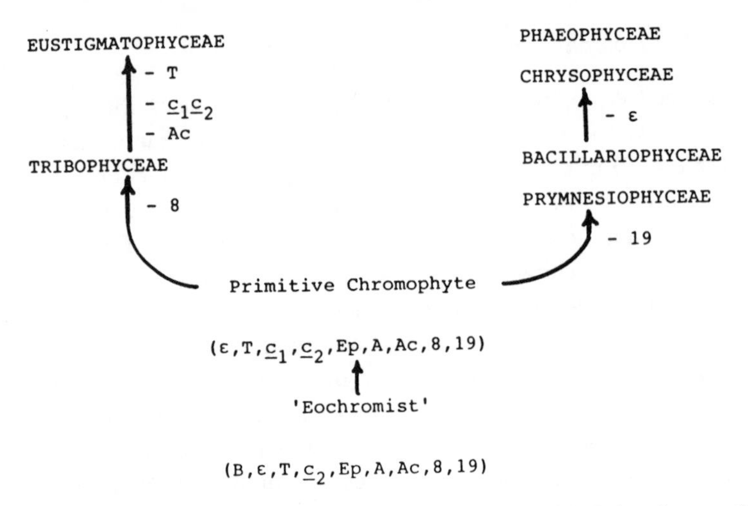

Fig. 7–3. Changes in pigments during the evolution of the classes of the Chromophyta from the eochromist proposed by Cavalier-Smith (1986), and the deletions of pathways of synthesis following evolution from the primitive Chromophyte. Abbreviations as in Fig. 7–2.

synthesis developed by the Chromophyta – in particular, for 8-keto formation, acetylation, and 19-hydroxylation – are common to the Chlorophyta; thus EM I, the primitive eomitozoan (Fig. 7–2) must have developed a complex series of pathways. Judged purely by pathways of synthesis, the most simple tree for the Chromophyta (sensu Cavalier-Smith) would be as in Fig. 7–3. The phylogenetic tree of Chromophyta (sensu Christensen) presented by Hibberd (1979) is based on both flagellar and chloroplast characters and does not fit the scheme shown in Fig. 7–3, which is based only on pathways of synthesis of pigments. His scheme shows that of the higher Chromophyta, the Prymesiophyceae are most primitive, whereas the other classes containing fucoxanthin evolved later than the group containing vaucheriaxanthin (Eustigmatophyceae and Tribophyceae), as judged by the secondary loss of the transitional helix in particular. However, the loss of pathways of synthesis of the vaucheriaxanthin group and the fucoxanthin group, as shown in Fig. 7–3, cannot allow a linear sequence, and, in addition, the scheme in Fig. 7–3 has the Eustigmatophyceae evolving after the Tribophyceae by secondary loss of several synthetic pathways. Accordingly, the best explanation, assuming the xenogenous origin of the chloroplasts, is that the pathways of pigment synthesis present will reflect changes in the chromophyte symbiont postulated in Fig. 7–3. Calculation of the homology between 5*S* rRNA has shown that the Chrysophyceae may have emerged before the Bacillariophyceae whereas the Phaeophyceae may have emerged after (Lim et al., 1986); this fits neither the scheme of Hibberd (1979), where the secondary loss of the photoreceptor apparatus from the Bacillariophyceae places this class later than the Phaeophyceae, nor the scheme in Fig. 7–3, where the secondary loss of the synthesis of ϵ-carotenoids from the Chrysophyceae places them later than the Bacillariophyceae. Lim et al. (1986) also found that sequence similarities between Chromophyta (sensu Cavalier-Smith, 1986) on the one hand, and Euglenophyceae, Cryptophyceae, Dinophyceae, and Rhodophyceae on the other hand were not high (65–69%), thus implying early divergence of these classes. However, as discussed in Section 7.1, using sequences of 5*S* rRNA extracted from whole cells may not be reliable when one is comparing algal classes.

7.4.5. *Chlorophyta*

Recent discussions of algal phylogeny have stressed that Prochlorophyceae such as *Prochloron* are likely symbionts giving rise to all classes of eukaryotes containing chlorophyll *b*. The Chlorophyta all have two or three (Euglenophyceae) membranes surrounding the chloroplast, and this again is taken as evidence for their symbiont origin (Whatley and Whatley, 1981). However, apart from possessing chlorophyll *b*, the Prochlorophyceae have few other attributes in common with the chloroplasts of the Chlorophyta (Sec. 7.3), and any symbiont is more likely to have evolved chlorophyll *b* independently of the Prochlorophyceae (Cavalier-Smith, 1982).

In addition to the pigments found in all higher plants (β-carotene, chlorophyll *a* and *b*, lutein, zeaxanthin, violaxanthin, and neoxanthin), the Chlorophyceae and Prasinophyceae sometimes contain siphonaxanthin and its ester, siphonein, loroxanthin, or prasinoxanthin and derivatives. These pigments have some taxonomic value (see Sec. 7.6.2), but possibly those present in the Prasinophyceae – the scaly flagellates believed to be primitive in the phylum (Stewart and Mattox, 1978; Taylor, 1978) – might help to indicate which are the most primitive orders within the class. Because they occur in the reputedly primitive Prasinophyceae, O'Kelly (1982) considered siphonaxanthin and siphonein primitive pigments when occurring in other taxa, and O'Kelly and Floyd (1984) showed that Dasycladales, the only order of the Ulvophyceae without siphonaxanthin, is at the head of the phylogenetic tree of the class Ulvophyceae. By the same argument, prasinoxanthin might be regarded as primitive, and, as it occurs in the Prasinophyceae and with similar chirality, it might share a common chemical precursor with siphonaxanthin (Foss et al., 1984). Loroxanthin, although occurring in the relatively primitive Ulotrichales (Yokohama, 1983; O'Kelly and Floyd, 1984), also occurs in the advanced Chlorococcales (Aitzetmuller et al., 1969), and thus may have little value in phylogeny. Although loroxanthin shares a C-19 hydroxyl group with siphonaxanthin and might also be a possible precursor for both siphonaxanthin and prasinoxanthin, it does not occur in the same species as either of those two xanthophylls. On the other hand, preliminary observations in this laboratory have shown that siphonaxanthin and prasinoxanthin may occur together in *Pterosperma* (Chapter 4, Sec. 4.1 4.). The combinations in which MgDVP, prasinoxanthin, and siphonaxanthin occur together may provide evidence for the taxonomy of the Prasinophyceae (Ricketts, 1967c).

The Euglenophyceae do not sit comfortably in the Chlorophyta; apart from chlorophyll *b*, they share few features with the rest of the phylum. The chloroplasts are surrounded by three membranes, not two, as in other classes, and their carotenoids are typical of the Chromophyta (diadinoxanthin and diatoxanthin), though earlier workers identified these incorrectly, an error repeated by Margulis (1981). Bjørnland (1982) and Fiksdahl and colleagues (Fiksdahl et al., 1984a; Fiksdahl and Liaaen-Jensen, 1988) have identified several other pigments not typical of the Chlorophyta (see Sec. 4.1.4B) in members of the Euglenophyceae. The third membrane has been identified as enclosing the chloroplast endoplasmic reticulum (CER), but Gibbs (1978) has argued that this is the plasma membrane of a chlorophyte symbiont. As mentioned above, the Euglenophyceae do not carry out the pathways of synthesis of the remaining Chlorophyta (Fig. 7–2). The close affinity of the Euglenophyceae and Dinophyceae has been postulated on their fine structure (Taylor, 1974, 1976, 1978; McQuade, 1983), though their mitochondrial cristae differ. Gibbs (1981) has proposed that the symbiont giving rise to the Euglenophyceae was a primitive member of the Chlorophyceae, but Cavalier-Smith (1986) considered that the Chlorophyta have evolved by mutation following the original symbiosis to form his eomitozoan (Fig. 7–2). The minimum number of mutations that I believe must occur without polyphyletic origins of xanthophylls is shown in Fig. 7–2, with the original eomitozoan (EMI) mutating to EMII and EMIII. Analysis of 5*S* rRNA has shown that the Euglenophyceae are closer to the Protozoa than other Chlorophyta (Delihas et al., 1981; Kumazaki, Hori, and Osawa, 1982a,b), and the euglenophytes separate clearly from other Viridiplantae in the cladogram generated by Bremer (1985). A cladogram generated from the 5*S* rRNA sequence data of Hori et al. (1985) by Bremer, Humphries, Mishler, and Churchill (1987) widely separates *Chlamydomonas* from five other genera of the Chlorophyta, though the analysis did not include *Euglena*. Homologies between base sequences of 5*S* rRNA from *Chlorella* and from *Synechococcus, Prochloron, Euglena,* and *Chlamydomonas* are low (Yamada and Shimaji, 1986), consistent with the data for the Chlorophyta of Bremer et al. (1987) mentioned above. However, homologies using base sequences of 5*S* rRNA may not be reliable (Sec. 7.1), and a cladistic analysis of 5*S* rRNA secondary and primary structures by Wolters and Erdmann (1988) has shown that plastids of *Euglena, Chlorella,* and *Chlamydomonas,* and the cyanelle of *Cyanophora* share a specific character and are closely related to *Synechococcus*-type cyanophytes.

7.5. Phylogenetic trees based on photosynthetic pigments

Many trees have been proposed (Sec. 7.1), but most have been based on anatomy and fine structure and usually take into account only chlorophylls and phycobilins. As many carotenoids are now found to harvest light for photosynthesis (Chapter 6), these should receive more weight than they usually do. Liaaen-Jensen (1977, 1978) has drawn a phylogenetic tree based on all photosynthetic pigments, but assuming independent origins of several pathways. Unfortunately, the details of the mechanisms of biosynthesis of many xanthophylls are still uncertain; more than one precursor has been suggested for the triple bond between C-7 and C-8, and for the keto group at C-8 (Liaaen-Jensen, 1978); this would reduce the number of deletions shown in Fig. 7–2, where a phylogeny based on photosynthetic pigments is shown. This tree is an expansion of that proposed by Cavalier-Smith (1982, 1986), taking into account the mutations giving rise to new pathways of synthesis of carotenoids without proposing polyphyletic origins of any pigments, except the chlorophyll *b* in the Prochlorophyceae. In the Chromophyta, new pathways of synthesis of xanthophylls developed, in particular those leading to fucoxanthin, and in the Chlorophyta, those leading to siphonaxanthin in the Caulerpales. In the Chromophyta, the most simple phylogeny based on carotenoids is shown in Fig. 7–3; as discussed above (Sec. 7.4.4), this is not consistent with the scheme proposed by Hibberd (1979) and the data of Lim et al. (1986).

7.6. Value of photosynthetic pigments in taxonomy within classes

In earlier sections in Chapters 3, 4, and 5, the importance of the pigments in assigning algae to the various classes was discussed. In this section, we consider their value in assigning algae to the various taxa within the classes.

7.6.1. *Chlorophylls*

As chlorophyll *b* is present in all Chlorophyta and Prochlorophyceae, it has no value in the taxonomy of these taxa and is merely a primitive character delimiting the Viridiplantae (sensu Cavalier-Smith, 1982) in the cladograms discussed by Bremer (1985) and Bremer et al. (1987). Chlorophylls c_1 and c_2 usually occur together in all classes of the Chromophyta, except the Cryptophyceae and Dinophyceae (chlorophyll c_2 only) and Eustigmatophyceae (no chlorophyll *c*). In the Chrysophyceae, Andersen and Mulkey (1983) have found only chlorophyll c_1 in marine strains from the family Synuraceae Bourrelly 1981 (Mallomonadaceae), and the lack of chlorophyll c_2 is of taxonomic value here. In his classification of the Kingdom Chromista, Cavalier-Smith (1981) has separated *Synura* from the Chrysophyceae, and Andersen (1987) has separated this family into a new class, the Synurophyceae. So far, the distribution of the new chlorophyll, c_3, confirms that the pigments of Bacillariophyceae and Prymnesiophyceae are similar (Stauber and Jeffrey, 1988), as both classes could be subdivided into three pigment groups: (1) fucoxanthin and chlorophyll $c_1 + c_2$; (2) fucoxanthin and chlorophyll $c_2 + c_3$; (3) fucoxanthin and derivatives, and chlorophyll $c_2 + c_3$. Group 1 contains the majority of species examined, and further minor variations were found among the diatoms, such as one with chlorophyll c_1, c_2, and c_3, and one with only chlorophyll c_2. The genus *Nitzschia* contributed species to each group, and deserves further taxonomic study. Whether chlorophyll *c* distribution will provide taxonomic markers for the families of diatoms requires further investigation (Stauber and Jeffrey, 1988). The distribution of chlorophyll *d* is rare and irregular (Chapter 3, Sec. 3.1) in the Rhodophyta and has no taxonomic value.

The pigment identified as MgDVP by Ricketts (1966a) or as chlorophyll c_1 by Wilhelm et al. (1986) and Wilhelm (1987) is restricted to some members of the Prasinophycae and is accompanied by one or more of the xanthophylls, siphonaxanthin, siphonein, or prasinoxanthin. Norris

(1980) has suggested that intergenetic variability in the pigments in the genus *Pyramimonas* (Ricketts, 1970) may help indicate that the genus is polyphyletic. In a cladistic study of the Viridiplantae (sensu Cavalier-Smith, 1981), Sluiman (1985) considered Prasinophyceae paraphyletic and unacceptable in the classification, because no clearly derived features were common to the whole class. Stewart and Mattox (1978, 1980; Mattox and Stewart, 1984) have also found the class Prasinophyceae unsatisfactory and have divided it into the Micromonadophyceae and the Pleurastrophyceae; they admit that the Micromonadophyceae is a heterogeneous group, shown by the unique LHCP in *Mantoniella* and *Micromonas* (Fawley, Stewart, and Mattox, 1986), which is absent in others of this class (Table 6–2). The distribution of MgDVP (or chlorophyll c_1) and of the xanthophylls siphonaxanthin, siphonein, and prasinoxanthin may well contribute to the classification of these difficult flagellates, whose relationships to higher Chlorophyta are discussed by Moestrup (1982).

7.6.2. Carotenoids

Presence or absence of certain xanthophylls is valuable as a taxonomic marker for assigning algae to classes (Chapter 4), but within classes seems of little utility unless a substantial number of species are examined. Jensen (1966) and Strain (1958, 1966) published lists of the carotenoids in the Phaeophyceae, and Goodwin (1980) collected valuable tables of xanthophylls in most classes of algae. Subsequently, Brown and McLachlan (1982) examined xanthophylls in the Gracilariaceae (Rhodophyceae), and Pennington et al. (1985) and Withers et al. (1981) published tables showing xanthophylls in seven species of Cryptophyceae and Chrysophyceae, respectively. Inspection of these tables demonstrates the infrequent value of carotenoid distribution in taxonomy within each class. In the Rhodophyceae, Brown and McLachlan (1982) found lutein replaced by β-cryptoxanthin in 22 strains of *Gracilaria* (Gracilariaceae: Gigartinales), suggesting that the distribution might have taxonomic value. The Porphyridiaceae (Porphyridiales) may also show this distribution, though only three species have been analyzed (Goodwin, 1980). The carotenoids of the Dinophyceae are usually similar in all orders, though the peridinin derivative, pyrroxanthin, is almost entirely restricted to the Glenodiniaceae, except for *Amphidinium carterae* in the Gymnodiniaceae.

Stauber and Jeffrey (1988) have found β-carotene, fucoxanthin, diadinoxanthin, and diatoxanthin in all of 71 isolates from 22 families of diatoms. Only one species contained a derivative of fucoxanthin, and any taxonomic value of pigments in this class rests with the distribution of the forms of chlorophyll *c*.

Few taxa of the Chlorophyta have been examined thoroughly, except for the Caulerpales (Ulvophyceae: sensu O'Kelly and Floyd, 1984) where siphonaxanthin and/or its ester, siphonein, are virtually always found, the one exception being *Caulerpa filiformis* (Strain, 1965). These pigments occur also in other orders (Ulotrichales and Siphonocladales: O'Kelly and Floyd, 1984; Ulvales: Yokohama et al., 1977; Kageyama and Yokohama, 1978; Yokohama 1981a,b; O'Kelly, 1982; Cladophorales: Weber and Czygan, 1972), but usually in specimens growing in low light intensity (O'Kelly and Floyd, 1984). Siphonaxanthin is absent from some Dichotomosiphonaceae that contain siphonein (Kleinig, 1969), apparently indicating active acetylation to form the ester from siphonaxanthin. The possible role of prasinoxanthin as an indicator of primitive members of the Chlorophyceae has been discussed in Section 7.4.5. The value of this pigment in assigning species to taxa must await further work on the relationships within the scaly flagellates.

7.6.3. Biliproteins

The original classification of biliproteins specific to the class Cyanophyceae (C-PE and C-PC) and the subclasses of the Rhodophyceae, the Bangiophycidae (B-PE) and the Florideophycidae (R-PE and R-PC), no longer hold, and, for example, the Bangiales usually contain R-PE (Honsell et al., 1984). Honsell et al. (1984) have provided a list of the identifications of the phycobiliproteins in the Rhodophyceae, mainly from Hirose and Kumano (1966) and from Glazer et al. (1982). Virtually all members of the Porphyridiaceae, Goniotrichaceae, and Erythropeltidaceae analyzed contain either B-PE I or II (Table 5–1), but B-PE occurs in some but not all Achrochaetiaceae and Rhodomelaceae; this distribution could thus be of taxonomic value, and Gabrielson, Garbary, and Scagel (1985) used it in a cladistic analysis of the Rhodophyceae, though the analysis did not segregate the Bangiophycidae from the Florideophycidae. This analysis showed that the Bangiophycidae was paraphyletic and recommended that it and the Florideophycidae be included in the single class Rhodophyceae. In a subsequent cladistic analysis of the Florideophycidae, Gabrielson and Garbary (1987) used presence of B-PE and absence of R-PE as primitive states; most orders were assigned derived states based on distribution of PE, but the Erythropeltidales, assigned primitive pigment states, appeared as the most primitive order in all six versions of the cladograms. They considered R-PE in the Bangiales to be a synapomorphy with the florideophytes. As the order Acrochaetidales contains both B- and R-PE, they suggested that it might not be monophyletic.

Chapman (1974) used presence of C-PC as one factor in assigning the paradoxical alga, *Cyanidium caldarium,* to the class Rhodophyceae, rather than to the Chlorophyceae or Cryptophyceae. Pan et al. (1986) studied PE in 30 species of Rhodophyceae, defining PE with two peaks as type I (i.e., B-PE) and as PE with three peaks type II (i.e., R-PE) (see Fig. 5–2). Although three out of 12 species with type I PE were florideophycean Rhodophyceae, they considered that the type of PE was of taxonomic significance.

The Cryptophyceae are a small class of the Eukaryota that have not yet been subdivided. Unlike the Rhodophyceae, a given species contains only one biliprotein, Cr-PC or Cr-PE, and none contains APC (Chapter 5). As both types of pigment occur within some genera as now defined, the type of biliprotein present has no taxonomic value (Klaveness, 1986). Each pigment occurs in three or four spectroscopic forms (Table 5–2), partly owing to the unusual phycobilins, CV, and the so-called bilin 697. Although the biliproteins are of no taxonomic value at present, pigments might prove a valuable characteristic if the class is subdivided. Hill (unpublished) has found that the type of Cr-PC present is the only characteristic separating a marine strain of a cryptophyte from a freshwater form. Hill and Rowan (1989) have suggested that because two different strains of *Hemiselmis virescens* contained different types of biliproteins, their identification should be examined again.

8

General summary

The physical position of the pigments and the ratios in which they occur in the components of the thylakoid membranes, and reconstitution of energy-transducing systems from components of similar or different classes of algae – in particular, in the classes containing phycobilisomes – is being actively investigated, as we have seen in Chapters 1, 5, and 7.

For several reasons, examining photosynthetic pigments in algae is valuable. Qualitative analysis indicates the taxon to which the species belongs, often to the level of the class, and certainly to the phylum (sensu Christensen), and, as we have seen in Chapter 7, distribution of pigments contributes to hypothetical schemes for evolution and phylogeny of the algae, though it tends to be neglected in favor of anatomy and cytology. Quantitative analysis is of two kinds: First, measuring pigments (usually chlorophyll *a*) in a crop of algae provides an estimate of biomass per unit area or volume; in Chapter 2, we saw how much effort has gone into these analyses. Second, quantitative analysis of the major pigments of all three types provides insight into the photosynthetic mechanism of a population in relation to the light intensity; thus these analyses are similar to studies of sun and shade plants on land, explaining the relative abundance of the two photosystems and their roles in energy transduction.

Chromatography is now regarded as the most reliable method for analysis of chlorophylls, although the relatively simple spectroscopic or fluorometric methods continue to attract workers in spite of errors due to the presence of chlorophyll derivatives and other pigments. TLC has been widely used since the late 1960s, but HPLC is now preferred, as seen by the number of recent papers listed in Table 2–6. Chromatography has always been the only effective method for analyzing carotenoids, and again HPLC is replacing older methods. Biliproteins are usually estimated by preliminary separation on column chromatography, although some polychromatic equations have been used.

The distribution of chlorophyll *b* requires little further study except when used as a marker for new members of the Prochlorophyceae (Burger-

Wiersma et al., 1986); it is still occasionally valuable to justify removal of algae from classes in the Chromophyta, as in the recent establishment of the new class Chlorarachnophyceae by Hibberd and Norris (1984). Absence of chlorophyll c_2 from marine Chrysophyceae has led Andersen (1987) to propose a new class, the Synurophyceae, for the genera *Synura* and *Mallomonas*. The modern methods of chemical analysis used at Trondheim under Liaaen-Jensen have identified a number of unusual carotenoids, and more work of this kind is needed to describe the xanthophylls remaining unidentified in the literature, including fucoxanthin-like pigments in the Dinophyceae, which cannot be identified by TLC alone. The variation in spectra of the biliproteins stems from the variation in type and ratio of the phycobilins in their subunits and sometimes reflects the ability of the organism to survive in green light at depth in water. In particular, the unusual pigments, PEC (containing CV) and CU–PE (containing PUB), found in some marine Cyanophyceae, can provide absorption in the green region of the spectrum, where other types of pigment do not absorb light effectively. The pigment composition is not rigidly determined genetically, as the proportion of PUB may be altered experimentally in the laboratory by manipulation of light quality. The importance of CU–PE was discussed in Section 5.3. The phycobilin CV also extends the range of absorption of light by the crytophyte pigments Cr–PC_{630} and Cr–PC_{645} into the green region of the spectrum. Many of the interesting variations in normal distribution and type of pigment occur in marine phytoplankton – for example, the members of the new class Synurophyceae and those containing the new chlorophyll c_3; further study of the pigments in these marine species would be desirable.

As shown in Table 6–2, many workers have examined pigment–protein complexes extracted from all classes of algae. In particular, the LHCP fractions specific for either PS I (LHC I) or PS II (LHC II) are being found, and further work in this area is required. The distribution and function of carotenoids other than fucoxanthin (and derivatives), peridinin, and β-carotene in these complexes also require further investigation, as few workers have analyzed the carotenoids present (Table 6–2), undoubtedly because of the difficulty of dealing with such small amounts of pigments.

Pigment distribution has provided evidence for phylogenetic schemes for the evolution of the algae; erecting schemes not involving polyphyletic origin of some pathways of carotenoid synthesis requires one to assume hypothetical precursors (the eomitozoan and eochromist of Cavalier-Smith;

see Fig. 7–2) containing combinations of pigments not found in modern algae. Monophyletic origin for the phycobilins implies that the eomitozoan must have contained biliproteins with all three of the common phycobilins (PCB, PEB, and PUB) and thus must have been derived from a cyanophyte containing CU–PE and C–PC.

References

Aasen, A. J., and Liaaen-Jensen, S. 1966. The carotenoids of the flexibacteria. II. A new xanthophyll from *Saphrospira grandis*. *Acta Chem. Scand.* 20: 811–19.

Abaychi, J. K., and Riley, J. P. 1979. The determination of phytoplankton pigments by high-performance liquid chromatography. *Anal. Chim. Acta* 64: 525–7.

Adams, S. M., Kao, O. H. W., and Berns, D. S. 1979. Psychrophile C-phycocyanin. *Plant Physiol.* 64: 525–7.

Aihara, M. S., and Yamamoto, H. Y. 1968. Occurrence of antheraxanthin in two Rhodophyceae, *Acanthophora specifera* and *Gracilaria lichenoides*. *Phytochemistry* 7: 497–9.

Aitzetmuller, K., Strain, H. H., Svec, W. A., Grandolfo, M., and Katz, J. J. 1969. Loroxanthin, a unique xanthophyll from *Scenedesmus obliquus* and *Chlorella vulgaris*. *Phytochemistry* 8: 1761–70.

Aitzetmuller, K., Svec, W. A., Katz, J. J., and Strain, H. H. 1968. Structure and chemical identity of diadinoxanthin and the principal xanthophyll of *Euglena*. *Chem. Commun.* 32–3.

Alberte, R. S., and Andersen, R. A. 1986. Antheraxanthin, a light-harvesting carotenoid found in chromophyte algae. *Plant Physiol.* 80: 583–7.

Alberte, R. S., Friedman, A. L., Gustafson, D. L., Rudnick, M. S., and Lyman, H. 1981. Light-harvesting systems of brown algae and diatoms. Isolation and characterization of chlorophyll *a/c* and chlorophyll *a*/fucoxanthin pigment–protein complexes. *Biochim. Biophys. Acta* 635: 304–16.

Alberte, R. S., Wood, A. M., Kursar, T. A., and Guillard, R. R. L. 1984. Novel phycoerythrins in marine *Synechococcus* spp. Characterization and evolutionary and ecological implications. *Plant Physiol.* 75: 732–9.

Allen, M. B. 1958. Possible functions of chlorophyll. Studies with green algae that lack chlorophyll *b*. *Brookhaven Symp. Biol.* 11: 339–42.

Allen, M. B., Dougherty, E. C., and McLaughlin, J. J. A. 1959. Chromoprotein–pigments of some cryptomonad flagellates. *Nature* 184: 1047–9.

Allen, M. B., French, C. S., and Brown, J. S. 1960. Native and extractable forms of chlorophyll in various algal groups. In *Comparative Biochemistry of Photoreactive Systems*, ed. M. B. Allen, pp. 33–52. New York: Academic Press.

Allen, M. B., Fries, L., Goodwin, T. W., and Thomas, D. M. 1964. The carotenoids of algae: Pigments from some cryptomonads, a heterokont and some Rhodophyceae. *J. Gen. Microbiol.* 34: 259–67.

Allen, M. B., Goodwin, T. W., and Phagpolngarm, S. 1960. Carotenoid distribution in certain naturally occurring algae and in some artifically induced mutants of *Chlorella pyrenoidosa*. *J. Gen. Microbiol.* 23: 93–103.

Andersen, R. A. 1987. Synurophyceae classis nov., a new class of algae. *Am. J. Bot.* 74: 337–53.

Andersen, R. A., and Mulkey, T. J. 1983. The occurrence of chlorophyll c_1 and c_2 in the Chrysophyceae. *J. Phycol.* 19: 289–94.

Anderson, J. M. 1980. P-700 content and polypeptide profile of chlorophyll–protein complexes of spinach and barley thylakoids. *Biochim. Biophys. Acta* 591: 113–26.

– 1983. Chlorophyll–protein complexes of a *Codium* species, including a light-harvesting siphonaxanthin–chlorophyll *a*/*b*–protein complex, an evolutionary relic of some Chlorophyta. *Biochim. Biophys. Acta* 724: 370–80.

– 1985. Chlorophyll–protein complexes of a marine green alga, *Codium* species (Siphonales). *Biochim. Biophys. Acta* 806: 145–53.

Anderson, J. M., and Barrett, J. 1986. Light-harvesting pigment–protein complexes of algae. *Enc. Plant Physiol.* (2nd ed.), 19: 269–85.

Anderson, J. M., Waldron, J. C., and Thorne, S. W. 1978. Chlorophyll–protein complexes of spinach and barley thylakoids. *FEBS Lett.* 92: 227–33.

– 1980. Chloroplast–protein complexes of a marine green alga, *Caulerpa cactoides*. *Plant Sci. Lett.* 17: 144–57.

Angapindu, A., Silberman, H., Tantnatana, P., and Kaplan, I. R. 1958. The separation of chlorophylls by paper and cellulose column chromatography. *Arch. Biochem. Biophys.* 75: 56–68.

Antia, N. J. 1965. The optical activity of fucoxanthin. *Can. J. Chem.* 11: 339–43.

Antia, N. J., Bisalputra, T., Cheng, J. Y., and Kalley, J. P. 1975. Pigment and cytological evidence for reclassification of *Nannochloris oculata* and *Monallantus salina*. *J. Phycol.* 11: 339–43.

Antia, N. J., and Cheng, J. Y. 1977. Reexamination of the carotenoid pigments of the unicellular blue-green alga, *Agmenellum quadruplicatum*. *J. Fish. Res. Board Can.* 34: 659–68.

– 1982. The keto-carotenoids of two marine coccoid members of the Eustigmatophyceae. *Br. Phycol. J.* 17: 39–50.

– 1983. Evidence for an anomalous xanthophyll composition in a clone of *Dunaliella tertiolecta* (Chlorophyceae). *Phycologia* 22: 235–42.

Araki, S., Oohusa, T., Omata, T., and Murata, N. 1984. Column chromatographic separation of chlorophyllide *a* and phaeophorbide *a*. *Plant Cell Physiol.* 25: 841–3.

Arnon, D. I. 1949. Copper enzymes in isolated chloroplasts. Polyphenol oxidase in *Beta vulgaris*. *Plant Physiol.* 24: 1–15.

Arpin, N., Svec, W. A., and Liaaen-Jensen, S. 1976. New fucoxanthin-related carotenoids from *Cocolithus huxleyi*. *Phytochemistry* 15:529–32.

Asami, M. 1952. On the paper chromatography of the leaf pigments. *Bot. Mag. Tokyo* 65: 771–2.

Augier, J. 1953. La constitution chimique de qualques floridées Rhodomelacées. *Rev. Gén. Bot.* 60: 257–83.

Bacon, M. F. 1966. Artifacts from chromatography of chlorophylls. *Biochem. J.* 101: 34–6C.

Bacon, M. F., and Holden, M. 1967. Changes in chlorophyll resulting from various chemical and physical treatments of leaves and leaf extracts. *Phytochemistry* 6: 193–210.

Balch, W. M., and Haxo, F. T. 1984. Spectral properties of *Noctiluca militaris* Suriray, a heterotrophic dinoflagellate. *J. Plankton Res.* 6: 515–25.

Barber, R. T., White, A. W., and Siegelman, A. W. 1969. Evidence for a cryptomonad symbiont in the ciliate *Cyclatrichium mainiere*. *J. Phycol.* 5: 86–8.

Barrett, J., and Anderson, J. M. 1980. The P-700–chlorophyll *a*–protein complex and two major light-harvesting complexes of *Acrocarpia paniculata* and other brown seaweeds. *Biochim. Biophys. Acta* 590: 309–23.

Barrett, J., and Jeffrey, S. W. 1964. Chlorophyllase and formation of an atypical chlorophyllide in marine algae. *Plant Physiol.* 33: 44–7.

– 1971. A note on the occurrence of chlorophyllase in marine algae. *J. Exp. Mar. Biol. Ecol.* 7: 255–62.

Bauer, L. 1952. Trennung der Karotinoide und Chlorophylle mit Hilfe der Paperchromatographie. *Naturwissenschaften* 39: 88.

Beale, S. I. 1984. Biosynthesis of photosynthetic pigments. In *Chloroplast Biosynthesis – Topics in Photosynthesis,* Vol. 5, ed. N. R. Baker and J. Barber, pp. 133–205. Amsterdam: Elsevier.

Beer, S., and Eshel, A. 1985. Determining phycoerythrin concentrations in aquatic crude extracts of red algae. *Aust. J. Mar. Freshwater Res.* 36: 785–92.

Beer, S., and Levy, I. 1983. Effects of photon fluence rate and light spectrum composition on growth, photosynthesis and pigment relations in *Gracilaria* sp. *J. Phycol.* 19: 516–32.

Bennett, A., and Bogorad, L. 1971. Properties of subunits and aggregates of blue-green algal biliproteins. *Biochemistry* 10: 3625–34.

– 1973. Complimentary chromatic adaptation in a filamentous blue-green alga. *J. Cell Biol.* 58: 419–35.

Benson, E. A., and Cobb, A. H. 1981. The separation, identification and quantitative determination of photopigments from the siphonaceous marine alga, *Codium fragile*. *New Phytol.* 88: 627–32.

– 1983. Pigment/protein complexes of the intertidal alga *Codium fragile* (Suringar) Hariot. *New Phytol.* 95: 581–94.

Berger, R., Liaaen-Jensen, S., McAlister, V., and Guillard, R. R. L. 1977. Carotenoids of Prymnesiophyceae (Haptophyceae). *Biochem. Syst. Ecol.* 5: 71–5.

Bernhard, K., Moss, G. F., and Toth, G. 1974. Sterioisomers of fucoxanthin. *Tetrahed. Lett.* No. 44: 3899–3902.

Berns, D. S. 1967. Immunochemistry of biliproteins. *Plant Physiol.* 42: 1569–86.

Bidigare, R. R., Kennicutt, M. C., and Brooks, J. M. 1985. Rapid determination of chlorophylls and their degradation products by high-performance liquid chromatography. *Limnol. Oceanogr.* 30: 432–5.

Bishop, J. E., Rapoport, H., Klotz, A. V., Chan, C. F., Glazer, A. N., Fuglistaller, P., and Zuber, H. 1987. Chromopeptides from phycoerythrocyanin. Structure and linkage of the three bilin groups. *J. Am. Chem. Soc.* 109: 875–81.

Bjørnland, T. 1982. Chlorophylls and carotenoids of the marine alga *Eutreptionella gymnastica*. *Phytochemistry* 21: 1715–19.

– 1983. Chlorophyll *a* and carotenoids of five isolates of the red alga *Antithamnion plumula*. *Biochem. Syst. Ecol.* 11: 73–6.

– 1984. Chlorophyll *a* and carotenoids of the red alga *Erythrotrichia carnea*. *Biochem. Syst. Ecol.* 12: 279–83.

Bjørnland, T., and Aguilar-Martinez, M. 1976. Carotenoids in red algae. *Phytochemistry* 15: 291–6.

Bjørnland, T., Borch, G., and Liaaen-Jensen, S. 1984. Configurational studies on red algal carotenoids. *Phytochemistry* 23: 1711–15.

Bjørnland, T., Fiksdahl, A., Haxo, F. T., and Liaaen-Jensen, S. 1985. Carotenoids of *Euglena gracilis* strain Ulex 753. In *Proc. 2nd Int. Phycol. Cong.*, Copenhagen, August 1985, p. 13.

Bjørnland, T., and Tangen, K. 1979. Pigmentation and morphology of a marine *Gyrodinium* (Dinophyceae) with a major carotenoid different from peridinin and fucoxanthin. *J. Phycol.* 15: 457–63.

Blinks, L. R. 1964. Accessory pigments and photosynthesis. In *Photophysiology*, Vol. 1, ed. A. C. Giese, pp. 199–221. New York: Academic Press.

Boardman, N. K., and Thorne, S. W. 1971. Sensitive fluorescence method for the determination of chlorophyll *a*/chlorophyll *b* ratios. *Biochim. Biophys. Acta* 253: 222–31.

Boczar, B. A., and Prézelin, B. B. 1986. Light- and $MgCl_2$-dependent characteristics for chlorophyll–protein complexes isolated from the marine dinoflagellate, *Glenodinium* sp. *Biochim. Biophys. Acta* 850: 300–9.

– 1987. Chlorophyll–protein complexes from the red-tide dinoflagellate, *Gonyaulax polyedra* Stein: Isolation, characterization and the effect of growth irradiance on chlorophyll distribution. *Plant Physiol.* 83: 805–12.

Boczar, B. A., Prézelin, B. B., Markwell, J. P., and Thornber, J. P. 1980. A chlorophyll *c*-containing pigment–protein complex from the marine dinoflagellate *Glenodinium* sp. *FEBS Lett.* 120: 243–7.

Bogorad, L. 1975. Phycobiliproteins and complementary chromatic adaptation. *Annu. Rev. Plant Physiol.* 26: 369–401.

Bonen, L., and Doolittle, W. F. 1975. On the prokaryotic nature of red algal chloroplasts. *Proc. Natl. Acad. Sci. U.S.A.* 72: 2301–4.

Boney, A. D. 1975. *Phytoplankton* (Studies in Biology No. 52). London: Arnold.

Boto, K. C., and Bunt, J. S. 1978. Selective exication fluorimetry for determination of chlorophylls and phaeophorbides. *Anal. Chem.* 50: 392–5.

Bowles, N. D., Pearl, H. W., and Tucker, J. 1985. Effective solvents and extraction periods employed in phytoplankton carotenoid and chlorophyll determination. *Can. J. Fish. Aquat. Sci.* 42: 1127–31.

Bramley, P. M. 1985. The in vitro biosynthesis of carotenoids. *Adv. Lipid Res.* 21: 243–79.

Brandt, P., Kaiser-Jarry, K., and Wiessner, W. 1982. Chlorophyll–protein complexes. Variability of CPI and the existence of two distinct forms of LHCP and one low-molecular-weight chlorophyll *a* protein. *Biochim. Biophys. Acta* 679: 404–9.

Brandt, P., Zufall, E., and Wiessner, W. 1983. Relation between the light-harvesting chlorophyll *a*–protein complex LHCPa and photosystem I in the alga *Chlamydomonas stillata. Plant Physiol.* 71: 128–31.

Braumann, T., and Grimme, H. L. 1979. Single step separation and identification of photosynthetic pigments by high-performance liquid chromatography. *J. Chromatogr.* 170: 264–8.

Braumann, T., and Grimme, H. L. 1981. Reverse-phase high-performance liquid chromatography of chlorophylls and carotenoids. *Biochim. Biophys. Acta* 637: 8–17.

Bremer, K. 1985. Summary of green plant phylogeny and classification. *Cladistics* 1: 369–85.

Bremer, K., Humphries, C. J., Mishler, B. D., and Churchill, S. P. 1987. On cladistic relationships in green plants. *Taxon* 36: 339–49.

Breton, J., and Katoh, S. 1987. Orientation of the pigments in photosystem II: Low temperature linear-dichroism study of a core particle and of its chlorophyll–protein subunits isolated from *Synechococcus* sp. *Biochim. Biophys. Acta* 892: 99–107.

Britton, G. 1983. *The Biochemistry of Natural Pigments.* Cambridge: Cambridge University Press.

Britton, G., and Goodwin, T. W. 1971. Biosynthesis of carotenoids. *Methods Enzymol.* 18C: 654–701.

Brooks, C., and Gantt, E. 1973. Comparisons of phycoerythrins (542, 566 nm) from Cryptophycean algae. *Arch. Mikrobiol.* 88: 193–204.

Brown, A. S., Foster, J. A., Voynow, P. V., Franzblau, C., and Troxler, R. F. 1975. Allophycocyanin from the filamentous Cyanophyte, *Phormidium luridum. Biochemistry* 14: 3581–8.

Brown, A. S., and Troxler, R. F. 1977. Bilin–apoprotein linkages in Rhodophyan phycobiliproteins. The role of cysteine. *FEBS Lett.* 82: 206–10.

Brown, J. M. A., Dromgoole, F. I., and Guest, P. M. 1980. The effect of spectrophotometer characteristics on the measurement of chlorophylls. *Aquat. Bot.* 9: 173–8.

Brown J. S. 1985. Three photosynthetic antenna porphyrins on a primitive green alga. *Biochim. Biophys. Acta* 807: 143–6.

– 1987. Functional organization of chlorophyll *a* and carotenoids in the alga, *Nannochloropsis salina. Plant Physiol.* 83: 434–7.

Brown, L. M., Hargrave, B. F., and MacKinnon, H. D. 1981. Analysis of chlorophyll *a* in sediments by high-pressure liquid chromatography. *Can. J. Fish. Aquat. Sci.* 38: 205–14.

Brown, L. M., and McLachlan, J. 1982. Atypical carotenoids for the Rhodophyceae in the genus *Gracilaria* (Gigartinales). *Phycologia* 21: 9–16.

Brown, S. R. 1968. Absorption coefficients of chlorophyll derivatives. *J. Fish. Res. Board Can.* 25: 523–40.

Brown, W. G. 1939. Micro separations by chromatographic adsorption on blotting paper. *Nature* 143: 377–8.

Bruinsma, J. 1961. A comment on the spectrophotometric determination of chlorophyll. *Biochim. Biophys. Acta* 52: 576–8.

– 1963. The quantitative analysis of chlorophylls *a* and *b* in plant extracts. *Photochem. Photobiol.* 2: 241–9.

Bryant, D. A. 1977. Ph.D. thesis, University of California, Berkeley.

– 1981. A photoregulated expression of multiple phycocyanin species. *Eur. J. Biochem.* 119: 425–9.

– 1982. Phycoerythrin and phycocyanin: Properties and occurrence in cyanobacteria. *J. Gen. Microbiol.* 128: 835–44.

Bryant, D. A., and Cohen-Bazire, G. 1981. Effects of chromatic illumination on cyanobacterial phycobilisomes. *Eur. J. Biochem.* 119: 415–24.

Bryant, D. A., Cohen-Bazire, G., and Glazer, A. N. 1981. Characterization of the biliproteins of *Gloeobacter violaceus. Arch. Microbiol.* 129: 190–8.

Bryant, D. A., Glazer, A. N., and Eiserling, F. A. 1976. Characterization and structural properties of the major biliproteins of *Anabaena* sp. *Arch. Microbiol.* 110: 61–75.

Bryant, D. A., Guglielmi, G., de Marsac, N. T., Castets, A.-M., and Cohen-Bazire, G. 1979. The structure of cyanobacterial phycobilisomes: A model. *Arch. Microbiol.* 123: 113–27.

Bryant, D. A., Hixson, C. S., and Glazer, A. N. 1978. Structural studies on phycobiliproteins. III. Composition of bilin-containing peptides from the β subunits of C-phycocyanin, R-phycocyanin and phycoerythrocyanin. *J. Biol. Chem.* 253:220–5.

Buchecker, R., and Liaaen-Jensen, S. 1977. Absolute configuration of heteroxanthin and diadinoxanthin. *Phytochemistry* 16: 729–33.

Buchecker, R., Liaaen-Jensen, S., Borch, G., and Siegelman, H. W. 1976. Carotenoids of *Anacystis nidulans,* structures of caloxanthin and nostoxanthin. *Phytochemistry* 15: 1015–18.

Buchecker, R., Marti, U., and Eugster, C. H. 1984. Synthese von optisch activen Carotinoiden mit 3, 5, 6-Trihydroxy-5, 6-dihydro-β-Endgruppen. *Helv. Chim. Acta* 67: 2043–56.

Budzikiewicz, H., and Taraz, K. 1971. Chlorophyll *c*. *Tetrahedron* 27: 1447–60.

Bullerjahn, G. S., Matthijs, H. C. P., Mur, L. R., and Sherman, L. A. 1987. Chlorophyll–protein composition of the thylakoid membrane from *Prochlorothrix hollandica,* a prokaryote containing chlorophyll *b*. *Eur. J. Biochem.* 168: 295–300.

Burger-Wiersma, T., Veenhuis, M., Korthals, H. F., Van de Wiel, C. C. M., and Mur, L. R. 1986. A new prokaryote containing chlorophyll *a* and *b*. *Nature* 320: 262–4.

Burnison, B. K. 1980. Modified dimethyl sulfoxide (DMSO) extraction for chlorophyll analysis of phytoplankton. *Can. J. Fish. Aquat. Sci.* 37: 729–33.

Camm, E. L., and Green, B. R. 1983. Isolation of PS II reaction centre and its relationship to the minor chlorophyll–protein complexes. *J. Cell. Biochem.* 23: 171–9.

Canaani, O. D., and Gantt, E. 1980. Circular dichroism and polarized fluorescence characteristics of blue-green algal allophycocyanins. *Biochemistry* 19: 2950–6.

– 1982. Formation of hybrid phycobilisomes by association of phycobiliproteins from *Nostoc* and *Fremyella*. *Proc. Natl. Acad. Sci. U.S.A.* 79: 5277–81.

– 1983. Native and in vitro-associated phycobilisomes of *Nostoc* sp.: Composition, energy transfer and effect of antibodies. *Biochim. Biophys. Acta* 723: 340–9.

Canaani, O. D., Lipschultz, C. A., and Gantt, E. 1980. Reassembly of phycobilisomes from allophycocyanin and a phycocyanin–phycoerythrin complex. *FEBS Lett.* 115: 225–9.

Carneto, J. E., and Catoggio, J. A. 1976. Variation in pigment contents of the diatom, *Phaeodactylum tricornutum* during growth. *Mar. Biol.* 36: 105–12.

Caron, L., and Brown, J. 1987. Chlorophyll–carotenoid–protein complexes from the diatom *Phaeodactylum tricornutum:* Spectrophotometric, pigment and polypeptide analyses. *Plant Cell Physiol.* 28: 775–85.

Caron, L., Dubacq, J. P., Berkaloff, C., and Jupin, H. 1985. Subchloroplast fractions from the brown alga, *Fucus serratus:* Phosphatidylglycerol contents. *Plant Cell Physiol.* 26: 131–9.

Caron, L., Jupin, H., and Berkaloff, C. 1983. Effects of light quality on chlorophyll-forms Ca 684, Ca 690 and Ca 699 of the diatom *Phaeodactylum tricornutum*. *Photosynth. Res.* 4: 21–33.

Carter, R. J., Flett, D. J., and Gibbs, C. F. 1988. Correction of baseline irregularities in the gradient elution of phytoplankton pigments by HPLC using methanol/water/ethyl acetate. *J. Chromatogr. Res.* 26: 121–4.

Castelfranco, P. A., and Beale, S. I. 1982. Chlorophyll biosynthesis: Recent advances and areas of current interest. *Annu. Rev. Plant Physiol.* 34: 241–78.

Cavalier-Smith, T. 1975. The origin of muclei and of eukaryotic cells. *Nature* 256: 463–8.

– 1981. Eukaryotic kingdoms: Seven or nine? *BioSystems* 14: 461–81.

– 1982. The origin of plastids. *Biol. J. Linn. Soc.* 17: 289–306.

– 1986. The kingdom Chromista: Origin and systematics. *Prog. Phycol. Res.* 4: 309–47.

– 1987a. The origin of eukaryotic and archebacterial cells. *Ann. N.Y. Acad. Sci.* 503: 17–54.

– 1987b. The simultaneous origin of mitochondria, chloroplasts, and microbodies. *Ann. N.Y. Acad. Sci.* 503: 55–71.

Cavalier-Smith, T., and Lee, J. J. 1985. Protozoa as hosts for endosymbioses and the conversion of symbionts into organelles. *J. Protozool.* 32: 376–9.

Chang, W. Y. B., and Rossmann, R. 1981. A numerical simulation of trichromatic equations on chlorophyll estimation using the spectrophotometric technique. *Hydrobiologia* 79: 265–70.

Chang, W. Y. B., and Rossmann, R. 1982. The influence of phytoplankton composition on the relative effectiveness of grinding and sonication for chlorophyll extraction. *Hydrobiologia* 88: 245–9.

Chapman, D. J. 1965. Studies on the carotenoids of the flagellate Cryptophyceae, and the chloroplast pigments of the endosymbiotic algae in *Cyanophora paradoxa* and *Glaucocystis nostochinearum*. Ph.D. thesis, University of California, San Diego.

– 1966a. Three new carotenoids isolated from algae. *Phytochemistry* 5: 1311–3.

– 1966b. The pigments of the symbiotic algae (Cyanomes) of *Cyanophora paradoxa* and *Glaucocystis nostochinearum* and two Rhodophyceae, *Porphyridium aerugeneum* and *Asterocytis ramosa*. *Arch. Mikrobiol.* 55: 17–25.

– 1974. Taxonomic status of *Cyanidium caldarium*, the Porphyridiales and Gonierichales. *Nova Hedwigia* 25: 673–82.

Chapman, D. J., Cole, W. J., and Siegelman, H. W. 1967. The structure of phycoerythrobilin. *J. Am. Chem. Soc.* 89: 5976–7.

Chapman, D. J., and Haxo, F. T. 1963. Identity of ϵ-carotene and ϵ_1-carotene. *Plant Cell Physiol.* 4: 59–63.

– 1966. Chloroplast pigments of Chloromonadophyceae. *J. Phycol.* 2: 89–91.

Cheng, J. Y., Don-Paul, M., and Antia, S. J. 1974. Isolation of an unusually stable cis-isomer of alloxanthin from a bleached autolysed culture of *Chroomonas salina* grown photoheterotrophically on glycerol. Observations of cis–trans isomerization of alloxanthin. *J. Protozool.* 21: 761–8.

Chu, Z.-X., and Anderson, J. M. 1985. Isolation and characterization of a siphonaxanthin–chlorophyll *a/b*–protein complex of photosystem I from a *Codium* species (Siphonales). *Biochim. Biophys. Acta* 806: 154–60.

Clement-Metral, J. D., Gantt, E., and Redlinger, T. 1985. A photosystem II–phycobilisome preparation from the red alga, *Porphyridium cruentum:* Oxygen evolution, ultrastructure, and polypeptide isolation. *Arch. Biochem. Biophys.* 238: 10–17.

Codgell, R. J. 1983. Photosynthetic reaction centers. *Annu. Rev. Plant Physiol.* 34: 21–45.

Cohen-Bazire, G., Beguin, S., Rimon, S., Glazer, A. N., and Brown, D. M. 1977. Physico-chemical and immunological properties of allophycocyanin. *Arch. Microbiol.* 111: 225–38.

Cohen-Bazire, G., and Bryant, D. A. 1982. Phycobilisomes: Composition and structure. In *The Biology of Cyanobacteria,* ed. N. G. Carr and B. A. Whitton, pp. 143–90. Oxford: Blackwell.

Colijn, F., and Dijkema, K. S. 1981. Species composition of benthic diatoms and distribution of chlorophyll *a* on an intertidal flat in the Dutch Wadden Sea. *Mar. Ecol. Prog. Ser.* 4: 9–21.

Comar, C. L., and Zscheile, F. P. 1942. Analysis of plant extracts for chlorophylls *a* and *b* by a photoelectric spectrophotometric method. *Plant Physiol.* 17: 198–209.

Conquist, A. 1981. Discussion paper. *Ann. N.Y. Acad. Sci.* 361: 500–4.

Cooper, R. D. G., Davies, J. B., Leftwick, A. P., Price, C., and Weedon, B. C. L. 1975. Carotenoids and related compounds. Pt. XXXII. Synthesis of astaxanthin, phoenicoxanthin, hydroxyechinenone and the corresponding diosphenols. *J. Chem. Soc.* (*Perkin Trans.*) 2195–2204.

Cox, G., Hiller, R. G., and Larkum, A. W. D. 1985. An unusual cyanophyte, containing phycourobilin and symbiotic with sponges and ascidians. *Mar. Biol.* 89: 149–63.

Chereskin, B. M., Castelfranco, P. A., Dallas, J. L., and Straub, K. M. 1983. Mg-2,4-divinyl pheoporphyrin a_5: The product of a reaction catalysed in vitro by developing chloroplasts. *Arch. Biochem. Biophys.* 226: 10–18.

Chereskin, B. M., Clement-Metral, J. D., and Gantt, E. 1985. Characterization of a purified photosystem II–phycobilisome particle preparation from *Porphyridium cruentum. Plant Physiol.* 77: 626–9.

Chereskin, B. M., Wong, Y.-S., and Castelfranco, P. A. 1982. In vitro synthesis of the chlorophyll isocyclic ring. *Plant Physiol.* 70: 982–93.

Chevolot-Magueur, A.-M., Cave, A., Potier, P., Teste, J., Chiaroni, A., and Riche, C. 1976. Composés bromés de *Rytiphlea tinctoria* (Rhodophyceae). *Phytochemistry* 15: 767–71.

Chevolot-Magueur, A.-M., Lavorel, J., and Potier, P. 1974. Mine en évidence de l'activité photosynthétique du pigment de *Rytiphlea tinctoria* (Clem) C. Ag. *C. R. Hebd. Séances Acad. Sci.* 278: 261–4.

Chiba, Y., Aiga, I., Idemori, M., Satoh, Y., Matsushita, K., and Sasa, T. 1967. Studies on chlorophyllase of *Chlorella protothecoides. Plant Cell Physiol.* 8: 623–35.

Cholnoky, K., Gyorgyev, K. D., Ronai, A., Szabolcs, J., Toth, G., Galasko, G., Mallams, A. K., Waight, E. S., and Weedon, B. C. L. 1969. Carotenoids and related compounds. Pt. XXI. Structure of neoxanthin (foliaxanthin). *J. Chem. Soc.* [*C*] 1256–63.

Christensen, T. 1962. Alger. In *Systematisk Botanik*, Vol. 2, 178 p., No. 2, ed. T. W. Böcher, M. Lange, and T. Sørensen, Copenhagen: Munksgaard.

– 1964. The gross classification of algae. In *Algae and Man,* ed. D. F. Jackson, pp. 59–64. New York: Plenum Press.

Craigie, J. S., Leigh, C., Chen, L. C., and McLachlan, J. 1971. Pigments, polysaccharides and photosynthetic products of *Phaeosaccion callensii. Can. J. Bot.* 49: 1067–74.

Cunningham, F. X., and Schiff, J. A. 1986a. Chlorophyll–protein complexes from *Euglena gracilis* and mutants deficient in chlorophyll *b*. I. Pigment composition. *Plant Physiol*. 80: 223–30.

– 1986b. Chlorophyll–protein complexes from *Euglena gracilis* and mutants deficient in chlorophyll *b*. II. Polypeptide composition. *Plant Physiol*. 80: 231–38.

Czeczuga, B. 1987. Phycobiliproteins in the phycobionts of the *Stereocaulon* species. *Biochem. Syst. Ecol*. 15: 15–7.

Czygan, F.-C., and Heumann, W. 1967. Die Zusammensetzung und Biogenese der Carotinoide im *Pseudomonas echinoides* und einegen Mutanten. *Arch. Mikrobiol*. 57: 123–34.

Dale, R. E., and Teale, F. W. J. 1970. Number and distribution of chromophore types in native phycobiliproteins. *Phytochem. Photobiol*. 12: 99–117.

Dales, R. P. 1960. On the pigments of the Chrysophyceae. *J. Mar. Biol. Assoc. U.K*. 39: 693–9.

Daley, R. J., and Brown, S. R. 1973. Experimental characterization of lacustrine chlorophyll diagenesis. *Arch. Hydrobiol*. 72: 277–304.

Daley, R. J., Gray, C. B. J., and Brown, S. R. 1973a. A quantitative, sensitive method for determining algal and sedimentary chlorophyll derivatives. *J. Fish. Res. Board. Can*. 30: 345–56.

– 1973b. Reverse-phase thin-layer chromatography of chlorophyll derivatives. *J. Chromatogr*. 76: 175–83.

Davidson, J. 1954. Procedure for the extraction, separation and estimation of the major fat-soluble pigments in hay. *J. Sci. Food Agric*. 5: 1–7.

Davies, A. J., Khare, A., Mallams, A. R., Massy-Westropp, R. A., Moss, G. P., and Weedon, B. C. L. 1984. Carotenoids and related compounds. Pt. 38. Synthesis of (3*RS*,3′*RS*)-alloxanthin and other acetylenes. *J. Chem. Soc*. (*Perkin Trans*.) 2747–57.

Davies, B. H. 1965. Analysis of carotenoid pigments. In *Chemistry and Biochemistry of Plant Pigments,* 1st ed., ed. T. W. Goodwin, pp. 489–532. London: Academic Press.

– 1976. Carotenoids. In *Chemistry and Biochemistry of Plant Pigments,* 2nd ed., Vol. 2, ed. T. W. Goodwin, pp. 38–165. London: Academic Press.

Davies, B. H., Matthews, S., and Kirk, J. T. O. 1970. The nature and biosynthesis of the carotenoids of different colour varieties of *Capsicum annum*. *Phytochemistry* 9: 797–805.

Davies, D., and Holdsworth, E. S. 1980. The use of high pressure chromatography for the identification and preparation of pigments concerned in photosynthesis. *J. Liq. Chromatogr*. 3: 123–32.

Davis, P. S. 1957. Method for determination of chlorophyll in seawater. *CSIRO Aust. Div. Fish. Oceanogr. Divl. Rep. No. 7*.

Dayhoff, M. O., and Schwartz, R. M. 1980. Prokaryote evolution and the symbiotic origin of eukaryotes. In *Endocytobiology, Endosymbiosis and Cell Biology,* ed. W. Schwemmler and H. E. A. Schenk, pp. 53–82. Berlin: de Gruyter.

Delihas, N., Andersen, J., Andresini, W., Kaufman, L., and Lyman, H. 1981. The 5*S* ribosomal RNA of *Euglena gracilis* cytoplasmic ribosomes closely homologous to the 5*S* RNA of trypanosomatid protozoa. *Nucleic Acids Res.* 9: 6627–83.

Diner, B. A. 1979. Energy transfer from the phycobilisomes to photosystem II reaction centers in wild type *Cyanidium caldarium*. *Plant Physiol.* 63: 30–4.

Diversé-Pierluissi, M., and Krogmann, D. W. 1988. A zeaxanthin protein from *Anacystis nidulans*. *Biochim. Biophys. Acta* 933: 372–7.

Dodge, J. D. 1971. A dinoflagellate with both a mesokaryotic and a eukaryotic nucleus. I. Fine structure of the nuclei. *Protoplasma* 73: 145–57.

– 1975. A survey of chloroplast ultrastructure in the Dinophyceae. *Phycologia* 14: 253–63.

Dolphin, D. 1983. Porphyrins and related tetrapyrrolic substances. In *Chromatography: Fundamentals and Applications of Chromatographic and Electrophoretic Methods,* Pt. B: *Applications,* ed. E. Heftmann, pp. B377–406. Amsterdam: Elsevier.

Donze, M., and de Groot, H. P. 1982. A cheap high-capacity continuous centrifuge. *J. Plankton Res.* 4: 187–8.

Doolittle, W. F., and Bonen, L. 1981. Molecular sequence data indicating an endosymbiotic origin of plastids. *Ann. N.Y. Acad. Sci.* 361: 248–58.

Dörnemann, D., and Senger, H. 1981. Isolation and partial characterization of a new chlorophyll associated with the reaction centre of photosystem I of *Scenedesmus*. *FEBS Lett.* 126: 323–7.

– 1986. The structure of chlorophyll RC I, a chromophore of the reaction center of photosystem I. *Photochem. Photobiol.* 43: 573–81.

Dougherty, R. C., Strain, H. H., Svec, W. A., Uphaus, R. A., and Katz, J. J. 1966. Structure of chlorophyll *c′*. *J. Am. Chem. Soc.* 88: 5037–8.

– 1970. The structure, properties and distribution of chlorophyll *c*. *J. Am. Chem. Soc.* 92: 2826–33.

Douglas, S. E., and Doolittle, W. F. 1984. Nucleotide sequence of the 5*S* rRNA gene and flanking regions in the cyanobacterium, *Anacystis nidulans*. *FEBS Lett.* 166: 307–10.

Dubinsky, Z., and Polna, M. 1976. Pigment composition during a *Peridinium* bloom in Lake Kenneret (Israel). *Hydrobiologia* 51: 239–43.

Duncan, M. J., and Harrison, P. J. 1982. Comparison of solvents for extracting chlorophylls from marine macrophytes. *Bot. Mar.* 25: 445–7.

Durand, M., and Berkaloff, C. 1985. Pigment composition and chloroplast organization of *Gambierdiscus toxicans* Adachi and Fukuyo (Dinophyceae). *Phycologia* 24: 217–23.

Dustan, P. 1979. Distribution of zooxanthellae and photosynthetic chloroplast pigments of the reef-building coral *Montastrea annularis* Ellis and Solander in relation to depth on a West Indian coral reef. *Bull. Mar. Sci.* 29: 79–95.

Duysens, L. N. M. 1951. Transfer of light energy within the pigment system present in photosynthesizing cells. *Nature* 168: 548–50.

Egger, K., Nitsche, H., and Kleinig, H. 1969. Diatoxanthin und Diadinoxanthin bestandteile der Xanthophyll-gumisches von *Vaucheria* und *Botrydium*. *Phytochemistry* 8: 1583–5.

Eidem, A., and Liaaen-Jensen, Ṡ. 1974. Synthesis of 11,11′-deuterated ϵ-carotene and lycopene. *Acta Chem. Scand.* B28: 273–6.

Elmorjani, K., Thomas, J.-C., and Sebban, P. 1986. Phycobilisomes of wild-type and pigment mutants of the cyanobacterium *Synechocystis* PCC 6803. *Arch. Microbiol.* 146: 186–91.

Emerson, R. 1957. Dependence of yield of photosynthesis in long-wave red on wavelength and intensity of supplementary light. *Science* 125: 746.

Emerson, R., and Arnold, W. 1932. A separation of the reactions in photosynthesis by means of intermittent light. *J. Gen. Physiol.* 15: 391–420.

Eschenmoser, W., and Eugster, C. H. 1978. Synthese und Chiralitat von (5*S*,6*R*)-5,6-Epoxy-5,6-dihydro-β,β-carotin und (5*R*,6R)-5,6-dihydro-β, β-carotin-5,6-diol unem Carotinoid mit ungewöhnlichen Eigenschaften. *Helv. Chim. Acta* 61: 822–31.

Esenbeck, N. von. 1836. Uber einen blau-rothen Farbstoff, der sich bei der Zersitsung von *Oscillatorien* beildet. *Ann. Chimie (Liebig)* 17: 75–82.

Eskins, K., and Dutton, H. J. 1979. Sample preparation for high-performance liquid chromatography of higher plants. *Anal. Chem.* 51: 1885–6.

Eskins, K., Schoefield, C. R., and Dutton, H. J. 1977. High-performance liquid chromatography of plant pigments. *J. Chromatogr.* 135: 219–20.

Establier, R., and Lubian, L. M. 1982. Pigment composition of *Nannochloris maculata* Butcher and *N. occulata* Droop (CCAP, 251/6). Taxonomic implications. *Inv. Pesq.* 46: 451–7.

Falkowski, F. G., and Owens, T. G. 1980. Light-shade adaption. Two strategies in marine phytoplankton. *Plant Physiol.* 66: 592–5.

Falkowski, P. G., and Sucher, J. 1981. Rapid, quantitative separation of chlorophylls and their degradation products by high-performance liquid chromatography. *J. Chromatogr.* 213: 349–51.

Farnham, W. F., Blunden, G., and Gordon, S. M. 1985. Occurrence and pigment analysis of the sponge endosymbiont *Microspora ficulina* (Chlorophyceae). *Bot. Mar.* 27: 78–81.

Faust, M. A., and Norris, K. H. 1982. Rapid in vitro spectro-photometric analysis of chlorophyll pigments in intact phytoplankton cultures. *Br. Phycol. J.* 17: 351–61.

Fawley, M. W. (1988) Separation of chlorophylls c_1 and c_2 from pigment extracts of *Pavlova gyrans* by reverse-phase high performance liquid chromatography. *Plant Physiol.* 86: 76–8.

Fawley, M. W., Morton, S. J., Stewart, K. D., and Mattox, K. R. 1987. Evidence for a common evolutionary origin of light-harvesting fucoxanthin–chlorophyll *a*/*c*-protein complexes of *Pavlova gyrans* (Prymnesiophyceae) and *Phaeodactylum tricornutum* (Bacillariophyceae). *J. Phycol.* 23: 377–81.

Fawley, M. W., Stewart, K. D., and Mattox, K. R. 1986. The novel light-harvesting pigment–protein complex of *Mantoniella squamata* (Chlorophyta): Phylogenetic implications. *J. Mol. Evol.* 23: 168–76.

Feldman, J., and Tixier, R. 1947a. Sur l'existance d'un nouveau pigment dans les plastès d'une Rhodophycee. *C. R. Hebd. Séances Acad. Sci.* 225: 201–2.

– 1947b. Sur la floridorubine, pigment rouge des plastes d'une Rhodophycee (*Rytiphlea tinctoria* (Clem.) C. Ag.). *Rev. Gén. Bot.* 54: 341–53.

Fiksdahl, A., Bjørnland, T., and Liaaen-Jensen, S. 1984a. Algal carotenoids with novel end groups. *Phytochemistry* 23: 649–55.

Fiksdahl, A., and Liaaen-Jensen, S. 1988. Diacetylenic carotenoids from *Euglena viridis*. *Phytochemistry* 27: 1447–50.

Fiksdahl, A., Liaaen-Jensen, S., and Siegelman, H. W. 1978a. Carotenoids of *Coccolithus pelagicus*. *Biochem. Syst. Ecol.* 7: 47–8.

Fiksdahl, A., Mortensen, J. T., and Liaaen-Jensen, S. 1978b. High-pressure liquid chromatography of carotenoids. *J. Chromatogr.* 157: 111–17.

Fiksdahl, A., Withers, N., Guillard, R. R. L., and Liaaen-Jensen, S. 1984b. Carotenes of the Raphidophyceae. A chemosystematic contribution. *Comp. Biochem. Physiol.* 78B: 265–71.

Fiksdahl, A., Withers, N., and Liaaen-Jensen, S. 1984c. Carotenoids of *Heterosigma akashiwo:* A chemosystematic contribution. *Biochem. Syst. Ecol.* 12: 355–6.

Foppen, F. H. 1971. Tables for the identification of carotenoid pigments. *Chromatogr. Rev.* 14: 133–298.

Ford, R. C. 1987. Investigation of highly stable photosystem I chlorophyll–protein complexes from the thermophilic cyanobacterium *Phormidium laminosum*. *Biochim. Biophys. Acta* 893: 115–25.

Fork, D. C. 1963. Observations on the function of chlorophyll *a* and accessory pigments in photosynthesis. In *Photosynthetic Mechanisms of Green Plants*. ed. B. Kok and A. T. Jagendorf, pp. 352–61. Publication 1145 NAS-NRC. Washington, DC.

Foss, P., Guillard, R. R. L., and Liaaen-Jensen, S. 1984. Prasinoxanthin. A chemosystematic marker for algae. *Phytochemistry* 23: 1629–33.

– 1986. Carotenoids from eucaryotic ultraplankton clones (Prasinophyceae). *Phytochemistry* 25: 118–24.

Foss, P., Lewin, R. A., and Liaaen-Jensen, S. 1987. Carotenoids of *Prochloron* sp. (Prochlorophyta). *Phycologia* 26: 142–4.

Fournier, R. O. 1978. Membrane filtering. In *Phytoplankton Manual,* ed. A. Sournia, pp. 108–14. Paris: UNESCO.

Fox, G. E., Stackebrandt, F., Hespell, R. B., Gibson, J., Maniloff, J., Dyer, T. A., Wolf, R. S., Balch, W. E., Tanner, R. S., Magrum, L. J., Zaglen, L. B., Blackmore, R., Gupta, R., Bonen, L., Lewis, B. J., Stahl, D. A., Luehrsen, K. R., Caen, K. N., and Woese, C. R. 1980. The phylogeny of prokaroytes. *Science* 209: 457–63.

Francis, G. W., and Halfen, L. N. 1972. Cyanophyta. Oscillatoriaceae. γ-Carotene and lycopene in *Oscillatoria princeps*. *Phytochemistry* 11: 2347–8.

Francis, G. W., Hertzberg, S., Andersen, K., and Liaaen-Jensen, S. 1970. New carotenoid glycosides from *Oscillatoria limosa*. *Phytochemistry* 9: 629–35.

Francis, G. W., Knutsen, G., and Lien, T. 1973. Loroxanthin from *Chlamydomonas reinhardii*. *Acta Chem. Scand.* 27: 3599–600.

Francis, G. W., Strand, L. P., Lien, T., and Knutsen, G. 1975. Variations in the carotenoid content of *Chlamydomonas reinhardii* throughout the cell cycle. *Arch. Microbiol.* 104: 249–54.

French, C. S. 1960. The chlorophylls in vivo and in vitro. *Enc. Plant Physiol.* (1st ed.) 5(1): 252–97.

French, C. S., Smith, J. H. C., Virgin, H. I., and Airth, R. L. 1956. Fluorescence spectrum curves of chlorophylls, pheophytins, phycoerythrins, phycocyanins and hypericin. *Plant Physiol.* 31: 360–74.

Friedman, A. L., and Alberte, R. S. 1984. A diatom light-harvesting pigment-protein complex: Purification and characterization. *Plant Physiol.* 76: 483–9.

Fritsch, F. E. 1935. The structure and reproduction of algae, Vol. 1. Cambridge: Cambridge University Press.

Fuglistaller, P., Mimuro, M., Suter, F., and Zuber, H. 1987. Allophycocyanin complexes of the phycobilisomes from *Mastigocladus laminosus*. Influence of the linker polypeptide $L_C^{8.9}$ on the spectral properties of the phycobiliprotein subunits. *Biol. Chem. Hoppe-Seyler* 368: 353–67.

Fuglistaller, F., Suter, F., and Zuber, H. 1986. Linker polypeptides of the phycobilisomes from the cyanobacterium, *Mastigocladus laminosus. Biol. Chem. Hoppe-Seyler* 367: 601–14.

Fuglistaller, P., Widmer, H., Sidler, W., Frank, G., and Zuber, H. 1981. Isolation and characterization of phycoerythrocyanin and chromatic adaptation of the thermophilic cyanobacterium *Mastigocladus laminosus. Arch. Microbiol.* 129: 268–74.

Fujita, Y., and Shimura, S. 1974. Phycoerythrin of the marine blue-green alga, *Trichodesmium thiebantii. Plant Cell Physiol.* 45: 934–42.

Fujiwara-Arasaki, T., Yamamoto, M., and Kakiuchi, K. 1985. C-Phycoerythrin from a red alga, *Porphyra tenera:* Subunit structure. *Biochim. Biophys. Acta* 828: 261–5.

Gabrielson, P. W., Garbary, D. J., and Scagel, R. F. 1985. The nature of the ancestral red alga: Inferences from a cladistic analysis. *BioSystems* 18: 335–46.

Gabrielson, P. W. and Garbary, D. J. 1987. A cladistic analysis of Rhodophyta: Florideophycean orders. *Br. Phycol. J.* 22: 125–38.

Gantt, E. 1977. Recent contributions in phycobiliproteins and phycobilisomes. *Photochem. Photobiol.* 26: 685–9.

– 1979. Phycobiliproteins of Cryptophyceae. In *Biochemistry and Physiology of Protozoa,* Vol. 1, 2nd ed., ed. M. Landowski and S. H. Hunter, pp. 121–37. New York: Academic Press.

– 1980. Structure and function of phycobilisomes: Light harvesting pigment complexes in red and blue-green algae. *Int. Rev. Cytol.* 66: 45–80.

– 1981. Phycobilisomes. *Annu. Rev. Plant Physiol.* 32: 327–47.

– 1986. Phycobilisomes. *Enc. Plant Physiol.* (2nd ed.) 19: 260–8.

Gantt, E., and Conti, S. F. 1966a. Granules associated with the chloroplast lamellae of *Porphyridium cruentum. J. Cell Biol.* 29: 423–34.

– 1966b. Phycobiliprotein localization in algae. *Brookhaven Symp. Biol.* 19: 393–405.

Gantt, E., Edwards, M. R., and Provasoli, L. 1971. Chloroplast structures of the Cryptophyceae. *J. Cell Res.* 48: 280–90.

Gantt, E., and Lipschultz, C. A. 1972. Phycobilisomes of *Porphyridium cruentum*. *J. Cell Biol.* 54: 313–24.

– 1973. Energy transfer in phycobilisomes from phycoerythrin to allophycocyanin. *Biochim. Biophys. Acta* 292: 858–61.

– 1974. Phycobilisomes of *Porphyridium cruentum*. Pigment analyses. *Biochemistry* 13: 2960–6.

– 1980. Structure and phycobiliprotein composition of phycobilisomes from *Griffithsia pacifica* (Rhodophyceae). *J. Phycol.* 16: 394–8.

Gantt, E., Lipschultz, C. A., and Zilinskas, B. 1976. Further evidence for a phycobilisome model from selective dissociation, fluorescence emission, immunoprecipitation and electron microscopy. *Biochim. Biophys. Acta* 430: 375–88.

Gantt, E., Scott, J., and Lipschultz, C. 1986. Phycobilin composition and chloroplast structure in the freshwater alga, *Compsopogon coeruleus* (Rhodophyta). *J. Phycol.* 22: 480–4.

Gardner, E. E., Stevens, S. E., and Fox, J. L. 1980. Purification and characterization of the C-phycocyanin from *Agmenellum quadruplicatum*. *Biochim. Biophys. Acta* 624: 187–95.

Garside, C., and Riley, J. P. 1968. The absorptivity of fucoxanthin. *Deep-Sea Res.* 15: 627.

– 1969. A thin-layer chromatographic method for the determination of plant pigments in sea water and cultures. *Anal. Chim. Acta* 46: 179–91.

Geisert, M., Rose, T., Bauer, W., and Zahn, R. K. 1987. Occurrence of carotenoids and sporopollenin in *Nannochlorium eukaryotum*, a novel, marine alga with unusual characteristics. *BioSystems* 20: 133–42.

Geiss, F. 1987. *Fundamentals of Thin Layer Chromatography*. Heidelberg: Hüthig.

Gibbs, C. F. 1979. Chlorophyll *b* interference in the fluorometric determination of chlorophyll *a* and phaeopigments. *Aust. J. Mar. Freshwater Res.* 30: 596–606.

Gibbs, S. P. 1962. Chloroplast development in *Ochromonas danica*. *J. Cell Biol.* 15: 343–61.

Gibbs, S. 1978. The chloroplasts of *Euglena* may have evolved from symbiotic green algae. *Can. J. Bot.* 56: 2883–9.

Gibbs, S. P. 1981. The chloroplasts of some algal groups may have evolved from endosymbiotic eukaryotic algae. *Ann. N.Y. Acad. Sci.* 361: 193–208.

Gibbs, S. P., Chu, L. L., and Magnussen, C. 1980. Evidence that *Olisthodiscus luteus* is a member of the Chrysophyceae. *Phycologia* 19: 173–7.

Giddings, T. H., Wasman, C., and Staehelin, L. A. 1983. Structure of the thylakoids and envelope membranes of the cyanelles of *Cyanophora paradoxa*. *Plant Physiol.* 71: 409–19.

Gieskes, W. W., and Kraay, G. W. 1983a. Unknown chlorophyll derivatives in the North Sea and the tropical Atlantic Ocean revealed by HPLC analysis. *Limnol. Oceanogr.* 28: 757–66.

Gieskes, W. W. C., and Kraay, G. W. 1983b. Dominance of Cryptophyceae during the plankton spring bloom in the central North Sea, detected by HPLC analysis of pigments. *Mar. Biol.* 75: 179–85.

Gieskes, W. W., and Kraay, G. W. 1986. Analysis of phytoplankton pigments by HPLC before, during and after the mass occurrence of the microflagellate, *Corymbellus aureus* during the spring bloom in the open northern North Sea in 1983. *Mar. Biol.* 92: 45–53.

Gillan, F. T., and Johns, R. B. 1980. Input and early diagenesis of chlorophylls in a temperate intertidal sediment. *Mar. Chem.* 9: 243–53.

Gillott, M. A., and Gibbs, S. P. 1980. The cryptomonad nucleomorph: Its ultrastructure and evolutionary significance. *J. Phycol.* 16: 558–68.

Gingrich, J. C., Lundell, D. J., and Glazer, A. N. 1983. Core structure in cyanobacterial phycobilisomes. *J. Cell. Biochem.* 22: 1–14.

Glazer, A. N. 1976. Phycocyanins. Structure and function. *Photochem. Photobiol. Rev.* 1: 71–114.

– 1977. Structure and molecular organization of the photosynthetic accessory pigments of cyanobacterial and red algae. *Mol. Cell. Biochem.* 18: 125–40.

– 1981. Photosynthetic accessory proteins with bilin prosthetic groups. In *The Biochemistry of Plants,* Vol. 8, ed. E. E. Conn and P. K. Stumpf, pp. 51–96. New York: Academic Press.

– 1982. Phycobilisomes: Structure and dynamics. *Annu. Rev. Microbiol.* 36: 173–98.

– 1983. Comparative biochemistry of photosynthetic light-harvesting systems. *Annu. Rev. Biochem.* 52: 125–57.

– 1984. Phycobilisome. A macromolecular complex optimized for light-energy transfer. *Biochim. Biophys. Acta* 768: 28–51.

– 1985. Light harvesting by phycobilisomes. *Annu. Rev. Biophys. Biophys. Chem.* 14: 47–77.

Glazer, A. N., and Apell, G. S. 1977. A common evolutionary origin for the biliproteins of cyanobacteria, Rhodophyta and Cryptophyta. *FEMS Lett.* 1: 113–6.

Glazer, A. N., Apell, G. S., Hixson, C. S., Bryant, D. A., Rimon, S., and Brown, D. M. 1976. Biliproteins of cyanobacteria and Rhodophyta: Homologous family of photosynthetic accessory pigments. *Proc. Natl. Acad. Sci. U.S.A.* 73: 428–51.

Glazer, A. N., and Bryant, D. A. 1975. Allophycocyanin B (λ_{max} 671, 618 nm). A new cyanobacterial phycobiliprotein. *Arch. Microbiol.* 104: 15–22.

Glazer, A. N., Chan, C., Williams, R. C., Yeh, S. N., and Clark, J. H. 1985b. Kinetics of energy flow in the phycobilisome core. *Science* 230: 1951–3.

Glazer, A. N., and Clark, J. H. 1986. Phycobilisomes: Macromolecular structure and energy flow dynamics. *Biophys. J.* 49: 115–16.

Glazer, A. N., and Cohen-Bazire, G. 1971. Subunit structure of the phycobiliproteins of blue-green algae. *Proc. Natl. Acad. Sci. U.S.A.* 68: 1398–1401.

– 1975. A comparison of cryptophytan phycocyanins. *Arch. Microbiol.* 104: 29–32.

Glazer, A. N., Cohen-Bazire, G., and Stanier, R. Y. 1971a. Comparative immunology of algal biliproteins. *Proc. Natl. Acad. Sci. U.S.A.* 68: 3005–8.

– 1971b. Characterization of phycoerythrin from a *Cryptomonas* sp. *Arch. Mikrobiol.* 80: 1–18.

Glazer, A. N., and Fang, S. 1973. Chromophore content of blue-green algal phycobiliproteins. *J. Biol. Chem.* 248: 659–62.

Glazer, A. N., Fang, S., and Brown, D. M. 1973. Spectroscopic properties of C-phycocyanin and of its α and β subunits. *J. Biol. Chem.* 248: 5677–85.

Glazer, A. N., and Hixson, C. S. 1975. Characterization of R-phycocyanin. *J. Biol. Chem.* 250: 5487–95.

– 1977. Subunit structures and chromophore composition of Rhodophycean phycoerythrins: *Porphyridium cruentum* B-phycoerythrin. *J. Biol. Chem.* 252: 36–42.

Glazer, A. N., Lundell, D. J., Yamanaka, G., and Williams, R. C. 1983. The structures of a "simple" phycobilisome. *Ann. Microbiol.* (*Inst. Pasteur*) 134B: 159–80.

Glazer, A. N., West, J. A., and Chan, C. 1982. Phycoerythrins as chemotaxonomic markers in red algae: A survey. *Biochem. Syst. Ecol.* 10: 203–15.

Glazer, A. N., Yeh, S. N., Webb, S. P., and Clark, J. H. 1985a. Disk-to-disk transfer as the rate-limiting step for energy flow in phycobilisomes. *Science* 227: 419–23.

Glick, R. E., and Zilinskas, B. A. 1982. Role of the colourless polypeptides in phycobilisome reconstruction from separated phycobiliproteins. *Plant Physiol.* 69: 991–7.

Goedheer, J. C. 1966. Visible absorption and fluorescence of chlorophyll and its aggregates in solution. In *The Chlorophylls,* ed. L. P. Vernon and C. R. Seely, pp. 147–85. New York: Academic Press.

Goedheer, J. C., and Birnie, F. 1965. Fluorescence polarization and location of fluorescence maxima of C-phycoerythrin. *Biochim. Biophys. Acta* 94: 579–81.

Goeyens, L., Post, E., Dehairs, F., Vandenhoudt, A., and Baeyens, W. 1982. The use of high-pressure liquid chromatography with fluorimetric detection for chlorophyll *a* determination in natural extracts of chloropigments and their degradation products. *Int. J. Environ. Anal. Chem.* 12: 51–63.

Goldsmith, T. H., and Krinsky, N. I. 1960. The epoxide nature of the carotenoid, neoxanthin. *Nature* 188: 491–3.

Golecki, J. R. 1979. Ultrastructure of cell wall and thylakoid membranes of the thermophilic cyanobacterium *Synechococcus lividus* under the influence of temperature shifts. *Arch. Microbiol.* 120: 125–33.

Golecki, J. R., and Drews, G. 1982. Supramolecular organization and composition of membranes. In *The Biology of Cyanobacteria,* ed. N. G. Carr and B. A. Whitton, pp. 125–41. Oxford: Blackwell.

Goodwin, T. W. 1955. Carotenoids. In *Modern Methods of Plant Analysis,* Vol. 3, ed. K. Paech and M. V. Tracey, pp. 272–311. Berlin: Springer-Verlag.

– 1965. *Chemistry and Biochemistry of Plant Pigments,* 1st ed. London: Academic Press.

– 1971a. Algal carotenoids. In *Aspects of Terpenoid Chemistry and Biochemistry,* ed. T. W. Goodwin, pp. 315–56. London: Academic Press.

– 1971b. Biosynthesis. In *Carotenoids,* ed. O. Isler, pp. 577–637. Basel: Birkhauser.

– 1973. Prochirality in biochemistry. *Essays in Biochemistry* 9: 103–60.

– 1974. Carotenoids and biliproteins. In *Algal Physiology and Biochemistry,* ed. W. P. D. Stewart, pp. 176–205. Oxford: Blackwell.

– 1976. *Chemistry and Biochemistry of Plant Pigments,* 2nd ed., Vols. 1 and 2. London: Academic Press.

– 1979. Isoprenoid distribution and biosynthesis in flagellates. In *Biochemistry and Physiology of Protozoa,* 2nd ed., Vol. 1, ed. M. Landowski and S. H. Hunter, pp. 91–119. New York: Academic Press.

– 1980. *The Biochemistry ofl the Carotenoids,* 2nd ed., Vol. 1. London: Chapman and Hall.

Goodwin, T. W., and Gross, J. A. 1958. The carotenoid distribution in various bleached sub-strains of *Euglena gracilis. J. Protozool.* 5: 292–6.

Govindjee, and Braun, B. Z. 1974. Light absorption, emission and photosynthesis. In *Algel Physiology and Biochemistry,* ed. W. P. D. Stewart, pp. 346–90. Oxford: Blackwell.

Govindjee, and Govindjee, R. 1974. The absorption of light in photosynthesis. *Sci. Am.* 231(6): 68–82.

– 1975. Introduction to photosynthesis. In *Bioenergetics of Photosynthesis,* ed. Govindjee, pp. 1–50. New York: Academic Press.

Gowen, R. J., Tett, P., and Wood, B. J. B. 1982. The problem of degradation products in the estimation of chlorophyll by fluorescence. *Arch. Hydrobiol. Beih. Ergebn. Limnol.* 16: 101–6.

Grabowski, J., and Gantt, E. 1978. Photophysical properties of phycobiliproteins from phycobilisomes: Fluorescence lifetimes, quantum yields, and polarization spectra. *Photochem. Photobiol.* 28:39–45.

Gray, B. H., Cosner, J., and Gantt, E. 1976. Phycocyanins with absorption maxima at 637 nm and 623 nm from *Agmanellum quadruplicatum. Photochem. Photobiol.* 24: 299–302.

Gray, B. H., and Gantt, E. 1975. Spectral properties of phycobilisomes and phycobiliproteins from the blue-green alga, *Nostoc* sp. *Photochem. Photobiol.* 21: 121–8.

Gray, M. W., and Doolittle, W. F. 1982. Has the endosymbiont hypothesis been proven? *Microbiol. Rev.* 46: 1–42.

Green, B. R., and Camm, E. L. 1982. The nature of the light-harvesting complex as defined by sodium dodecyl sulfate polyacrylamide gel electrophoresis. *Biochim. Biophys. Acta.* 681: 256–62.

Green, B. R., Camm, E. L., and Van Hooten, J. 1982. The chlorophyll-protein complexes of *Acetabularia*. A novel chlorophyll *a/b* complex which forms oligomers. *Biochim. Biophys. Acta* 681: 248–55.

Greenwood, A. D., Griffiths, H. B., and Santore, U. J. 1977. Chloroplasts and cell compartments in Cryptophyceae. *Br. Phycol. J.* 12: 119.

Gross, J. A., Stroz, R. J., and Britton, G. 1975. The carotenoid hydrocarbons of *Euglena gracilis* and derived mutants. *Plant Physiol.* 55: 175–7.

Guard-Friar, D., Eisenberg, B. L., Edwards, M. R., and MacColl, R. 1986. Immunochemistry on cryptomonad biliproteins. *Plant Physiol.* 80: 38–42.

Guard-Friar, D., and MacColl, R. 1984. Spectroscopic properties of tetrapyrroles on denatured biliproteins. *Arch. Biochem. Biophys.* 230: 300–5.

– 1986. Sub-unit separation (α,α',β) of cryptomonad biliproteins. *Photochem. Photobiol.* 43: 81–5.

Guglielmi, G., Cohen-Bazire, G. and Bryant, D. A. 1981. The structure of *Gloeobacter violaceus* and its phycobilisomes. *Arch. Microbiol.* 129: 181–9.

Gugliemelli, L. A. 1984. Isolation and characterization of pigment-protein particles from the light-harvesting complex of *Phaeodactylum tricornutum*. *Biochim. Biophys. Acta* 766: 45–50.

Gugliemelli, L. A., Dutton, H. J., Jursinic, P. A., and Siegelman, H. W. (1981). Energy transfer in a light-harvesting carotenoid-chlorophyll *c*-chlorophyll *a*-protein of *Phaeodactylum tricornutum*. *Photochem. Photobiol.* 33: 903–7.

Guikema, J. A., and Sherman, L. A. 1983. Chlorophyll-protein organization of membranes from the cyanobacterium *Anacystis nidulans*. *Arch. Biochem. Biophys.* 220: 155–66.

Guillard, R. R. L., and Lorenzen, C. J. 1972. Yellow-green algae with chlorophyllide *c*. *J. Phycol.* 8: 10–14.

Guillard, R. R. L., Murphy, L. S., Foss, P., and Liaaen-Jensen, S. 1985. *Synechococcus* spp. as likely zeaxanthin-dominant ultraplankton in the North Atlantic. *Limnol. Oceanogr.* 30: 412–14.

Gutman, A. L. 1984. Catabolism of δ-aminolaevulinic acid in etiolated *Euglena gracilis*. Phytochemistry 23: 2773–5.

Gysi, J. R., and Chapman, D. J. 1982. Phycobilins and phycobiliproteins of algae. In *CRC Handbook of Biosolar Resources,* Vol. 1, Pt. 1, ed. O. R. Zaborsky, pp. 83–102. Boca Raton, FL: CRC Press.

Gysi, J. R., and Zuber, H. 1979. Properties of allophycocyanin II and its α- and β-subunits from the themophile blue-green alga *Mastigocladus laminosus*. *Biochem. J.* 181: 577–83.

Hager, A. 1955. Chloroplasten-Farbstoffe, ihre paperchromatographische trennung und ihre Veränderungen durch Aussenfaktorem. *Z. Naturforsch.* 10b: 310–12.

– 1980. The reversible, light-induced conversions of xanthophylls in the chloroplast. In *Pigments in Plants,* 2nd ed., ed. F.-C. Czygan, pp. 57–79. Stuttgart: Fischer.

Hager, A., and Meyer-Bertenrath, T. 1966. Die Isolierung und quantitative Bestemmung der Carotenoids und Chlorophylle von Blattern, Algen und isolierten Chloroplasten mit Hilfe dünnschichtenchromatographischen Methoden. *Planta* 69: 198–217.

– 1967a. Die Identifizierung der an Dünnschichten getrennten Carotinoide grüner Blätten und Algen. *Planta* 76: 149–68.

– 1967b. Beziehungen zwischen Absorptionsspectrum und Konstitution bei Carotenoidem von Algen und Höhern Pflanzen. *Ber. Dtsch. Bot. Ges.* 80: 426–36.

Hager, A., and Stransky, H. 1970a. Das carotinoidmuster und die Verbreitung des lichtenduzierten Xanthophyllcyclus in verschiedenen algenklassen. I. Methoden zen Identifizierung der Pigmente. *Arch. Mikrobiol.* 71: 132–63.

– 1970b. Das carotinoidmuster und die Verbreitung des lichtenduzierten Xanthophyllcyclus in verschiedenen algenklassen. III. Grünalgen. *Arch. Mikrobiol.* 72: 68–83.

– 1970c. Das carotinoidmuster und die Verbreitung des lichtenduzierten Xanthophyllcyclus in verschiedenen algenklassen. V. Einzelne Vertreter der Cryptophyceae, Euglenophyceae, Bacillariophyceae, Chrysophyceae und Phaeophyceae. *Arch. Mikrobiol.* 73: 77–89.

Halfen, L. N., and Francis, G. W. 1972. The influence of culture temperature on the carotenoid composition of the blue-green alga, *Anacystis nidulans*. *Arch. Mikrobiol.* 81: 25–35.

Hallegraeff, G. M. 1976. Pigment diversity in freshwater phytoplankton. I A comparison of spectrophotometric and paper chromatographic methods. *Int. Rev. Ges. Hydrobiol.* 61: 149–68.

– 1977. Pigment diversity in freshwater phytoplankton. II. Summer-succession in three Dutch lakes with different trophic characteristics. *Int. Rev. Ges. Hydrobiol.* 62: 19–39.

– 1981. Seasonal study of phytoplankton pigments and species at a coastal station off Sydney: Importance of diatoms and nanoplankton. *Mar. Biol.* 61: 107–18.

Hallegraeff, G. M., and Jeffrey, S. W. 1984. Tropical phytoplankton species and pigments of continental shelf waters of North and North-West Australia. *Mar. Ecol. Prog. Ser.* 20: 59–74.

– 1985. Description of new chlorophyll *a* alteration products in marine phytoplankton. *Deep-Sea Res.* 32: 697–705.

Hallenstvet, M., Liaaen-Jensen, S., and Skulberg, O. M. 1979. Carotenoids of *Oscillatoria bornetii* f. *tenuis*. *Biochem. Syst. Ecol.* 7: 1–2.

Hanes, C. S., and Isherwood, F. A. 1949. Separation of the phosphoric esters on the filter paper chromatogram. *Nature* 164: 1107–12.

Harris, D. G., and Zscheile, F. P. 1943. Effects of solvent upon absorption spectra of chlorophylls *A* and *B;* their ultraviolet absorption spectra in ether solution. *Bot. Gaz.* 104: 515–27.

Hattori, A., Crespi, H. L., and Katz, J. J. 1965. Association and dissociation of phycocyanin and the effects of deuterium substitution on the process. *Biochemistry* 4: 1225–38.

Hattori, A., and Fujita, Y. 1959. Spectroscopic studies on the phycobilin pigments obtained from blue-green and red algae. *J. Biochem.* (*Tokyo*) 46: 903–9.

Hausmann, P., Falk, H., and Sitte, P. 1985. DNA in the nucleomorph of *Cryptomonas* demonstrated by DAPI fluorescence. *Z. Naturforsch.* 40c: 933–5.

Haxo, F. T. 1985. Photosynthetic action spectrum of the coccolithophorid *Emiliania huxleyi* (Haptophyceae): 19′-Hexanoyloxyfucoxanthin as antenna pigment. *J. Phycol.* 21: 282–7.

Haxo, F. T., and Blinks, L. R. 1950. Photosynthetic action spectra of marine algae. *J. Gen. Physiol.* 33: 389–422.

Haxo, F. T., and Fork, D. C. 1959. Photosynthetically active accessory pigments of Cryptomonads. *Nature* 184: 1051–2.

Haxo, F. T., O'hEocha, C., and Norris, P. 1955. Comparative studies of chromatographically purified phycoerythrins and phycocyanis. *Arch. Biochem.* 54: 162–73.

Healey, F. P. 1968. The carotenoids of four blue-green algae. *J. Phycol.* 4: 126–9.

Hellebust, J. A., and Craigie, J. S. 1978. *Physiological and Biochemical Methods: Handbook of Phycological Methods.* Cambridge: Cambridge University Press.

Herrin, D. L., Plumley, F. G., Ikeuchi, M., Michaels, A. S., and Schmidt, G. W. 1987. Chlorophyll antenna proteins of photosystem I: Topology, synthesis and regulation of the 20-kDa subunit of *Chlamydomonas* light-harvesting complex of photosystem I. *Arch. Biochem. Biophys.* 254:397–408.

Hertzberg, S., and Liaaen-Jensen, S. 1966a. The carotenoids of blue-green algae. I. The carotenoids of *Oscillatoria rubescens* and an *Athrospera* sp. *Phytochemistry* 5: 557–63.

– 1966b. The carotenoids of blue-green algae. II. The carotenoids of *Aphanezamenon flos-aqua*. *Phytochemistry* 5: 565–70.

– 1966c. Bacterial carotenoids. XIX. The carotenoids of *Myobacterium phlei* strain Vera. 1. The structures of the minor carotenoids. *Acta Chem. Scand.* 20: 1187–94.

– 1967. The carotenoids of blue-green algae. III. A comparative study of mutatochrome and flavacin. *Phytochemistry* 6: 1119–26.

– 1969a. The structure of myxoxanthophyll. *Phytochemistry* 8: 1259–80.

– 1969b. The structure of oscillaxanthin. *Phytochemistry* 8: 1281–92.

– 1971. The constitution of aphanizophyll. *Phytochemistry* 10: 3251–2.

Hertzberg, S., Mortensen, T., Borch, G., Siegelman, H. W., and Liaaen-Jensen 1977. On the absolute configuration of 19′-hexanoyloxy fucoxanthin. *Phytochemistry* 16: 587–90.

Hibberd, D. J. 1976. The ultrastructure and taxonomy of the Chrysophyceae and Prymnesiophyceae (Haptophyceae): A survey with some new observations on the ultrastructure of the Chrysophyceae. *Bot. J. Linn. Soc.* 72: 55–80.

– 1979. The structure and phylogenetic significance of the flagellar transition region in the chlorophyll *c*-containing algae. *BioSystems* 11: 243–61.

– 1981. Notes on the taxonomy and nomenclature of the algal classes, Eustigmatophyceae and Tribophyceae (synonym Xanthophyceae). *Bot. J. Linn. Soc.* 82: 93–119.

Hibberd, D. J., and Chretiennot-Dinet, M. J. 1979. The ultrastructure and taxonomy of *Rhizochromulina marina* gen. et sp. nov.: An amoeboid, marine chrysophyte. *J. Mar. Biol. Assoc. U.K.* 59: 179–93.

Hibberd, D. J., and Leedale, G. F. 1970. Eustigmatophyceae – A new algal class with unique organization of the motile cell. *Nature* 225: 758–60.

– 1971. A new algal class – The Eustigmatophyceae. *Taxon* 20: 523–5.

– 1972. Observations on the cytology and ultrastructure of the new algal class, Eustigmatophyceae. *Ann. Bot.* 36: 49–71.

Hibberd, D. J., and Norris, R. E. 1984. Cytology and ultrastructure of *Chlorarachnion reptans* (Chlorarachniophyta divisio nova, Chlorarachniophyceae classis nova). *J. Phycol.* 20: 310–30.

Hill, D. R. A., and Rowan, K. S. 1989. The distribution of biliproteins in the Cryptophyceae. *Phycologia,* in press.

Hill, R., and Bendall, F. 1960. Function of the two cytochrome components in chloroplasts: A working hypothesis. *Nature* 186: 136–7.

Hiller, R. G., and Larkum, A. W. D. 1985. The chlorophyll-protein complexes of *Prochloron* sp. (Prochlorophyta). *Biochim. Biophys. Acta* 806: 107–15.

Hiller, R. G., Larkum, A. W. D., and Wrench, P. M. 1988. Chlorophyll proteins of the prymnesiophyte *Pavlova lutheri* (Droop) comb. nov.: Identification of the major light-harvesting complex. *Biochim. Biophys. Acta* 932: 223–31.

Hiller, R. G., and Martin, C. D. 1987. Multiple forms of a type I phycoerythrin from a *Chroomonas* sp. (Cryptophyceae) varying in subunit composition. *Biochim. Biophys. Acta* 923: 98–102.

Hiller, R. G., Post, A., and Stewart, A. C. 1983. Isolation of intact detergent-free phycobilisomes by trypsin. *FEBS Lett.* 156: 180–4.

Hirayama, O. 1967. Lipids and lipoprotein complex in photosynthetic tissue. II. Pigments and lipids in blue-green alga, *Anacystis nidulans. J. Biochem. (Japan)* 61: 179–85.

Hirose, H., and Kumano, S. 1966. Spectroscopic studies on the phycoerythrins from rhodophycean algae with special reference to their phylogenetical relations. *Bot. Mag. (Tokyo)* 79: 105–13.

Hirose, H., Kumano, S., and Madono, K. 1969. Spectroscopic studies on phycoerythrins from Cyanophycean and Rhodophycean algae with special reference to their phylogenetical relations. *Bot. Mag. (Tokyo)* 82: 197–203.

Hixson, C. S. 1976. Characterization of Rhodophytan biliproteins. Ph.D. dissertation, University of California, Los Angeles.

Hiyama, T., Nishimura, M., and Chance, B. 1969. Determination of carotenes by thin-layer chromatography. *Anal. Biochem.* 29: 339–50.

Hladik, J., Pancoska, P., and Sofoua, D. 1982. The influence of the carotenoids on the conformation of chlorophyll-protein complexes isolated from the cyanobacterium *Plectonema boryanum*. *Biochim. Biophys. Acta* 681: 263–72.

Hoffman, P., and Werner, D. 1966. Spectrophotometric chlorophyll determination having special regard to various types of equipment. *Jena Rev.* 7: 300–3.

Holden, M. 1962. Separation by paper chromatography of chlorophylls *a* and *b* and some of their breakdown products. *Biochim. Biophys. Acta* 56: 378–9.

– 1965. Chlorophylls. In *Chemistry and Biochemistry of Plant Pigments,* 1st ed., ed. T. W. Goodwin, pp. 461–88. London: Academic Press.

– 1976. Chlorophylls. In *Chemistry and Biochemistry of Plant Pigments,* 2nd ed., Vol.. 2, ed. T. W. Goodwin, pp. 1–37. London: Academic Press.

Holm, G. 1954. Chlorophyll mutations in barley. *Acta Agric. Scand.* 4: 457–71.

Holm-Hansen, O., Lorenzen, C. J., Holmes, R. W., and Strickland, J. D. H. 1965. Fluorometric determination of chlorophyll. *J. Cons. Perm. Int. Explor. Mer.* 30: 3–15.

Holm-Hansen, O., and Riemann, B. 1978. Chlorophyll *a* determination: Improvement in methodology. *Oikos* 30: 438–47.

Holt, A. S. 1961. Further evidence of the relation between 2-desvinyl-2-formyl-chlorophyll *a* and *d*. *Can. J. Bot.* 39: 327–31.

– 1965. Nature, properties and distribution of chlorophylls. In *Chemistry and Biochemistry of Plant Pigments,* 1st ed., ed. T. W. Goodwin, pp. 3–28. London: Academic Press.

Holt, A. S., and Jacobs, E. E. 1954. Spectroscopy of plant pigments. 1. Ethyl chlorophyllides and their pheophorbides. *Am. J. Bot.* 41: 710–22.

Holt, A. S., and Morley, H. V. 1959. A proposed structure for chlorophyll *d*. *Can. J. Chem.* 37: 507–14.

Holt, T. K., and Krogmann, D. W. 1981. A carotenoid-protein from cyanobacteria. *Biochem. Biophys. Acta* 63: 408–14.

Holzwarth, A. R., Wendler, J., and Wehrmeyer, W. 1983. Studies in chromophore coupling in isolated phycobilisomes. 1. Picosecond fluorescence kinetics of energy transfer in phycocyanin from *Chroomonas* sp. *Biochim. Biophys. Acta* 724: 388–95.

Honsell, E., Ghirardelli, L. A., and Avanzini, A. 1985. Relationship between phycobilisome formation and temperature in *Nilophyllum punctatum* chloroplasts. In *Proc. 2nd Int. Phycol. Cong.*, Copenhagen, Aug. 1985, p. 69.

Honsell, E., Kosovel, V., and Talarico, L. 1984. Phycobilin distribution in Rhodophyta: Studies and interpretation on the basis of their absorption spectra. *Bot. Mar.* 27: 1–16.

Hori, H., Lim, B.-L., and Osawa, S. 1985. Evolution of green plants as deduced from 5*S* rRNA sequences. *Proc. Natl. Acad. Sci. U.S.A.* 82: 820–3.

Hsu, B.-D., and Lee, J.-Y. 1987. Orientation of pigments and pigment-protein complexes in the diatom *Cylindrotheca fusiformis*. A linear dichroism study. *Biochim. Biophys. Acta* 893: 572–7.

Hu, H., Yu, M., and Zhang, X. 1980. Discovery of phycobilin in *Gymnodinium cyaneum* Hu sp. nov. and its phylogenetic significance. *Kexue Tongbao* 25: 882–4.

Huang, C., Berns, D. S., and Guarino, D. U. 1984. Characterization of components of P-700-chlorophyll *a*-protein complex from a blue-green alga, *Phormidium luridum*. *Biochim. Biophys. Acta* 765: 21–9.

Humphrey, G. F. 1960. The concentration of phytoplankton pigments in Australian waters. *CSIRO Div. of Fish. and Oceanogr. Tech. Paper No. 9*, 27 p.

– 1978. The recalculation of marine chlorophyll concentrations with special reference to Australian waters. *Aust. J. Mar. Freshwater Res.* 29: 409–16.

Hushovd, O. T., Gulliksen, O. M., and Nordley, Ø. 1982. Isolation of chloroplast membranes and electrophoretic separation of chlorophyll-containing proteins from *Ulva mutabilis* Føya. *Bot. Mar.* 25: 155–61.

Ikamori, M., and Arasaki, S. 1977. Photosynthetic pigments in marine algae. I. Two-dimensional paper chromatographic separation of chlorophylls and carotenoids from green algae and sea grasses. *Bull. Jpn. Soc. Phycol.* 25: 58–66.

Ingram, K., and Hiller, R. G. 1983. Isolation and characterization of a major chlorophyll a/c_2 light-harvesting protein from a *Chroomonas* species (Cryptophyceae). *Biochim. Biophys. Acta* 722: 310–19.

Isaksen, M., and Francis, G. W. 1986. Reverse-phase thin-layer chromatography of carotenoids. *J. Chromatogr.* 355: 358–62.

Ish-Shalom, D., and Ohad, I. 1983. Organization of chlorophyll-protein complexes of photosystem I in *Chlamydomonas reinhardii*. *Biochim. Biophys. Acta* 722: 398–50.

Isler, O. 1971. *Carotenoids*. Basel: Birkhauser.

Isono, T., and Katoh, T. 1987. Subparticles of *Anabaena* phycobilisomes. II. Molecular assembly of allophycocyanin cores in reference to 'anchor' protein. *Arch. Biochem. Biophys.* 256: 317–24.

Itagaki, T., Nakayama, K., and Okada, M. 1986. Chlorophyll-protein complexes associated with photosystem I isolated from the green alga, *Bryopsis maxima*. *Plant Cell Physiol.* 27: 1241–7.

Jackson, A. H. 1976. Structure, properties and distribution of chlorophylls. In *Chemistry and Biochemistry of Plant Pigments*, 2nd. ed., Vol. 1, ed. T. W. Goodwin, pp. 1–63. London: Academic Press.

Jacobsen, T. R. 1978. A quantitative method for the separation of chlorophylls *a* and *b* from phytoplankton pigments by high pressure liquid chromatography. *Mar. Sci. Commun.* 4: 33–47.

– 1982. Comparison of chlorophyl *a* measurements by fluorimetric, spectrophotometric and high pressure liquid chromatographic methods in aquatic environments. *Arch. Hydrobiol. Beih. Ergebn. Limnol.* 16: 35–46.

Jaspers, E. M. W. 1965. Pigmentation of tobacco crown-gall tissues cultured in vitro in dependence of the composition of the medium. *Physiol. Plant.* 18: 933–40.

Jeffrey, S. W. 1961. Paper chromatographic separation of chlorophylls and carotenoids from marine algae. *Biochem. J.* 80: 336–42.

– 1962. Purification of chlorophyll *c* from *Sargassum flavicans*. *Nature* 194: 600.

– 1963. Purification and properties of chlorophyll *c* from *Sargassum flavicans*. *Biochem. J.* 86: 313–18.

– 1965. Paper chromatographic separation of pigments in marine phytoplankton. *Aust. J. Mar. Freshwater Res.* 16: 307–13.

– 1968a. Quantitative thin-layer chromatography of chlorophylls and carotenoids from marine algae. *Biochim. Biophys. Acta* 162: 271–83.

– 1968b. Two spectrally distinct components of preparations of chlorophyll *c*. *Nature* 220: 1032–3.

– 1968c. Pigment composition of siphonales algae in the brain coral *Favia*. *Biol. Bull.* 135: 141–8.

– 1968d. Photosynthetic pigments of the phytoplankton of some coral reef waters. *Limnol. Oceanogr.* 13: 350–55.

– 1969. Properties of two spectrally different components in chlorophyll *c* preparations. *Biochem. Biophys. Acta* 177: 456–67.

– 1972. Preparation and some properties of crystalline chlorophyll c_1 and c_2 from marine algae. *Biochim. Biophys. Acta* 279: 15–33.

– 1974. Profiles of photosynthetic pigments in the ocean using thin-layer chromatography. *Mar. Biol.* 26: 101–10.

– 1976a. The occurrence of chlorophyll c_1 and c_2 in algae. *J. Phycol.* 12: 249–54.

– 1976b. A report of green algal pigments in the central North Pacific Ocean. *Mar. Biol.* 37: 33–7.

– 1981. An improved thin-layer chromatographic technique for marine phytoplankton pigments. *Limnol. Oceanogr.* 26: 191–7.

Jeffrey, S. W., and Allen, M. B. 1964. Pigments, growth and photosynthesis in cultures of two chrysomonads, *Coccolithus huxleyi* and a *Hymenomonas* sp. *J. Gen. Microbiol.* 36: 277–88.

– 1967. A paper chromatographic method for the separation of phytoplankton pigments at sea. *Limnol. Oceanogr.* 12: 533–7.

Jeffrey, S. W., Dounce, R., and Benson, A. A. 1974. Carotenoid transformations in the chloroplast envelope. *Proc. Natl. Acad. Sci. U.S.A.* 71: 807–10.

Jeffrey, S. W., and Hallegraeff, G. M. 1980a. Studies on phytoplankton species and photosynthetic pigments in a warm core eddy of the East Australian current. I. Summer populations. *Mar. Ecol. Prog. Ser.* 3: 265–94.

– 1980b. Studies on phytoplankton species and photosynthetic pigments in a warm-core eddy of the East Australian current. II. A note on pigment methodology. *Mar. Ecol. Prog. Ser.* 3: 295–301.

– 1987a. Chlorophyllase distribution in marine phytoplankton: A problem for chlorophyll analysis. *Mar. Ecol. Prog. Ser.* 35: 293–304.

– 1987b. Plankton pigments, species and light climate in a complex warm-core eddy of the East Australian Current. *Deep-Sea Res.* 34: 649–73.

Jeffrey, S. W., and Haxo, F. T. 1968. Photosynthetic pigments of dinoflagellates (Zooxanthellae) from corals and clams. *Biol. Bull.* 135: 149–65.

Jeffrey, S. W., and Humphrey, G. F. 1975. New spectroscopic equations for determining chlorophylls *a, b,* c_1 and c_2 in higher plants, algae and natural phytoplankton. *Biochem. Physiol. Pflanzen.* 167: 191–4.

Jeffrey, S. W., and Shibata, K. 1969. Some spectral characteristics of chlorophyll *c* from *Tridacna crocea* zooxanthellae. *Biol. Bull.* 136: 54–62.

Jeffrey, S. W., Sielicki, M., and Haxo, F. T. 1975. Chloroplast pigment patterns in dinoflagellates. *J. Phycol.* 11: 374–84.

Jeffrey, S. W., and Stauber, J. L. 1985. Photosynthetic pigments in diatoms. In *Proc. 2nd Int. Phycol. Cong.,* Copenhagen, August 1985, p. 74.

Jeffrey, S. W., and Vesk, M. 1977. Effect of blue-green light on photosynthetic pigments and chloroplast structure in the marine diatom, *Stephanopyxus turnis. J. Phycol.* 13: 271–9.

Jeffrey, S. W., and Wright, S. W. 1987. A new spectrally distinct component in preparations of chlorophyll *c* from the micro-alga, *Emiliania huxleyi* (Prymnesiophyceae). *Biochim. Biophys. Acta* 894: 180–8.

Jensen, A. 1959. Quantitative determination of carotene by paper chromatography. *Acta Chem. Scand.* 13: 1259–60.

– 1960. Chromatographic separation of carotenes and other chloroplast pigments on aluminium oxide-containing paper. *Acta Chem. Scand.* 14: 2051.

– 1961. Algal carotenoids. Fucoxanthin monoacetate. *Acta Chem. Scand.* 15: 1604.

– 1963. Paper chromatography of carotene and carotenoids. In *Carotine und Carotenoide,* ed. K. Lang, pp. 119–28. Darmstadt: Stemkopff.

– 1966. Carotenoids of Norwegian brown seaweeds and of seaweed meals. *Norw. Inst. Seaweed Res. Trondheim Rep. No. 31.*

– 1978. Chlorophylls and carotenoids. In *Handbook of Phycological Methods: Physiological and Biochemical Methods,* ed. J. A. Hellebust and J. S. Cragie, pp. 59–70. Cambridge: Cambridge University Press.

Jensen, A., and Aasmundrud, O. 1963. Paper chromatographic characterization of chlorophylls. *Acta Chem. Scand.* 17: 907–12.

Jensen, A., and Liaaen-Jensen, S. 1959. Quantitative paper chromatography of carotenoids. *Acta Chem. Scand.* 13: 1863–8.

Jensen, A., and Sakshaug, E. 1973. Studies on the phytoplankton ecology of the Trondheimsfjord II. Chloroplast pigments in relation to abundance and physiological state of the phytoplankton. *J. Exp. Mar. Biol. Res.* 11: 137–55.

Jerlov, N. G. 1976. *Marine Optics.* Amsterdam: Elsevier.

Jiang, L., Ma, J., He, H., Zeng, F., Yang, Z., and Liu, Y. 1983. Some properties of R-phycoerythrin. In *Proc. Joint China-U.S. Phycol. Symp.,* ed. C. K. Tsing, pp. 365–78. Beijing, China: Science Press.

Johannes, B., Brezezinka, H., and Budzeikiewicz, H. 1971. Photosynthesis on green plants, VI. Isolation of diadinoxanthin from *Euglena gracilis. Z. Naturforsch* 26b: 277–8.

Johansen, J. E., and Liaaen-Jensen, S. 1974. Synthesis of 7,7′-deuterated rhodopsin and 7-deuterated apo-8-lycopenal. *Acta Chem. Scand.* [*B*] 28: 301–7.

Johansen, J. E., Svec, W. A., Liaaen-Jensen, S., and Haxo, F. T. 1974. Carotenoids of Dinophyceae. *Phytochemistry* 13: 2261–72.

Jones, I. D., Butler, L. S., Gibbs, E., and White, R. C. 1972. An evaluation of reversed phase partition for thin-layer chromatographic identification of chlorophylls and derivatives. *J. Chromatogr.* 70: 87–98.

Jones, O. T. C. 1963a. The inhibition of bacteriochlorophyll biosynthesis in *Rhodopseudomonas speroides* by 8-hydroxyquinoline. *Biochem. J.* 88: 335–43.

– 1963b. Magnesium, 2,4-divinylphaeoporphyrin a_5 monomethyl ester, a protochlorophyll-like pigment produced by *Rhodospeudomonas spheroides. Biochem. J.* 89: 182–9.

Jung, J., Song, P.-S., Paxton, R. J., Edelstein, M. S., Swanson, R., and Hazen, E. E. 1980. Molecular topology of the phycocyanin photoreceptor from *Chroomonas* species. *Biochemistry* 19: 24–32.

Jungalwala, F. B., and Cama, H. R. 1962. Carotenoids in *Delonix regia* (Gul Mohr) flower. *Biochem. J.* 85: 1–8.

Kageyama, A., and Yokohama, Y. 1978. The function of siphonein in a siphonous green alga, *Dichotomosiphon tuberosus. Jpn. J. Phycol.* 26: 151–5.

Kan, K.-S., and Thornber, J. P. 1976. Light-harvesting chlorophyll *a*/*b*-protein complex of *Chlamydomonas reinhardii. Plant Physiol.* 57: 47–52.

Kaplan, S., and Arntzen, C. J. 1982. Photosynthetic membrane structure and function. In *Photosynthesis,* Vol. 1: Energy Conversion by Plants and Bacteria, ed. Govindjee, pp. 65–151. New York: Academic Press.

Karrer, P., Fatzer, W., Favarger, M., and Jucker, E. 1943. Die Antheridienfarbstoffe von *Chara*-Arten (Armlenchlergewächse). *Helv. Chim. Acta* 26: 2121–2.

Karrer, P., and Jucker, E. 1948. *Carotenoide*. Basel: Berkhauser. (Trans. E. A. Brande, New York: Elsevier, 1950.)

Katoh, T., Dörnemann, D., and Senger, H. 1985. Evidence for chlorophyll RC I in Cyanobacteria. *Plant Cell Physiol.* 26: 1583–6.

Ke, B., Imsgard, F., Kjøsen, H., and Liaaen-Jensen, S. 1970. Electronic spectra of carotenoids at 77° K. *Biochim. Biophys. Acta* 210: 139–52.

Keast, J. F., and Grant, B. R. 1976. Chlorophyll *a*:*b* ratios in some siphonous green algae in relation to species and environment. *J. Phycol.* 12: 328–31.

Keegstra, K., Werner-Washburn, M. N., Cline, K., and Andrews, J. (1984). The chloroplast envelope. Is it homologous with the double membranes of mitochondria and gram-negative bacteria? *J. Cell. Biochem.* 24: 55–68.

Kikuchi, R., Ashida, K., and Hirao, S. 1979. Phycobilins of different color types of *Porphyra yezoensis* Ueda. *Bull. Jpn. Soc. Sci. Fish.* 45: 1461–4.

Killilea, S. D., O'Carra, P., and Murphy, R. F. 1980. Structures and apoprotein linkages of phycoerythrobilin and phycocyanobilin. *Biochem. J.* 187: 311–20.

Kirilovsky, D., Kessel, M., and Ohad, I. 1983. In vitro reassociation of phycobiliproteins and membranes to form functional membrane-bound phycobilisomes. *Biochim. Biophys. Acta* 724: 416–26.

Kirilovsky, D., and Ohad, I. 1986. Functional assembly in vitro of phycobilisomes with isolated photosystem II particles of eukaryotic chloroplasts. *J. Biol. Chem.* 261: 12317–23.

Kirk, J. T. O., and Tilney-Bassett, R. A. E. 1978. *The Plastids: Their Chemistry, Structure, Growth and Inheritance,* 2nd ed. Amsterdam: Elsevier.

Kite, G. C., and Dodge, J. D. 1985. Structural organization of plastid DNA in two anomalously pigmented dinoflagellates. *J. Phycol.* 21: 50–6.

Kjøsen, H., and Liaaen-Jensen, S. 1972. Carotenoids of higher plants. 6. Total synthesis of lycoxanthin and lycophyll. *Acta Chem. Scand.* 26: 4121–9.

Kjøsen, H., Norgard, S., Liaaen-Jensen, S., Svec, W. A., Strain, H. H., Wegfahrt, P., Rapoport, H., and Haxo, F. T. 1976. Algal carotenoids. XV. Structural studies on peridinin. Part 2. Supporting evidence. *Acta Chem. Scand.* [*B*] 30: 157–64.

Klaveness, D. 1986. Classic and modern criteria for determining species of Cryptophyceae. *Bull. Plankton Soc. Japan* 32: 111–28.

Kleinig, H. 1969. Carotenoids of siphonous green algae: A chemotaxonomic study. *J. Phycol.* 5: 281–4.

Kleinig, H., and Egger, K. 1967. Carotenoide der *Vaucheriales Vaucheria* und *Botrydium* (Xanthophyceae). *Z. Naturforsch.* 22b: 868–72.

Kleppel, G. S., and Pieper, R. E. 1984. Phytoplankton pigments in the gut of planktonic copepods from coastal waters off southern California. *Mar. Biol.* 78: 193–8.

Klimov, V. V., Klevanik, A. V., Shuvalov, V. A., and Krasnovsky, A. A. 1977. Reduction of pheophytin in the primary light reaction of photosystem II. *FEBS Lett.* 82: 183–6.

Klotz, A. V., and Glazer, A. N. 1985. Characterization of the bilin attachment sites in R-phycoerythrin. *J. Biol. Chem.* 260: 4856–63.

Knight, R., and Mantoura, R. F. C. 1985. Chlorophyll and carotenoid pigments in Foramenifera and their symbiotic algae: Analysis by high performance liquid chromatography. *Mar. Ecol. Prog. Ser.* 23: 241–9.

Knox, R. S. 1975. Excitation energy transfer and migration: Theoretical considerations. In *Bioenergetics of Photosynthesis,* ed. Govindjee, pp. 183–221. New York: Academic Press.

– 1977. Photosynthetic efficiency and exciton transfer and trapping. In *Primary Processes in Photosynthesis,* ed. J. Barker, pp. 55–97. Amsterdam: Elsevier.

Koller, K. P., and Wehrmeyer, W. 1975. B-Phycoerythrin from *Rhodella violaceae*. *Arch. Microbiol.* 104: 255–61.

Korthals, H. J., and Steenbergen, C. L. M. 1985. Separation and quantification of pigments from natural phototrophic microbial populations. *FEMS Microbiol. Ecol.* 31: 177–85.

Kost, H.-P., 1988. *CRC Handbook of Chromatography: Plant Pigments*. Boca Raton, FL: CRC Press.

Kremer, B. P., Kies, L., and Rostami-Rabet, A. 1979. Photosynthetic performance of cyanelles in the endocyanomes *Cyanophora, Glaucosphaera, Glocochate* and *Glaucocystis*. *Z. Pflanzenphysiol*. 92: 303–17.

Krinsky, N. I. 1963. A relationship between partition coefficients of carotenoids and their functional groups. *Anal. Biochem*. 6: 293–302.

Krinsky, N. I., and Goldsmith, T. H. 1960. The carotenoids of the flagellated alga, *Euglena gracilis*. *Arch. Biochem. Biophys*. 91: 271–9.

Krinsky, N. I., and Welankiwar, S. 1984. Assay of carotenoids. *Methods Enzymol*. 105: 155–62.

Kumano, M., Tomioka, N., and Sugiura, M. 1983. The complete nucleotide sequence of a 23*S* rRNA gene from a blue-green alga, *Anacystis nidulans*. *Gene* 24: 219–25.

Kumazaki, T., Hori, H., and Osawa, S. 1982a. Nucleotide sequences of cytoplasmic 5*S* ribosomal RNA from *Euglena gracilis*. *J. Mol. Evol*. 18: 293–6.

– 1982b. The nucleotide sequence of 5*S* ribosomal RNA from a protozoan species *Chilomonas paramecium* belonging to the class Phytomastigophorea. *FEBS Lett*. 149: 281–4.

Kursar, T. A., and Alberte, R. S. 1983. Photosynthetic unit organization in a red alga. Relationships between light-harvesting pigments and reaction centers. *Plant Physiol*. 72: 409–14.

Kursar, T. A., Swift, H., and Alberte, R. S. 1981. Morphology of a novel cyanobacterium and characterization of light-harvesting complexes from it: Implications for biliprotein evolution. *Proc. Natl. Acad. Sci. U.S.A*. 78: 6888–92.

Kursar, T. A., Van der Meer, J., and Alberte, R. S. 1983a. Light-harvesting system of the red alga *Gracilaria tikvahiae*. I. Biochemical analysis of pigment mutations. *Plant Physiol*. 73: 353–60.

– 1983b. Light-harvesting system of the red alga *Gracilaria tikvahiae*. II. Phycobilisome characteristics of pigment mutants. *Plant Physiol*. 73: 361–9.

Kutzing, F. T. 1843. *Phycologia Generalis*. Leipzig: F. A. Brockhaus.

Larkum, A. W. D., and Barrett, J. 1983. Light-harvesting processes in algae. *Adv. Bot. Res*. 10: 1–219.

Larkum, A. W. D., Cox, G. C., Hiller, R. G., Parry, D. L., and Dibbayawan, T. P. 1987. Filamentous cyanophytes containing phycourobilin and in symbiosis with sponges and an ascidian of coral reefs. *Mar. Biol*. 95: 1–14.

Larsen, J. 1988. An ultrastructural study of *Amphidinium poecilochoum* (Dinophyceae), a phagotrophic dinoflagellate feeding on small species of cryptophytes. *Phycologia* 27: 366–77.

Leclerc, J.-C. 1983. Absorption and emission spectra of the phycobiliproteins from the cyanobacterium *Pseudoanabaena* 7409. *Photosynthetica* 17: 431–41.

Lee, R. E. 1972. Origin of plastids and the phylogeny of algae. *Nature* 237: 44–5.

Lee, S. H. Y., Gilbert, C. W., and Buetow, D. E. 1985. Temporal appearance of chlorophyll-protein complexes and the *N,N'*-dicyclohexyl carbodiimide-binding coupling factor-subunit III in forming thylakoid membranes of *Euglena gracilis*. *J. Plant Physiol.* 118: 7–21.

Lefort-Tran, M., Cohen-Bazire, G., and Pouphile, M. 1973. Les membranes photosynthétique des algues à biliproteines observées après cryodésapage. *J. Ultrastruct. Res.* 44: 199–207.

Lemberg, R., and Legge, J. W. 194. *Hematin Compounds and Bile Pigments*. New York: Interscience.

Lenz, J., and Fritsche, P. 1980. The estimation of chlorophyll *a* in water samples: A comparative study on retention in a glass-fibre and membrane filter and on the reliability of the storage methods. *Arch. Hydrobiol. Beih. Ergebn. Limnol.* 19: 46–57.

Lenz, J., and Zeitzchel, B. 1968. Zur Bestimmung des Extinktionskoeffizienten für Chlorophyll *a* in Methanol. *Kieler Meeresforsch.* 24: 41–50.

Lewin, R. A. 1976. Prochlorophyta as a proposed new division of algae. *Nature* 261: 697–8.

– 1977. *Prochloron:* Type genus of the Prochlorophyta. *Phycologia* 16: 217.

– 1983. The problems of *Prochloron*. *Ann. Microbiol.* (*Inst. Pasteur*) 134B: 37–41.

– 1984. *Prochloron* – A status report. *Phycologia* 23:203–8.

Lewin, R. A., Norris, R., Jeffrey, S. W., and Pearson, B. E. 1977. An aberrant chrysophycean alga *Pelagococcus subviridis* gen. nov. et sp. nov. from the South Pacific Ocean. *J. Phycol.* 13: 259–66.

Ley, A. C., Butler, W. L., Bryant, D. A., and Glazer, A. N. 1977. Isolation and function of allophycocyanin B from *Porphyridium cruentum*. *Plant Physiol.* 59: 974–80.

Liaaen, S., and Sørensen, N. A. 1956. Postmortal changes in the carotenoids of *Fucus vesiculosus*. In *Second Int. Seaweed Symp. Trondheim,* ed. T. Braaruth and N. A. Sorensen, pp. 22–31. London: Pergamon.

Liaaen-Jensen, S. 1971. Isolation reactions. In *Carotenoids,* ed. O. Isler, pp. 68–188. Basel: Birkhäuser.

– 1977. Algal carotenoids and chemosystematics. In *Marine Natural Products Chemistry: Nato Special Program Panel on Marine Sciences,* pp. 239–59. New York: Plenum Press.

– 1978. Marine carotenoids. In *Marine Natural Products: Chemical and Biological Perspectives,* Vol. 2, ed. P. J. Scheuer, pp. 1–73. New York: Academic Press.

– 1979. Carotenoids: A chemosystematic approach. Pure Appl. Chem. 51: 661–75.

– 1980. Stereochemistry of naturally occurring carotenoids. *Prog. Chem. Org. Nat. Products* 39: 123–72.

– 1985. Carotenoids of lower plants – Recent progress. *Pure Appl. Chem.* 57: 649–58.

Liaaen-Jensen, S., and Andrewes, A. G. 1985. Analyses of carotenoids and related pigments. *Methods Microbiol.* 18: 235–55.

Liaaen-Jensen, S., and Jensen, A. 1971. Quantitative determination of carotenoids in photosynthetic tissue. *Methods Enzymol.* 23: 586–602.

Lichtlé, C., Duval, J. C., and Lemoine, Y. 1987. Comparative biochemical, functional and ultrastructural studies of photosynthetic particles from a Cryptophycea: *Cryptomonas rufescens:* Isolation of an active phycoerythrin particle. *Biochim. Biophys. Acta* 894: 76–90.

Lichtle, C., Jupin, H., and Duval, J. C. 1980. Energy transfer from photosystem II to photosystem I in *Cryptomonas rufescens* (Cryptophyceae). *Biochim. Biophys. Acta* 591: 104–12.

Lichtlé, C., and Thomas, J. C. 1976. Étude ultrastructurale des thylakoides des algues à phycobiliproteines: Comparaison des resultats obtenus par fixation classique et cryodécapage. *Phycologia* 15: 393–404.

Liebezeit, G. 1980. Chlorophyll *a* in marine phytoplankton: Separation by HPLC and specific fluorimetric detection. *J. High Res. Chromatogr. Commun.* 3: 531–3.

Lim, B. L., Kawai, H., Hori, H., and Osawa, S. 1986. Molecular evolution of 5*S* ribosomal RNA from red and brown algae. *Jpn. J. Genet.* 61: 169–76.

Lind, E. F., Lane, H. C., and Gleason, L. S. 1953. Partial separation of the plastid pigments by paper chromatography. *Plant Physiol.* 28: 325–8.

Lipschultz, C. A., and Gantt, E. 1981. Association of phycoerythrin and phycocyanin: In vitro formation of a functional energy transferring phycobilisome complex of *Porphyridium cruentum*. *Biochemistry* 20: 3371–6.

Loeber, D. E., Russell, S. H., Touse, F. P., Weedon, B. C. L., and Diment, T. 1971. Carotenoids and related compounds XXVIII. Synthesis of zeaxanthin, β-cryptoxanthin and zeinoxanthin (α-cryptoxanthin). *J. Chem. Soc.* [*C*] 404–8.

Loeblich, A. R. 1976. Dinoflagellate evolution: Speculation and evidence. *J. Protozool.* 23: 13–28.

Loeblich, L. A., and Loeblich, A. R. 1973. A search for binucleate or chrysophyte-containing dinoflagellates. *J. Protozool.* 20: 518.

Loeblich, L. A., and Smith, V. E. 1968. Chloroplast pigments of the marine dinoflagellate *Gyrodinium resplendens*. *Lipids* 3: 5–13.

Loftus, M. E., and Carpenter, J. H. 1971. A fluorimetric method for determining chlorophylls *a, b,* and *c*. *J. Mar. Res.* 29: 319–338.

Long, E. B., and Cooke, G. D. 1971. A quantitative comparison of pigment extraction by membrane and glass-fibre filters. *Limnol. Oceanogr.* 16: 990–2.

Loomis, W. E. 1960. Historical introduction. *Enc. Plant Physiol.* (1st ed.) 5(1): 85–114.

Lorenzen, C. J. 1967. Determination of chlorophyll and pheo-pigments: Spectrophotometric equations. *Limnol. Oceanogr.* 12: 343–6.

Lorenzen, C. J., and Jeffrey, S. W. 1980. Determination of chlorophyll in seawater – Report of intercalibration tests. *UNESCO Tech. Papers in Mar. Sci. 35,* 20 pp.

Lu, R.-Z., Wang, S.-Z., and Yu, Y.-L. 1983. Studies on the fluorescence spectral properties of chlorophyll *d* in different solvents. *Bot. Res.: Cont. Inst. Bot. Academica Sinica* 1: 197–205.

Lu, R.-Z., and Yu, Y.-L. 1986. Reconstitution of phycobilisomes from dissociated phycobilisomes of *Anabaena varibalis*. *Photosynthetica* 20: 397–400.

Lubian, L. M., and Establier, R. 1982. Comparative study of pigment composition of some *Nannochloris* strains (Eustigmatophyceae). *Inv. Pesq.* 46: 379–89.

Ludwig, M., and Gibbs, S. P., 1985. DNA is present in the nucleomorph of *Cryptomonas*. Further evidence that the chloroplast evolved from a eukaryotic endosymbiont. *Protoplasma* 127: 9–20.

Lundell, D. J., and Glazer, A. N. 1983a. Molecular architecture of a light-harvesting antenna. Structure of the 18*S* core-rod subassembly of the *Synechococcus* 6301 phycobilisome. *J. Biol. Chem.* 258: 894–901.

– 1983b. Molecular architecture of a light-harvesting antenna. Core substructures in *Synechococcus* 6301 phycobilisomes: Two new allophycocyanin and allophycocyanin B complexes. *J. Biol. Chem.* 258: 902–8.

– 1983c. Molecular architecture of a light-harvesting antenna. Quarternary interactions in the *Synechococcus* 6301 phycobilisome core as revealed by partial tryptic digestion and circular dichroism studies. *J. Biol. Chem.* 258: 8708–13.

Lundell, D. J., Glazer, A. N., DeLange, R. J., and Brown, D. M. 1984. Bilin attachment sites in the α and β subunits of B-phycoerythrin: Amino acid sequence studies. *J. Biol. Chem.* 259: 5472–80.

Lundell, D. J., Glazer, A. N., Melis, A., and Malkin, R. 1985. Characterization of a cyanobacterial photosystem I complex. *J. Biol. Chem.* 260: 646–54.

Lundell, D. J., Williams, R. C., and Glazer, A. N. 1981a. Molecular architecture of a light-harvesting antenna: In vitro assembly of the rod substructure of *Synechococcus* 6301 phycobilisomes. *J. Biol. Chem.* 256: 3580–92.

Lundell, D. J., Yamanaka, G., and Glazer, A. N. 1981b. A terminal energy acceptor of the phycobilisome: The 75,000-dalton polypeptide of *Synechococcus* 6301 phycobilisomes – A new biliprotein. *J. Cell Biol.* 91: 315–19.

MacColl, R. 1982. Phycobilisomes and biliproteins. *Photochem. Photobiol.* 35: 899–904.

MacColl, R., and Berns, D. S. 1978. Energy transfer studies on cryptomonad biliproteins. *Photochem. Photobiol.* 27: 343–9.

MacColl, R., Berns, D. S., and Gibbons, O. 1976. Characterization of cryptomonad phycoerythrin and phycocyanin. *Arch. Biochem. Biophys.* 177: 265–75.

MacColl, R., Csatorday, K., Berns, D. S., and Traeger, E. 1980. Chromophore interactions in allophycocyanin. *Biochemistry* 19: 2817–20.

MacColl, R., Edwards, M. R., and Haaksma, C. 1978. Some porperties of allophycocyanin, from a thermophilic blue-green alga. *Biophys. Chem.* 8: 369–76.

MacColl, R., and Guard-Friar, D. 1983a. Phycocyanin 612: A biochemical and photochemical study. *Biochemistry* 22: 5568–72.

– 1983b. Phycocyanin 645. The chromophore assay of phycocyanin 645 from the cryptomonad protozoa *Chroomonas* sp. *J. Biol. Chem.* 258: 14327–9.

– 1987. *Phycobiliproteins*. Boca Raton, FL: CRC Press.

MacColl, R., Guard-Friar, D., and Csatorday, K. 1983. Chromatographic and spectroscopic analysis of phycoerythrin 545 and its sub-units. *Arch. Microbiol.* 135: 194–8.

MacColl, R., Habig, W., and Berns, D. S. 1973. Characterization of phycocyanin from *Chroomonas* species. *J. Biol. Chem.* 248: 7080–6.

MacColl, R., O'Connor, G., Crofton, G., and Csatorday, K. 1981. Phycoerythrocyanin: Its spectroscopic behaviour and properties. *Photochem. Photobiol.* 34: 719–23.

Machold, O., Simpson, D. J., and Møller, B. L. 1979. Chlorophyll-proteins of thylakoids from wild-type and mutants of barley (*Hordeum vulgare* L.). *Carlsberg Res. Commun.* 44: 235–54.

MacKay, R. M., Salgado, D., Bonen, L., Stackebrandt, E., and Doolittle, W. F. (1982) The 5*S* ribosomal RNAs of *Paracoccus denitrificans* and *Prochloron*. *Nucleic Acids Res.* 10: 2963–70.

MacKinney, G. 1940. Criteria for purity of chlorophyll preparations. *J. Biol. Chem.* 132: 91–109.

– 1941. Absorption of light by chlorophyll solutions. *J. Biol. Chem.* 140: 315–22.

MacLachlan, S., and Zalik, S. 1963. Plastid structure, chlorophyll concentration and free amino acid composition of a chlorophyll mutant of barley. *Can. J. Bot.* 41: 1053–62.

Mandelli, E. F. 1968. Carotenoid pigments of the dinoflagellate, *Glenodinium foliaceum* Stein. *J. Phycol.* 4: 347–8.

Mann, J. E., and Myers, J. 1968. On pigment, growth, and photosynthesis of *Phaeodactylum tricornutum*. *J. Phycol.* 4: 349–55.

Manning, W. M., and Strain, H. H. 1943. Chlorophyll *d*, a green pigment in red algae. *J. Biol. Chem.* 151: 1–19.

Manodori, A., Alhadeff, M., Glazer, A. N., and Melis, A. 1984. Photochemical apparatus organization in *Synechococcus* 6301 (*Anacystis nidulans*). Effect of phycobilisome mutation. *Arch. Microbiol.* 139: 117–23.

Manodori, A., and Melis, A. 1984. Photochemical organization in *Anacystis nidulans*. Effect of CO_2 concentration during cell growth. *Plant Physiol.* 74: 67–71.

Mantoura, R. F. C., and Llewellyn, C. A. 1983. The rapid determination of algal chlorophyll and carotenoid pigments and their breakdown products in natural waters by reverse phase high-performance liquid chromatography. *Anal. Chim. Acta* 151: 297–314.

Marchand, P. S., Reegg, R., Schwieter, C., Siddons, P. T., and Weedon, B.

C. L. 1965. Carotenoids and related compounds. XI. Synthesis of δ carotene and ϵ carotene. *J. Chem. Soc.* 2019–26.

Margalef, R. 1968. *Perspectives in Ecological Theory*. Chicago: University of Chicago Press.

Margulis, L. 1970. *Origin of Eukaryotic Cells*. New Haven: Yale University Press.

– 1975. *Symbiosis and Cell Evolution*. San Francisco: W. H. Freeman.

– 1981. *Symbiosis in Evolution*. San Francisco: Freeman.

Marker, A. H. F. 1972. The use of acetone and methanol in the extraction of chlorophyll in the presence of phaeophytin. *Freshwater Biol.* 2: 361–85.

– 1977. Some problems arising from the estimation of chlorophyll *a* and phaeophytin *a* in methanol. *Limnol. Oceanogr.* 22: 578–9.

– 1980. A note on the extraction of chlorophyll from benthic algae using methanol. *Arch. Hydrobiol. Beih. Ergebn. Limnol.* 14: 88–90.

Marker, A. H. F., Crowther, C. A., and Gunn, R. J. M. 1980a. Methanol and acetone as solvents for estimating chlorophyll *a* and phaeopigments by spectrophotometry. *Arch. Hydrobiol. Beih. Ergebn. Limnol.* 14: 52–69.

Marker, A. F. H., and Jinks, S. 1982. The spectroscopic analysis of chlorophyll *a* and phaeopigments in acetone, ethanol and methanol. *Arch. Hydrobiol. Beih. Ergebn. Limnol.* 16: 3–17.

Marker, A. F. H., Nusch, E. A., Rai, H., and Riemann, B. 1980b. The measurement of photosynthetic pigments in freshwaters and standardization of methods: Conclusions and recommendations. *Arch. Hydrobiol. Beih. Ergebn. Limnol.* 14: 91–106.

Marki-Fischer, E., Bütikofer, F. A., Buchecker, R., and Eugster, C. H. 1983. Absolute Konfiguration von Loroxanthin (=(3*R*, 3′*R*, 6′*R*)-β,ϵ-carotin-3,19,3′-triol). *Helv. Chim. Acta* 66: 1175–82.

Markwell, J. P., Thornber, J. P., and Boggs, R. T. 1979. Higher plant chloroplasts: Evidence that all the chlorophyll exists as chlorophyll-protein complexes. *Proc. Natl. Acad. Sci. U.S.A.* 76:1233–5.

Martin, A. J. P., and Synge, E. L. W. 1941. A new form of chromatogram employing two liquid phases. *Biochem. J.* 35: 1358–68.

Martin, C. D., and Hiller, R. G. 1987. Subunits and chromophores of a type I phycoerythrin from a *Chroomonas* sp. (Cryptophyceae). *Biochim. Biophys. Acta* 923: 88–97.

Mattox, K. R., and Stewart, K. D. 1984. Classification of green algae: A concept based on comparative cytology. In *Systematics of Green Algae*, ed. D. E. G. Irvine and D. M. John, pp. 29–72. London: Academic Press.

McFadden, G. I., Hill, D. R. A., and Wetherbee, R. 1986. A study of the genus *Pyramimonas* (Prasinophyceae) from southeastern Australia. *Nord. J. Bot.* 6: 209–34.

McLean, R. J. 1967. Pigments of the chrysophyte *Apistonema*. *Nova Hedwigia* 17: 369–72.

McQuade, A. B. 1983. Origins of the nucleate organism. II. *BioSystems* 16: 39–55.

Meeks, J. C. 1974. Chlorophylls. In *Algal Physiology and Biochemistry*, ed. W. D. P. Stewart, pp. 161–75. Ocford: Blackwell.

Melnikov, S. S., and Yevstigneyev, V. B. 1964. Isolation and spectral properties of chlorophyll *c*. *Biofizika* 9: 414–21.

Meyer, S. R., and Pienaar, R. N. 1984. The microanatomy of *Chroomonas africana* sp. nov. (Cryptophyceae). *S. Afr. J. Bot.* 3: 306–19.

Michel, H. P., Schneider, E., Tellenbach, M., and Boschetti, A. 1981. Intrinsic membrane proteins of the thylakoids of *Chlamydomonas reinhardii*. *Photosynth. Res.* 2: 203–12.

Michel-Wolwertz, M.-R., Sironval, C., and Goedheer, J. C. 1965. Presence of a chlorophyll *d*-like pigment in *Chlorella*. *Biochim. Biophys. Acta* 94: 584–5.

Miller, E. C. 1931. *Plant Physiology*. New York: McGraw-Hill.

Moed, J. R., and Hallegraeff, G. M. 1978. Some problems in the estimation of chlorophyll-*a* and phaeopigments from pre- and post-acidification spectrophotometric measurements. *Int. Rev. Ges. Hydrobiol.* 63: 787–800.

Moestrup, Ø. 1982. Phycological reviews, 7. Flagellar structure in algae: A review, with new observations, particularly on the Chrysophyceae, Phaeophyceae (Fucophyceae), Euglenophyceae and *Reckertia*. *Phycologia* 21: 427–528.

Molisch, H. 1894. Das Phycoerythrin, seine Krystallisier-barkeit und chemische Natur. *Bot. Z.* 52: 177–86.

Moran, R., and Porath, D. 1980. Chlorophyll determination in intact tissue using *N,N'*-dimethylformamide. *Plant Physiol.* 65: 478–9.

Mörschel, E., Koller, K. P., Wehrmeyer, W., and Schneider, H. 1977. Biliprotein assembly in the disc-shaped phycobilisomes of *Rhodella violacea*. I. Electron microscopy of phycobilisomes in situ and analyses of their architecture after isolation and negative staining. *Cytobiologie* 16: 118–29.

Mörschel, E., and Wehrmeyer, W. 1975. Cryptomonad biliprotein: Phycocyanin-645 from a *Chroomonas* species. *Arch. Microbiol.* 105: 153–8.

– 1977. Multiple forms of phycoerythrin-545 from *Cryptomonas maculata*. *Arch. Microbiol.* 113: 83–9.

– 1979. Electromicroscopical analysis of native biliprotein aggregates and their spatial assembly. *Ber. Dtsch. Bot. Ges.* 92: 393–402.

Moss, B. 1967. A note on the estimation of chlorophyll *a* in freshwater algal communities. *Limnol. Oceanogr.* 12: 340–2.

Moss, G. P., and Weedon, B. C. L. 1976. Chemistry of the carotenoids. In *Chemistry and Biochemistry of Plant Pigments,* 2nd ed., Vol. 1, ed. T. W. Goodwin, pp. 149–224. London: Academic Press.

Muckle, G., and Rüdiger, W. 1977. Chromophore content of C-phycoerythrin from various cyanobacteria. *Z. Naturforsch.* 32c: 957–62.

Mues, R., Edelbluth, E., and Zinsmeister, H. D. 1973. Das Carotenoid-muster von *Lophocolea bidentata* (L.) Dum. *Österr. Bot. Z.* 122: 177–84.

Murakami, A., Mimuro, M., Okki, K., and Fujita, Y. 1981. Absorption spectrum of allophycocyanin isolated from *Anabaena cylindrica:* Variation of the absorption spectrum induced by changes of the physico-chemical environment. *J. Biochem. (Tokyo)* 89: 79–86.

Murata, N., Sato, N., Omata, T., and Kuwabara, T. 1981. Separation and characterization of thylakoid and cell envelope of the blue-green alga (Cyanobacterium) *Anacystis nidulans*. *Plant Cell Physiol*. 22: 855–66.

Murray, A. P., Gibbs, C. F., Longmore, A. R., and Flett, D. J., 1986. Determination of chlorophyll in marine waters: Intercomparison of a rapid HPLC method with full HPLC, spectrophotometric and fluorometric methods. *Mar. Chem*. 19: 211–27.

Nakayama, K., Itagaki, T., and Okada, M. 1986. Pigment comparison of chlorophyll-protein complexes isolated from the green alga *Bryopsis maxima*. *Plant Cell Physiol*. 27: 311–17.

Namba, O., and Satoh, K. 1987. Isolation of a photosystem II reaction center consisting of D-1 and D-2 polypeptides and cytochrome *b*-559. *Proc. Natl. Acad. Sci. U.S.A*. 84: 109–12.

Nells, H. J. C. F., and DeLeenheer, A. P. 1983. Isocratic nonaqueous reverse-phase liquid chromatography of carotenoids. *Anal. Chem*. 55: 270–5.

Nelson, D. J. 1960. Improved chlorophyll extraction method. *Science* 132: 351.

Nelson, J. W., and Livingstone, A. L. 1967. Stabilization of xanthophyll and carotene by ethoxyquin during thin-layer chromatography. *J. Chromatogr*. 28: 465–7.

Neufeld, G. J., and Riggs, A. F. 1969. Aggregation properties of C-phycocyanin from *Anacystis nidulans*. *Biochim. Biophys. Acta* 181: 234–43.

Nies, M., and Wehrmeyer, W. 1980. Isolation and biliprotein characterization of phycobilisomes from the thermophilic cyanobacterium *Mastigocladus laminosus* Cohn. *Planta* 150: 330–7.

– 1981. Biliprotein assembly in the hemidiscoidal phycobilisomes of the thermophilic cyanobacterium *Mastigocladus laminosus* Cohn. Characterization of dissociation products with specific reference to the peripheral phycoerythrocyanin-phycocyanin complexes. *Arch. Microbiol*. 129: 374–9.

Nitsche, H. 1973. Heteroxanthin in *Euglena gracilis*. *Arch. Mikrobiol*. 90: 151–5.

Norgard, S., Svec, W. A., Liaaen-Jensen, S., Jensen, A., and Guillard, R. R. L. 1974a. Chloroplast pigments and algal systematics. *Biochem. Syst. Ecol*. 2: 3–6.

– 1974b. Algal carotenoids and chemotaxonomy. *Biochem. Syst. Ecol*. 2: 7–9.

Norris, R. E. 1980. Prasinophytes. In *Phytoflagellates,* ed. E. R. Cox, pp. 85–146. Amsterdam: Elsevier.

Nusch, E. A. 1980. Comparison of different methods for chlorophyll and phaeopigments determination. *Arch. Hydrobiol. Beih. Ergebn. Limnol*. 14: 14–36.

Nybom, N. 1955. The pigment characteristics of chlorophyll mutations in barley. *Hereditas* 41: 483–98.

Nybraaten, G., and Liaaen-Jensen, S. 1974. Algal carotenoids. XI. New carotenoid epoxides from *Trentipohlia iolithus*. *Acta Chem. Scand*. 28B: 483–5.

O'Carra, P. 1965. Purification and N-terminal analysis of algal biliproteins. *Biochem. J*. 94: 171–4.

– 1970. Algal biliproteins. *Biochem. J*. 119: 2–3P.

O'Carra, P., Murphy, R. F., and Killilea, S. D. 1980. The native forms of the phycobilin chromatophores of algal biliproteins. *Biochem. J.* 187: 303–9.

O'Carra, P., and O'hEocha, C. 1976. Algal biliproteins and phycobilins. In *Chemistry and Biochemistry of Plant Pigments,* 2nd ed., Vol. 1, ed. T. W. Goodwin, pp. 329–76. London: Academic Press.

O'Carra, O., O'hEocha, C., and Carroll, D. M. 1964. Spectral properties of the phycobilins. II. Phycoerythrobilin. *Biochemistry* 3: 1343–50.

Ogawa, T., and Shibata, K. 1965. A sensitive method for determining chlorophyll *b* in plant extracts. *Photochem. Photobiol.* 4: 193–200.

O'hEocha, C. 1958: Comparative biochemical studies of the phycobilins. *Arch. Biochem. Biophys.* 73: 207–19.

– 1961. Spectrophotometric studies of some red algal constituents. *Centre Natl. Rech. Sc. Colloq. Int.* 103: 121–33.

O'hEocha, C. 1962. Phycobilins. In *Physiology and Biochemistry of Algae,* ed. R. A. Lewin, pp. 421–35. London: Academic Press.

– 1963. Spectral properties of the phycobilins. I. Phycocyanobilin. *Biochemistry* 2: 375–82.

– 1965a. Phycobilins. In *Chemistry and Biochemistry of Plant Pigments,* 1st ed., ed. T. W. Goodwin, pp. 175–96. London: Academic Press.

– 1965b. Biliproteins in algae. *Annu. Rev. Plant Physiol.* 16: 415–32.

– 1971. Pigments of the red algae. *Oceanogr. Mar. Biol. Annu. Rev.* 9: 61–82.

O'hEocha, C., and O'Carra, P. 1961. Spectral studies of denatured phycoerythrin. *J. Am. Chem. Soc.* 83: 1091–3.

O'hEocha, C., O'Carra, P. O., and Mitchell, D. 1964. Biliproteins of cryptomonad algae. *Proc. R. Irish Acad.* 63B: 191–200.

O'hEocha, C., and Raftery, M. 1959. Phycoerythrins and phycocyanins of cryptomonads. *Nature* 184: 1049–51.

Okamura, M. Y., Feher, G., and Nelson, N. 1982. Reaction centers. In *Photosynthesis,* Vol. 1: *Energy Conversion by Plants and Bacteria,* ed. Govindjee, pp. 197–272. New York: Academic Press.

O'Kelly, C. J. 1982. Chloroplast pigments in selected marine Chaetophoraceae and Chaetosiphonaceae (Chlorophyta): The occurrence and significance of siphonaxanthin. *Bot. Mar.* 25: 133–7.

O'Kelly, C. J., and Floyd, G. E. 1984. Correlations among patterns of sporangial structures and development, life-histories and ultrastructural features in the Ulvophyceae. In *Systematics of the Green Algae,* ed. D. E. G. Irvine and D. M. John, pp. 121–56. London: Academic Press.

Olson, R. J., Chisholm, S. W., Zettler, E. R., and Armbrust, E. V. 1988. Analysis of *Synechococcus* pigment types in the sea using single and dual beam flow cytometry. *Deep Sea Res.* 35A: 425–40.

Omata, T., and Murata, N. 1983a. Preparation of chlorophyll *a*, chlorophyll *b* and bacteriochlorophyll *a* by column chromatography with DEAE-Sepharose CL-6B and Sepharose CL-6B. *Plant Cell Physiol.* 24: 1093–1100.

– 1983b. Isolation and characterization of the cytoplasmic membranes from the blue-green alga (Cyanobacterium) *Anacystis nidulans. Plant Cell Physiol.* 24: 1101–12.

Omata, T., and Murata, N. 1984. Isolation and characterization of three types of membranes from the cyanobacterium (blue-green alga) *Synechocystis* PCC 6714. *Arch. Microbiol.* 139: 113–6.

Ong, L. J., and Glazer, A. N. 1987. R-Phycocyanin II, a new phycocyanin occurring in marine *Synechococcus* species. *J. Biol. Chem.* 262: 6323–27.

Ong, L. J., Glazer, A. N., and Waterbury, J. B. 1984. An unusual pigment from a marine cyanobacterium. *Science* 224: 80–2.

Oquist, G., Fork, D. C., Schoch, S., and Malmberg, G. 1981. Solubilization and spectral characteristics of chlorophyll–protein complexes isolated from the thermophilic blue-green alga, *Synechococcus lividus*. *Carnegie Inst. Year Book* 80: 47–9.

Ortiz, W., and Stutz, E. 1980. Synthesis of the polypeptides of the chlorophyll–protein complexes in isolated chloroplasts of *Euglena gracilis*. *FEBS Lett.* 116: 298–302.

Otsuki, A., Watanabe, M. M., and Sugahara, K. 1987. Chlorophyll pigments in methanol extracts from ten axenic cultured diatoms and three green algae as determined by reverse phase HPLC with fluorometric detection. *J. Phycol.* 23: 406–14.

Owens, T. G., and Wold, E. R. 1986. Light-harvesting function in the diatom *Phaeodactylum tricornutum*. *Plant Physiol.* 80: 732–6.

Padgett, M. P., and Krogmann, D. W. 1987. Large scale preparation of pure phycobiliproteins. *Photosynth. Res.* 11: 225–35.

Paerl, H. W. 1984. Cyanobacterial carotenoids: Their roles in maintaining optimal photosynthetic production among aquatic bloom forming genera. *Oecologia* 61: 143–9.

Paerl, H. W., Lewin, R. A., and Cheng, L. 1984. Variations in chlorophyll and carotenoid pigmentation among *Prochloron* (*Prochlorophyta*) symbionts in diverse marine ascidians. *Bot. Mar.* 27: 257–64.

Paerl, H. W., Tucker, J., and Bland, P. T. 1983. Carotenoid enhancement and its role in maintaining blue-green algal (*Microcystis aerugenoia*) surface blooms. *Limnol. Oceanogr.* 28: 847–57.

Palla, J.-C., Mille, G., and Busson, F. 1970. Etude composée des carotenoides de *Spirulina platensis* (Gom.) Geitter et de *Spirulina geitieri* J. de Toni (Cyanophyceae). *C. R. Hebd. Séances Acad. Sc.* 270D: 1038–41.

Pan, Z., and Tseng, C. K. 1985. Studies of phycobiliproteins of *Porphyra* from China. *Oceanol. Limnol. Sinica* 16: 417–20.

Pan, Z., Zhou, B., Zeng, C., and Tseng, C. K. 1987. Comparative studies on spectral properties of R-phycoerythrin from the red seaweeds from Qingdao (China). *Chin. J. Oceanol. Limnol.* 4: 353–9.

Parry, D. L. 1984. Cyanophytes with R-phycoerythrin in association with seven species of ascidians from the Great Barrier Reef. *Phycologia* 23: 503–5.

Parsons, T. R. 1961. On the pigment composition of eleven species of marine phytoplankters. *J. Fish Res. Board Can.* 18: 1071–25.

– 1963. A new method for the microdetermination of chlorophyll in sea water. *J. Mar. Res.* 21: 164–71.

Parsons, T. R., and Strickland, J. D. H. 1963. Discussion of spectrophotometric determination of marine plant pigments with revised equations for ascertaining chlorophylls and carotenoids. *J. Mar. Res.* 11: 155–63.

Patterson, G. M. L., and Withers, N. W. 1982. Laboratory cultivation of *Prochloron,* a tryptophan autotroph. *Science* 217: 1034–5.

Pearlstein, R. M. 1982. Chlorophyll singlet excitons. In *Photosynthesis,* Vol. 1: *Energy Conversion by Plants and Bacteria,* ed. Govindjee, pp. 293–330. New York: Academic Press.

Pechar, L. 1987. Use of an acetone:methanol mixture for the extraction and spectrophotometric determination of chlorophyll *a* in phytoplankton. *Arch. Hydrobiol.* [*Suppl.*] 781: 99–117.

Pedersen, M., Saenger, P., Rowan, K. S., and Hofsten, W. A. 1979. Bromine, bromphenols and floridorubin in the red alga, *Lenormandia prolifera. Physiol. Plant.* 46: 121–6.

Pennington, F. C., Haxo, F. T., Borch, G., and Liaaen-Jensen, S. 1985. Carotenoids of cryptophyceae. *Biochem. Syst. Ecol.* 13: 215–19.

Pennington, F. C., Strain, H. H., Svec, W. A., and Katz, J. J. 1964. Preparation and properties of pyrochlorophyll *a*, methyl pyrochlorophyllide *a*, pyropheophytin *a* and methyl pyropheophorbide *a* derived from chlorophyll by decarbmethoxylation. *J. Am. Chem. Soc.* 86: 1418–26.

Perkins, H. J., and Roberts, D. W. A. 1964. On the relative intensities of the "blue" and "red" absorption bands of chlorophyll *a*. *Biochim. Biophys. Acta* 79: 20–9.

Petracek, F. J., and Zechmeister, L. 1956. Determination of partition coefficients of carotenoids as a tool in pigment analysis. *Anal. Chem.* 28: 1484–5.

Peyriere, M. 1968. Les problèmes cytologiques de *Rytiphlea tinctoria* algue rouge à floridorubine. *C. R. Hebd. Séances Acad. Sc. Paris* 226D: 2253–5.

Pigott, G. H., and Carr, N. G. 1972. Homology between nucleic acids of blue-green algae and chloroplasts of *Euglena gracilis. Science* 175: 1259–61.

Pineau, B., Dubertret, G., and Schantz, R. 1985. Functional and structural organization of chlorophyll in the developing photosynthetic membranes of *Euglena gracilis* Z. V. Separation and characterization of pigment–protein complexes of the differentiated thylakoids. *Photosynth. Res.* 6: 159–74.

Porra, R. J., and Grimme, L. H. 1974. A new procedure for the determination of chlorophylls *a* and *b* and its application to normal and regreening *Chlorella. Anal. Biochem.* 57: 255–67.

Prézelin, B. B. 1976. The role of peridinin–chlorophyll *a*–proteins in the photosynthetic light adaption of the marine dinoflagellate *Glenodinium* sp. *Planta* 130: 225–33.

Prézelin, B. B., and Alberte, R. S. 1978. Photosynthetic characteristics and organization of chlorophyll in marine dinoflagellates. *Proc. Natl. Acad. Sci. U.S.A.* 75: 1801–4.

Prézelin, B. B., and Haxo, F. T. 1976. Purification and characterization of peridinin–chlorophyll *a*–proteins from the marine dinoflagellates *Glenodinium* sp. and *Gonyaulux polyhedra. Planta* 128: 133–41.

Priestle, J. R., Rhyne, R. H., Salmon, J. B., and Hackert, M. L. 1982. Phycobiliproteins: Comparison of solution and single crystal fluorescence for C-phycocyanin and B-phycoerythrin. *Photochem. Photobiol.* 35: 827–34.

Rabinowitch, E. I. 1945. *Photosynthesis and Related Processes,* Vol. 1. New York: Interscience.

Ragan, M. A., and Chapman, D. J. 1978. *A Biochemical Phylogeny of the Protista.* New York: Academic Press.

Ramus, J. 1983. A physiological test of the theory of complementary chromatic adaption. II. Brown, green and red seaweeds. *J. Phycol.* 19: 173–8.

Ramus, J., Beale, S. I., Mauzerall, D., and Howard, K. L. 1976. Changes in photosynthetic pigment concentration in seaweeds as a function of water depth. *Mar. Biol.* 37: 223–9.

Ramus, J., Lemons, F., and Zimmerman, C. 1977. Adaptation of light-harvesting pigments to downwelling light and the consequent photosynthetic performance of the eulittoral rockweeds *Ascophyllum nodosum* and *Fucus vesiculosus. Mar. Biol.* 38: 293–303.

Ramus, J., and van der Meer, J. P. 1983. A physiological test of the theory of complementary chromatic adaption. I. Color mutants of a red seaweed. *J. Phycol.* 19: 86–91.

Randerath, K. 1968. *Thin-Layer Chromatography,* trans. D. D. Libman, 2nd ed. New York: Academic Press.

Raven, P. 1970. A multiple origin of plastids and mitochondria. *Science* 169: 641–6.

Rebeiz, C. A., Wu, S. M., Kuhadja, M., Daniell, H., and Perkins, E. J. 1983. Chlorophyll *a* biosynthetic routes and chlorophyll *a* chemical heterogeneity in plants. *Mol. Cell. Biochem.* 57: 97–125.

Redlinger, T., and Gantt, E. 1981. Phycobilisome structure of *Porphyridium cruentum.* Polypeptide composition. *Plant Physiol.* 68: 1375–9.

– 1982. A M 95,000 polypeptide in *Porphyridium cruentum* phycobilisomes and thylakoids. Possible function of linkage of phycobilisomes to thylakoids and in energy transfer. *Proc. Natl. Acad. Sci. U.S.A.* 79: 5542–6.

– 1983. Photosynthetic membranes of *Porphyridium cruentum.* An analysis of chlorophyll–protein complexes and heme-binding proteins. *Plant. Physiol.* 73: 36–40.

Reger, B. J., and Krauss, R. W. 1970. The photosynthetic response to a shift in the chlorophyll *a* to chlorophyll *b* ratio of *Chlorella. Plant Physiol.* 46: 568–75.

Reinman, S., and Thornber, J. P. 1979. The electrophoretic isolation and partial characterization of three chlorophyll–protein complexes from blue-green algae. *Biochim. Biophys. Acta* 547: 188–97.

Renstrom, B., Borch, G., Skulberg, O. M., and Liaaen-Jensen, S. 1981. Optical purity of (3*S*,3'*S*)-astaxanthin from *Haematococcus fluvialis. Phytochemistry* 20: 2561–4.

Resch, C. M., and Gibson, J. 1983. Isolation of the carotenoid-containing cell wall of three unicellular cyanobacteria. *J. Bacteriol.* 155: 345–50.

Reith, A., and Sagromsky, H. 1972. Chlorophylls in species of the Botrydiales, *Arch. Protistenk.* 114: 46–50.

Rhiel, E., Mörschel, E., and Wehrmeyer, W. 1987. Characterization and structural analysis of a chlorophyll *a/c* light harvesting complex and of photosystem I particles isolated from the thylakoid membranes of *Cryptomonas maculata* (Cryptophyceae). *Eur. J. Cell Biol.* 43: 82–92.

Richards, F. A., and Thompson, T. G. 1952. The estimation and characterization of plankton populations by pigment analysis. II. A spectrographic method for the estimation of plankton pigments. *J. Mar. Res.* 11: 156–72.

Ricketts, T. R. 1965. Chlorophyll *c* in some members of the Chrysophyceae. *Phytochemistry* 4: 725–30.

– 1966a. Magnesium 2,4-divinylphaeoporphyrin a_5 monomethyl ester, a protochlorophyll-like pigment present in some unicellular flagellates. *Phytochemistry* 5: 223–9.

– 1966b. The carotenoids of the phytoflagellate *Micromonas pusilla*. *Phytochemistry* 5: 571–80.

– 1967a. The pigments of the phytoflagellates *Pedinomonas minor* and *Pedinomonas tuberculata*. *Phytochemistry* 6: 19–24.

– 1967b. The pigment composition of some flagellates possessing scaly flagella. *Phytochemistry* 6: 669–76.

– 1967c. Further investigations into the pigment comparison of green flagellates possessing scaly flagella. *Phytochemistry* 6: 1375–86.

– 1970. The pigments of the Prasinophyceae and related organisms. *Phytochemistry* 9: 1835–43.

– 1971a. The structure of siphonein and siphonaxanthin from *Codium fragile*. *Phytochemistry* 10: 155–60.

– 1971b. Identification of xanthophylls KI and KIS of the Prasinophyceae as siphonein and siphonaxanthin. *Phytochemistry* 10: 161–4.

Riethman, H. C., Mawhinney, T. P., and Sherman, L. A. 1987. Phycobilisome-associated glycoproteins in the cyanobacterium *Anacystis nidulans* R2. *FEBS Lett.* 215: 209–14.

Riemann, B. 1978a. Carotenoid interference in the spectrophotometric determination of chlorophyll degradation products from natural populations of phytoplankton. *Limnol. Oceangr.* 23: 1059–66.

– 1978b. Absorption coefficients for chlorophylls *a* and *b* in methanol and a comment on interference of chlorophyll *b* in determinations of chlorophyll *a*. *Vatten* 3: 187–94.

Riley, J. P., and Segar, D. A. 1969. The pigments of some further marine phytoplankton species. *J. Mar. Biol. Assoc. U.K.* 49: 1047–56.

Riley, J. P., and Wilson, T. R. S. 1965. The use of thin-layer chromatography for the separation of phytoplankton pigments. *J. Mar. Biol. Assoc. U.K.* 45: 583–91.

– 1967. The pigments of some marine phytoplankton species. *J. Mar. Biol. Assoc. U.K.* 47: 351–62.

Rippka, R., Waterbury, J., and Cohen-Bazire, G. 1974. A cyanobacterium which lacks thylakoids. *Arch. Microbiol.* 100: 419–36.

Rodriguez, D., Tanaka, Y., Katayama, T., Simpson, K. L., Lee, T.-C., and Chichester, C. O. 1976. Hydroxylation of β-carotene on MicroCel C. *J. Agric. Food Chem.* 24: 819–22.

Romeo, A. 1981. The chlorophyll–protein complexes of the prymnesiophyte, *Pavlova lutheri. XIII Int. Bot. Cong. Abstr.*, p. 224. Canberra: Aust. Acad. Sci.

Ronneberg, H., Foss, P., Ramdahl, T., Borch, G., Skulberg, O. M., and Liaaen-Jensen, S. 1980. Occurrence and chirality of oscillaxanthin. *Phytochemistry* 19: 2167–70.

Rosenberg, G., and Ramus, J. 1982. Ecological growth strategies in the seaweeds *Gracilaria foliifera* (Rhodophyceae) and *Ulva* sp. (Chlorophyceae): Photosynthesis and antenna composition. *Mar. Ecol. Prog. Ser.* 8: 233–41.

Rosinski, J., Hainfeld, J. F. Rigbi, M., and Siegelman, H. W. 1981. Phycobilisome ultrastructure and chromatic adaptation in *Fremyella diplosiphon*. *Ann. Bot.* 47: 1–42.

Round, F. E. 1980. The evolution of pigmented and unpigmented unicells: A reconsideration of the Protista. *BioSystems* 12: 61–70.

Rowan, K. S. 1981. Photosynthetic pigments. In *Marine Biology: An Australian Perspective,* ed. M. C. Clayton and R. J. King, pp. 355–68. Melbourne: Longman-Cheshire.

Roy, S. 1987. High-performance liquid chromatographic analysis of chloropigments. *J. Chromatogr.* 391: 19–34.

Rüdiger, W. 1971. Gallenfarbstoff und Biliproteide. *Fortschr. Chemie Organ. Natur.* 29: 60–139.

– 1980. Plant biliproteins. In *Pigments in Plants,* ed. F.-C. Czygan, pp. 314–51. Stuttgart: Fischer.

Rüdiger, W., and Benz, J. 1984. Synthesis of chloroplast pigments. In *Chloroplast Biosynthesis,* ed. R. J. Ellis, pp. 225–44. Cambridge: Cambridge University Press.

Rüdiger, W., Wagenmann, R., and Muckle, G. 1980. Investigation of C-phycoerythrin from the cyanobacterium *Pseudanabaena* W 1174. *Arch. Microbiol.* 127: 253–7.

Rumbeli, R., Wirth, M., Suter, F., and Zuber, H. 1987. The phycobiliprotein $\beta^{16.2}$ of the allophycocyanin core from the cyanobacterium *Mastigocladus laminosus*. Characterization and complete amino-acid sequence. *Biol. Chem. Hoppe-Seyler* 368: 1-9.

Ruskowski, M., and Zilinskas, B. A. 1982. Allophycocyanin I and the 95 kilodalton polypeptide. *Plant Physiol.* 70: 1055–9.

Rutter, L. 1948. A modified technique in filter-paper chromatography. *Nature* 161: 435–6.

Saenger, P. 1970. Floridorubin and the taxonomy of the *Amansieae*. Ph.D. dissertation, University of Melbourne, Victoria, Australia.

Saenger, P., Ducker, S. C., and Rowan, K. S. 1971. Two species of *Ceramiales* from Australia and New Zealand. *Phycologia* 10: 105–11.

Saenger, P., Pedersen, M., and Rowan, K. S. 1976. Bromo-compounds of the red alga, *Lenormandia prolifera*. *Phytochemistry* 15: 1957–8.

Saenger, P., Rowan, K. S., and Ducker, S. C. 1969. The water-soluble pigments of the red alga, *Lenormandia prolifera*. *Phycologia* 7: 59–64.

Sagan, L. 1967. On the origin of mitosing cells. *J. Theoret. Biol.* 14: 225–74.

Sagromsky, H. 1962. Chlorophyllestimmungen an verschiedenen Entwicklungstedien von *Vaucheria*. *Ber. Deutsch. Bot. Ges.* 75: 345–8.

Sahlberg, I., and Hynninen, P. H. 1984. Thin-layer chromatography of chlorophylls and their derivatives on sucrose layers. *J. Chromatogr* 291: 331–8.

Sakshaug, E., and Holm-Hansen, O. 1986. Photoadaption in Antarctic phytoplankton: Variations in growth rate, chemical composition and *P* versus *I* curves. *J. Plankton Res.* 8: 459–73.

Sand-Jensen, K. 1976. A comparison of chlorophyll *a* determinations of unstored and stored plankton filters extracted by methanol and acetone. *Vatten* 4: 337–41.

Santore, U. J. 1982. Comparative ultrastructure of two members of the Cryptophyceae assigned to the genus *Chroomonas* – With comments on their taxonomy. *Arch. Protistenk.* 125: 5–29.

Sardana, R. K., and Mehkotra, R. S. 1979. Change in pigment composition during a *Gymnodinium* bloom in the Brahmsarovar tank at Kurukshetra. *Proc. Indian Acad. Sci.* 88B (Pt. II): 407–12.

Sartory, D. P. 1985. The determination of algal chlorophyllous pigments by high performance liquid chromatography and spectrophotometry. *Water Res.* 19: 605–10.

Sartory, D. P., and Grobbelaar, J. U. 1984. Extraction of chlorophyll *a* from freshwater phytoplankton for spectrophotometric analysis. *Hydrobiologia* 114: 177–87.

Satoh, K. 1986. Chlorophyll–protein complexes. *Photosynth. Res.* 10: 181–7.

– 1988. Reality of P-680 chlorophyll protein – Identification of the site of primary photochemistry in oxygenic photosynthesis. *Physiol. Plant.* 72: 209–12.

Saxena, A. M. 1988. Phycocyanin aggregation. A small angle neutron scattering and size exclusion chromatographic study. *J. Mol. Biol.* 200: 579–91.

Schaltegger, K. H. 1965. Dunnschichtchromatographische Bestimmung von Chlorophyllen und Carotenoiden aus Kirschblaumblättern. *J. Chromatogr.* 19: 75–80.

Schanz, F. 1982. A fluorometric method for determining chlorophyll *a* and phaeophytin *a* concentrations. *Arch. Hydrobiol. Beih. Ergebn. Limnol.* 16: 91–100.

Scheer, H. 1981. Biliproteins. *Angew. Chem. Int. Ed. Engl.* 20: 241–61.

– 1982. Phycobiliproteins: Molecular aspects of photosynthetic antenna system. In *Light Reaction Path of Photosynthesis*, ed. F. K. Fong, pp. 7–45. Berlin: Springer Verlag.

Schertz, F. M. 1925. The quantitative determination of xanthophyll by means of the spectrophotometer. *J. Agric. Res.* 30: 253–61.

Schimmer, B. P., and Krinsky, N. I. 1966. Reduction of carotenoid epoxides with lithium aluminium hydride. *Biochemistry* 5: 2649–57.

Scholz, B., and Ballschmiter, K. 1981. Preparation and reverse-phase high-performance chromatography of chlorophylls. *J. Chromatogr.* 208: 148–55.

Schuler, F., Brandt, P., and Wiessner, W. 1982. Isolation of PS II-particles with in vivo characteristics from *Euglena gracilis* strain Z. *Z. Naturforsch.* 37c: 256–9.

Schuster, G., Nechustai, R., Nelson, N., and Ohad, I. 1985. Purification and composition of photosystem I reaction centre of *Prochloron* sp., an oxygen-evolving prokaryote containing chlorophyll *b*. *FEBS Lett.* 191: 29–33.

Schutt, F. 1888a. Ueber das Phycoerythrin. *Ber. Dtsch. Bot. Ges.* 6: 36–51.

– 1888b. Weitere Beiträge zur Kenntniss des Phycoerythrins. *Ber. Dtsch. Bot. Ges.* 6: 305–35.

Schutte, H.-R. 1983. Secondary plant substances. Aspects of carotenoid biosynthesis. *Prog. Bot.* 45: 120–35.

Schwartz, K. D., and Dayhoff, M. D. 1981. Chloroplast origins: Inferences from protein and nucleic acid sequences. *Ann. N.Y. Acad. Sci.* 361: 260–72.

Schwartz, S. J., and von Elbe, J. H. 1982. High performance liquid chromatography – A review. *J. Liq. Chromatogr.* 5 (Suppl. 1): 43–73.

SCOR-UNESCO 1966. *Monogr. Oceanogr. Methodol.* 1: 11–18.

Searle, G. F. W., Barber, J., Porter, G., and Tredwell, C. J. 1978. Picosecond time-resolved energy transfer in *Porphyridium cruentum*. Part II. In the isolated light-harvesting complex (phycobilisomes). *Biochim. Biophys. Acta* 501: 246–56.

Seely, G. R. 1966. The structure and chemistry of functional groups. In *The Chlorophylls,* ed. L. P. Vernon and G. R. Seely, pp. 67–109. New York: Academic Press.

Seely, G. R., Duncan, M. J., and Vidaver, W. E. 1972. Preparative and analytical extraction of pigments from brown algae with dimethylsulfoxide. *Mar. Biol.* 12: 184–8.

Seely, G. R., and Jensen, A. 1965. Effect of solvent on the spectrum of chlorophyll. *Spectrochim. Acta* 21: 1835–45.

Seewalt, E., and Stackebrandt, E. 1982. Partial sequence of 16*S* ribosomal RNA and the phylogeny of *Prochloron*. *Nature* 295: 618–20.

Seibert, M., and Connolly, J. S. 1984. Flurorescence properties of C-phycocyanin isolated from a thermophilic cyanobacterium. *Photochem. Photobiol.* 40: 267–71.

Sestak, Z. 1958. Paper chromatography of chloroplast pigments. *J. Chromatogr.* 1: 263–308.

– 1965. Paper chromatography of chloroplast pigments (chlorophylls and carotenoids) – Pt. 2. *Chromatogr. Rev.* 7: 65–97.

– 1967. Thin layer chromatography of chlorophyll. *Photosynthetica* 1: 267–92.

– 1980. Paper chromatography of chloroplast pigments (chlorophylls and carotenoids). Pt. 3 *Photosynthetica* 14: 239–70.

– 1982. Thin layer chromatography of chlorophylls. 2. *Photosynthetica* 16: 568–617.

Setif, P., and Mathis, P. 1986. Photosystem I reaction center and its primary electron transfer reactions. *Enc. Plant Physiol. (N.S.)* 19: 476–86.

Seybold, A., and Egle, K. 1937. Lichtfeld and Blattfarbstoffe I. *Planta* 26: 491-515.

Seybold, A., Egle, K., and Hulsbruch, W. 1941. Chlorophyll- und Carotenoid-bestimmungen von Susswasseralgen. *Botan. Arch.* 42: 239–53.

Sheldon, R. W. 1972. Size separation of marine seston by membrane and glass-fiber filters. *Limnol. Oceanogr.* 17: 494–8.

Sherma, J. 1971. Chromatography of leaf pigments on silica gel and aluminium hydroxide loaded papers. *J. Chromatogr.* 61: 203–4.

Sherma, J., and Latta, M. 1978. Reverse-phase thin-layer chromatography of chloroplast pigments on chemically bonded C_{18} plates. *J. Chromatogr.* 154: 73–5.

Sherma, J., and Lipstone, G. S. 1969. Chromatography of chlorplast pigments on preformed thin layers. *J. Chromatogr.* 41: 220–7.

Sherma, J., and Strain, H. H. 1968. Chromatography of leaf chloroplast pigments on ion-exchange papers. *Anal. Chim. Acta* 40: 155–9.

Shibata, K. 1958. Spectrophotometry of intact biological materials. Absolute and relative measurements of their transmission reflection and absorption spectra. *J. Biochem.* (Tokyo) 45: 599–623.

Shioi, Y., Fukae, R., and Sasa, T. 1983. Chlorophyll analysis by high-performance liquid chromatography. *Biochim. Biophys. Acta* 722: 72–9.

Shoaf, W. T. 1978. Rapid method for the separation of chlorophyll *a* and *b* by high-pressure liquid chromatography. *J. Chromatogr.* 152: 247–9.

Shoaf, W. T., and Lium, B. W. 1976. Improved extraction of chlorophyll *a* and *b* from algae using dimethyl sulfoxide. *Limnol. Oceanogr.* 21: 926–8.

Siefermann-Harms, D. 1985. Carotenoids in photosynthesis. 1. Location in photosynthetic membranes and light-harvesting function. *Biochim. Biophys. Acta* 811: 325–55.

– 1987. The light-harvesting and protective functions of carotenoids in photosynthetic membranes. *Physiol. Plant.* 69: 561–8.

Siegelman, H. W., and Firer, E. M. 1964. Purification of phytochrome from oat seedlings. *Biochemistry* 3: 418–23.

Siegelman, H. W., and Kycia, J. H. 1978. Algal biliproteins. In *Handbook of Phycological Methods: Physiological and Biochemical Methods*, ed. J. A. Hellebust and J. S. Cragie, pp. 72–79. Cambridge: Cambridge University Press.

Simpson, C. F. 1976. *Practical High Performance Liquid Chromatography*. London: Heyden.

Skjenstad. T., Haxo, F. T., and Liaaen-Jensen, S. 1984. Carotenoids of clam, coral and nudibranch zooxanthellae in aposymbiotic culture. *Biochem. Syst. Ecol.* 12: 149–53.

Sluiman, H. J. 1985. A cladistic evaluation of the lower and higher green plants (Veridiplantae). *Plant Syst. Evol.* 149: 217–32.

Smallidge, R. L., and Quackenbush, F. W. 1973: β,β-Carotene-2,3,3′-triol: A new carotenoid in *Anacystis nidulans*. *Phytochemistry* 12: 2481–2.

Smith, D. C., and Douglas, A. E. 1987. *The Biology of Symbiosis*. London: Arnold.

Smith, J. C., Platt, T., Li, W. K. W., Horne, E. P. W., Harrison, W. G., Subba

Rao, D. V., and Irwin, B. D. 1985. Arctic marine photoanlotrophic picoplankton. *Mar. Ecol. Prog. Ser.* 20: 207–20.

Smith, J. H. C., and Benitez, A. 1955. Chlorophylls: Analysis in plant materials. In *Modern Methods of Plant Analysis,* Vol. 4, ed. K. Paech and M. V. Tracey, pp. 142–90. Berlin: Springer Verlag.

Snyder, L. R., and Kirkland, J. J. 1979. *Introduction to Modern Liquid Chromatography,* 2nd ed. New York: Wiley.

Sorby, H. C. 1873. On comparative vegetable chromatology. *Proc. R. Soc. (Lond.)* 21: 144–83.

– 1877. On the characteristic colouring-matter of the red groups of algae. *Bot. J. Linn. Soc.* 15: 34–40.

Spencer, K. G., Yu, M.-H., West, J. A., and Glazer, A. N. 1981. Phycoerythrin and interfertility patterns in *Callithamnion* (Rhodophyta) isolates. *Br. Phycol. J.* 16: 331–43.

Speziale, B. J., Schreiner, S. P., Giammatteo, P. A., and Schindler, J. E. 1984. Comparison of *N,N*-dimethylformamide, dimethyl sulfoxide, and acetone for extraction of phytoplankton chlorophyll. *Can. J. Fish. Aquat. Sci.* 41: 1519–22.

Spoehr, H. A., Smith, J. H. C., Strain, H. H., and Milner, H. N. 1936. Biochemical investigations. Carotenoid pigments. *Carnegie Inst. Year Book* 35: 198–201.

Spurgeon, S. I., and Porter, J. W. 1980. Carotenoids. In *The Biochemistry of Plants.* Vol. 4, ed. P. K. Stumpf and E. E. Conn, pp. 419–83. New York: Academic Press.

Stadnichuk, I. N., Odintsova, T. I., and Strongin, A. Y. 1984. Molecular organization and the pigment composition of R-phycoerythrin from the red alga *Callithamnion rubosum. Mol. Biol.* (Engl. trans.) 18: 272–7.

Stadnichuk, I. N., Romanova, N. I., and Selyakh, I. O. 1985. A phycourobilin-containing phycoerythrin from the cyanobacterium *Oscillatoria* sp. *Arch. Microbiol.* 143: 20–5.

Staehelin, L. A. 1976. Reversible particle movements associated with unstacking and restacking of chloroplast membranes in vitro. *J. Cell. Biol.* 71: 136–58.

Staehelin, L. A., Armond, P. A., and Miller, K. R. 1977. Chloroplast membrane organization at the supramolecular level and its functional implications. *Brookhaven Symp. Biol.* 28: 278–315.

Staehelin, L. A., and DeWit, M. 1984. Correlation of structure and function of chloroplast membranes at the supramolecular level. *J. Cell. Biochem.* 24: 261–9.

Stahl, E. 1965. *Thin-layer chromatography: A Laboratory Handbook.* Berlin: Springer Verlag.

Stanier, R. V., and Cohen-Bazire, G. 1977. Phototropic prokaryotes: The Cyanobacteria. *Annu. Rev. Microbiol.* 31: 225–74.

Stanier, R. V., Kunisawa, R., Mandel, M., and Cohen-Bazire, G. 1971. Purification and properties of unicellular blue-green algae (Order Chroococcales). *Bact. Rev.* 35: 171–205.

Stanier, R. V., and Van Neil, C. B. 1962. The concept of a bacterium. *Arch. Mikrobiol.* 42: 17–35.
Stauber, J. L., and Jeffrey, S. W. 1988. Photosynthetic pigments in fifty-one species of marine diatoms. *J. Phycol.* 24: 158–72.
Stauffer, R. E., Lee, G. F., and Armstrong, D. E. 1979. Estimating chlorophyll extraction biases. *J. Fish. Res. Board Can.* 36: 152–7.
Steidinger, K. A., and Cox, E. R. 1980. Free-living dinoflagellates. In *Phytoflagellates,* ed. E. R. Cox, pp. 401–32. New York: Elsevier.
Stewart, A. C. 1980. The chlorophyll–protein complexes of a thermophilic blue-green alga. *FEBS Lett.* 114: 67–72.
Stewart, D. E., and Farmer, F. H. 1984. Extraction, identification, and quantitation of phycobiliprotein pigments from phototrophic plankton. *Limnol. Oceanogr.* 29: 392–7.
Stewart, K. D., and Mattox, K. R. 1978. Structural evolution in the flagellated cells of green algae and land plants. *BioSystems* 10: 145–52.
– 1980. Phylogeny of phytoflagellates. In *Phytoflagellates,* ed. E. R. Cox, pp. 433–62. New York: Elsevier.
Stewart, W. D. P. 1974. *Algal Physiology and Biochemistry.* London: Blackwell.
Stiles, W. 1925. *Photosynthesis: The Assimilation of Carbon by Green Plants.* London: Longmans.
– 1936. *An Introduction to the Principles of Plant Physiology.* London: Methuen.
Stokes, G. G. 1852. On the change of refrangibility of light. *Philos. Trans. R. Soc.* 463–562.
– 1864. On the supposed identity of biliverdin with chlorophyll, with remarks on the constitution of chlorophyll. *Proc. R. Soc. (Lond.)* 13: 144–5.
Strain, H. H. 1934. Carotene. VIII: Separation of carotenes by adsorption. *J. Biol. Chem.* 105: 523–35.
– 1936. Leaf xanthophylls. *Science* 83: 241–2.
– 1938. *Leaf xanthophylls.* Publ. No. 490 Carnegie Inst. Washington.
– 1949. Functions and properties of the chloroplast pigments. In *Photosynthesis in Plants,* ed. J. Franck. and W. E. Loomis, pp. 133–69. Ames: Iowa State College Press.
– 1953. Paper chromatography of chloroplast pigments: Sorption at a liquid–liquid interface. *J. Phys. Chem.* 57: 638–40.
– 1954. Leaf xanthophylls: The action of acids on violaxanthin, violeoxanthin, taraxanthin and tareoxanthin. *Arch. Biochem. Biophys.* 48: 458–68.
– 1958. *Chloroplast Pigments and Chromatographic Analysis,* 32nd Annual Priestley Lecture. University Park, PA: Penn State University Press.
– 1965. Chloroplast pigments and the classification of some siphonalean green algae of Australia. *Biol. Bull.* 129: 366–70.
– 1966. Fat-soluble chloroplast pigments: Their identification and distribution in various Australian plants. In *Biochemistry of Chloroplasts,* Vol. 1, ed. T. W. Goodwin, pp. 387–406. London: Academic Press.
Strain, H. H., Benton, F., Grandolfo, M. C., Aitzetmuller, K., Svec, W. A.,

and Katz, J. J. 1970. Heteroxanthin, diatoxanthin and diadinoxanthin from *Tribonema aequale*. *Phytochemistry* 9: 2561–5.

Strain, H. H., Cope, B. T., McDonald, G. N., Svec, W. A., and Katz, J. J. 1971a. Chlorophylls c_1 and c_2. *Phytochemistry* 10: 1109–14.

Strain, H. H., Cope, B. T., and Svec, W. A. 1971b. Analytical procedures for the isolation, identification and investigation of the chlorophylls. *Methods Enzymol.* 23: 452–76.

Strain, H. H., and Manning, W. M. 1942. Chlorofucine (chlorophyll γ), a green pigment of diatoms and brown algae. *J. Biol. Chem.* 144: 625–36.

Strain, H. H., Manning, W. M., and Hardin, G. 1943. Chlorophyll *c* (chlorofucine) of diatoms and dinoflagellates. *J. Biol. Chem.* 145: 655–68.

– 1944. Xanthophylls and carotenes of diatoms, brown algae, dinoflagellates and sea anemones. *Biol. Bull.* 86: 169–89.

Strain, H. H., and Sherma, J. 1969. Modifications of solution chromatrography illustrated with chloroplast pigments. *J. Chem. Educ.* 46: 476–83.

– 1972. Investigations of the chloroplast pigments of higher plants, green algae and brown algae and their influence upon the invention, modifications and applications of Tswett's chromatographic method. *J. Chromatogr.* 73: 371–97.

Strain, H. H., Sherma, J., Benton, F. L., and Katz, J. J. 1965a. One-way paper chromatography of the chloroplast pigments of leaves. *Biochim. Biophys. Acta* 109: 1–15.

– 1965b. Two-way paper chromatography of the chloroplast pigments of leaves. *Biochim. Biophys. Acta* 109: 16–22.

– 1965c. Radial paper chromatography and columnar chromatography of the chloroplast pigments of leaves. *Biochim. Biophys. Acta* 109: 23–32.

Strain, H. H., Sherma, J., and Grandolfo, M. 1967. Alteration of chloroplast pigments by chromatography with sileceous adsorbents. *Anal. Chem.* 39: 926–32.

– 1968. Comparative chromatography of the chloroplast pigments. *Anal. Biochem.* 24: 54–69.

Strain, H. H., and Svec, W. A. 1966. Extraction, separation, estimation and isolation of the chlorophylls. In *The Chlorophylls,* ed. L. P. Vernon and G. R. Seeley, pp. 22–66. New York: Academic Press.

– 1969. Some procedures for the chromatography of the fat-soluble chloroplast pigments. *Adv. Chromatogr.* 8: 119–76.

Strain, H. H., Svec, W. A., Wegfahrt, P., Rapoport, H., Haxo, F. T., Norgard, S., Kjøsen, H., and Liaaen-Jensen, S. 1976. Algal carotenoids. XIV. Structural studies on peridinin. Pt. I. Structural elucidation. *Acta Chem. Scand.* 30B: 109–20.

Strain, H. H., Thomas, M. R., Crespi, H. L., Blake, M. I., and Katz, J. J. 1960. Chloroplast pigments and photosynthesis in deuterated green algae. *Ann. N.Y. Acad. Sci.* 84: 617–33.

Strain, H. H., Thomas, M. R., and Katz, J. J. 1963. Spectral properties of ordinary and fully deuterated chlorophylls *a* and *b*. *Biochim. Biophys. Acta* 75: 306–11.

Stransky, H. 1978. A method for the quantitative estimation of chloroplast pigments in the picomole-range by use of a isocratic HPLC-system. *Z. Naturforsch.* 33c: 836–40.

Stransky, H., and Hager, A. 1970a. Das Carotinoidmuster und die Verbreitung des lichten duzierten Xanthophyllcyclus in verschieden Algenklassen. II. Xanthophyceae. *Arch. Mikrobiol.* 71: 84–96.

– 1970b. Das Carotinoidmuster und die Verbreitung des lichten duzierten xanthophyllcyclus in verschieden Algenklassen. IV. Cyanophyceae und Rhodophyceae. *Arch. Mikrobiol.* 72: 84–96.

– 1970c. Das Carotinoidmuster und die verbreitung des lichten duzierten xanthophyllcyclus in verschieden Algenclassen. VI. Chemosystematische Betrachtung. *Arch. Mikrobiol.* 73: 315–23.

Straub, E. 1980. Two chromatographic separation methods for chlorophyl analysis: TLC and its transformation to modern HPLC. *Arch. Hydrobiol. Beih. Ergebn. Limnol.* 14: 79–80.

Staub, O. 1976. *Key to carotenoids: Lists of natural carotenoids*. Basel: Birkhaüser.

Strickland, J. D. H., and Parsons, T. R. 1968. A practical handbook of seawater analysis. *Fish. Res. Board Can. Bull.* 167: 1–311.

– 1972. A practical handbook of seawater analysis. *Fish. Res. Board Can. Bull.* (2nd ed.) 167: 1–326.

Sullivan, C. M., Entwisle, T., and Rowan, K. S. (1989). The identification of chlorophyll *c* in the Tribophyceae (Syn. Xanthophyceae) using spectrophotofluorometry. *Phycologia* (submitted).

Suzuki, N., Saitoh, K., and Adachi, K. 1987. Reverse-phase high-performance thin-layer chromatography and column liquid chromatography of chlorophylls and their derivatives. *J. Chromatogr.* 408: 181–90.

Svec, W. A. 1978. The isolation, preparation, characterization and estimation of the chlorophylls and the bacteriochlorophylls. In *The Porphyrins,* Vol. 5: *Physical Chemistry,* Part C, ed. D. Dolphin, pp. 341–99. New York: Academic Press.

Swift, I. E., Milborrow, B. V., and Jeffrey, S. W. 1982. Formation of neoxanthin, diadinoxanthin and peridinin from [^{14}C]zeaxanthin by a cell-free system from *Amphidinium carterae*. *Phytochemistry* 21: 2859–64.

Takahashi, Y., Koike, H., and Katoh, S. 1982. Multiple forms of chlorophyll–protein complexes from a thermophilic cyanobacterium *Synechococcus* sp. *Arch. Biochem. Biophys.* 219: 209–18.

Takaiwa, F., Kusuda, M., Saga, N., and Sugiura, M. 1982. The nucleotide sequence of 5*S* rRNA from a red alga, *Porphyra yezoensis*. *Nucleic Acids Res.* 10: 6037–40.

Tandeau de Marsac, N. 1977. Occurrence and nature of chromatic adaptation in Cyanobacteria. *J. Bacteriol.* 130: 82–91

Tandeau de Marsac, N., and Cohen-Bazire, G. 1977. Molecular composition of cyanobacterial phycobilisomes. *Proc. Natl. Acad. Sci. U.S.A.* 74: 1635–9.

Tangen, K., and Bjørnland, T. 1981. Observations on pigments and morphology of *Gyrodinium aureolum* Hulburt, a marine dinoflagellate containing 19′-hexanoyloxyfucoxanthin as the main carotenoid. *J. Plankton Res.* 3: 389–401.

Taylor, D. L. 1967. The pigments of the zooxanthellae symbiotic with the intertidal anemone. *Anemonia sulcata. J. Phycol.* 3: 238–40.

Taylor, F. J. R. 1974. Implications and extensions of the serial endosymbiosis theory of the origin of eukaryotes. *Taxon* 23: 229–58.

– 1976. Flagellate phylogeny: A study in conflicts. *J. Protozool.* 23: 28–40.

– 1978. Problems in the development of an explicit hypothetical phylogeny of the lower eukaryotes. *BioSystems* 10: 67–89.

– 1983. Some eco-evolutionary aspects of intra-cellular symbioses. *Int. Rev. Cytol.* [Suppl.] 14: 1–28.

Teale, F. W., and Dale, R. E. 1970. Isolation and spectral characterization of phycobiliproteins. *Biochem. J.* 116: 161–9.

Tett, P., Kelly, M. G., and Hornberger, G. M. 1975. A method for the spectrophotometric measurement of chlorophyll *a* and phaeophytin *a* in benthic microalgae. *Limnol. Oceanogr.* 20: 887–96.

Tett, P., and Wallis A. 1978. The general annual cycle of chlorophyll standing crop in Loch Creran. *J. Ecol.* 66: 227–39.

Thomas, D. M., and Goodwin, T. W. 1965. Nature and distribution of carotenoids in the Xanthophyta (Heterokontae). *J. Phycol.* 1: 118–21.

Thomas, M. 1935. *Plant Physiology,* 1st ed. London: Churchill.

Thornber, J. P. 1986. Biochemical characterization and structure of pigment–proteins of photosynthetic organisms. *Enc. Plant Physiol.* (2nd ed.) 19: 98–142.

Thorne, S. W., Newcombe, E. H., and Osmond, C. B. 1977. Identification of chlorophyll *b* in extracts of prokaryotic algae by fluorescence spectroscopy. *Proc. Natl. Acad. Sci. U.S.A.* 74: 575–8.

Throndsen, J. 1978. Centrifugation. In *Phytoplankton Manual,* ed. A. Sournia, pp. 98–103. Paris: UNESCO.

Tomas, R. N., and Cox, E. R. 1973a. Observations on the symbiosis of *Peridinium balticum* and its intracellular alga. I. Ultrastructure. *J. Phycol.* 9: 304–23.

– 1973b. The symbiosis of *Peridinium balticum* (Dinophyceae). I. Ultrastructure and pigment analysis. *J. Phycol.* [*Suppl.*] 9: 16.

Tomas, R. N., Cox, E. R., and Steidinger, K. A. 1973. *Peridinium balticum* (Lavender) Lommerman, an unusual dinoflagellate with a mesocaryotic and an eucaryotic nucleus. *J. Phycol.* 7: 91–8.

Tomioka, N., and Sugiura, M. 1983. The complete nucleotide sequence of a 16*S* ribosomal RNA gene from a blue-green alga, *Anacystis nidulans. Mol. Gen. Genet.* 191: 46–50.

Trees, C. C., Bidigare, R. R., and Brooks, J. M. 1986. Distribution of chlorophylls and phaeopigments in the Northwestern Atlantic Ocean. *J. Plankton Res.* 8: 447–58.

Trench, R. K. 1982. Physiology, biochemistry and ultrastructure of cyanellae. *Prog. Physiol. Res.* 1: 259–88.

Trench, R. K., and Ronzio, G. S. 1978. Aspects of the relation between *Cy-*

anophora paradoxa (Korschikoff) and its endosymbiotic cyanelles *Cyanocyta Korschikoffiana* (Hall and Claus). II. The photosynthetic pigments. *Proc. R. Soc. Lond.* [*B.*] 202: 445–62.

Troxler, R. F., and Brown, A. S. 1975. Metabolism of δ-aminolevulinic acid in red and blue-green algae. *Plant Physiol.* 55: 463–7.

Troxler, R. F., and Dokos, J. M. 1973. Formulation of carbon monoxide and bile pigment in red and blue-green algae. *Plant Physiol.* 51: 72–5.

Troxler, R. F., Foster, J. A., Brown, A. S., and Franzblau, C. 1975. *Cyanidium caldarum:* Properties and amino acid sequences at the amino terminus. *Biochemistry* 14: 268–74.

Troxler, R. F., Greenwald, L. S., and Zilinskas, B. A. 1980. Allophycocyanin from *Nostoc* sp. phycobilisomes. *J. Biol. Chem.* 255: 9380–7.

Tswett, M. 1906a. Physikalisch-chemische Studem über das Chlorophyll. Die Absorptionen. *Ber. Deutsch. Bot. Ges.* 24: 316–23.

– 1906b. Adsorptionsanalyse und Chromatographische Methode. Anwendung auf die Chemie des Chlorophylls. *Ber. Deutsch. Bot. Ges.* 24: 384–93.

– 1911. Über den makro- und microchemischem Naturweis des Carotins. *Ber. Deutsch. Bot. Ges.* 29: 630–6.

Tyagi, V. V., Mayne, B. C., and Peters, G. A. 1980. Purification and initial characterization of biliproteins from the endophytic cyanobacterium of *Azolla. Arch. Microbiol.* 128: 41–4.

Turner, M. F., and Gowen, R. K. 1984. Some aspects of the nutrition and taxonomy of fourteen small green and yellow-green algae. *Bot. Mar.* 27: 249–55.

Uzzell, T., and Spolsky, C. 1981. Two data sets: Alternative explanations and interpretations. *Ann. N.Y. Acad. Sci.* 361: 481–99.

Van der Velde, H. H. 1973a. The natural occurrence in red algae of two phycoerythrins with different molecular weight and spectral properties. *Biochim. Biophys. Acta* 303: 246–57.

– 1973b. The use of phycoerythrin absorption spectra in the classification of red algae. *Acta Bot. Neerl.* 22: 92–9.

Van Dorssen, R. J., Breton, J., Plijter, J. J., Satoh, K., Van Gorkom, H. J., and Amesz, J. 1987. Spectroscopic properties of the reaction center and of the 47 kDa chlorophyll protein of photosystem II. *Biochim. Biophys. Acta* 893: 267–74.

Venrick, E. L. 1987. On fluorometric determinations of filter-retained pigments. *Limnol. Oceanogr.* 32: 492–3.

Vermans, W. F. T., and Govindjee. 1981. The acceptor side of photosystem II in photosynthesis. *Photochem. Photobiol.* 34: 755–93.

Vernet, M., and Lorenzen, C. J. 1987. The relative abundance of pheophorbide *a* and pheophytin *a* in temperate marine waters. *Limnol. Oceanogr.* 32: 352–8.

Vernon, L. P. 1960. Spectrophotometric determination of chlorophylls and phaeophytins in plant extracts. *Anal. Chem.* 32: 1144–50.

Vernon, L. P., and Seely, G. R. 1966. *The Chlorophylls.* New York: Academic Press.

Vesk, M., and Jeffrey, S. W. 1977. Effect of bluegreen light on photosynthetic pigments and chloroplast structure in unicellular marine algae from six classes. *J. Phycol.* 13: 280–88.

– 1987. Ultrastructure and pigments of two strains of the picoplanktonic alga, *Pelagacoccus subviridis* (Chrysophyceae). *J. Phycol.* 23: 322–36.

Volk, S. L., and Bishop, N. E. 1968. Photosynthetic efficiency of a phycocyaninless mutant of *Cyandium*. *Photochem. Photobiol.* 8: 213–21.

Waaland, J. R., Waaland, S. D., and Bates, C. 1974. Chloroplast structure and pigment composition in the red alga, *Griffithsia pacifica:* Regulation by light intensity. *J. Phycol.* 10: 193–9.

Walton, T. J., Britton, G., Goodwin, T. W., Diner, B., and Moshier, S. 1970. The structure of siphonaxanthin. *Phytochemistry* 9: 2545–52.

Wanner, G., and Kost, H.-P. 1980. Investigations on the arrangement and fine structure of *Porphyridium cruentum* phycobilisomes. *Protoplasm* 102: 97–109.

Wartenberg, D. E. 1978. Spectrophotometric equipment technique. *Limnol. Oceanogr.* 23: 566–70.

Wasley, J. W. F., Scott, W. T., and Holt, A. S. 1970. Chlorophyllides *c*. *Can. J. Biochem.* 48: 376–83.

Watanabe, M. M., Takeda, Y., Sasa, T., Inouye, I., Suda, S., Sawaguchi, T., and Chiara, M. 1987. A green dinoflagellate with chlorophylls *A* and *B:* Morphology, fine structure of the chloroplast and chlorophyll composition. *J. Phycol.* 23: 382–9.

Watanabe, T., Hongu, A., Honda, K., Nakazato, M., Konno, M., and Saitoh, S. 1984. Preparation of chlorophylls and pheophytins by isocratic liquid chromatography. *Anal. Chem.* 56: 251–6.

Watson, B. A., Waaland, S. D., and Waaland, J. R. 1986. Phycocyanin from the red alga. *Anotrichium tenue:* Modification of properties by a colorless polypeptide (M, 30000). *Biochemistry* 25: 4583–7.

Weber, A. 1969. Über die Chloroplastenfarbstoffe einiger Conjugaten. *Flora* 160A: 457–73.

Weber, A., and Czygan, F. C. 1972. Chlorophylls and carotenoids of the Chaetophoraceae. 1. Siphonaxanthin in *Microthamnion keutzingianum* Hageli. *Arch. Mikrobiol.* 84: 243–53.

Weber, A., and Wettern, M. 1980. Some remarks on the usefulness of algal carotenoids as chemotaxonomic markers. In *Pigments in Plants,* 2nd ed., ed. F.-C. Czygan, pp. 104–16. Stuttgart: Fischer.

Weinstein, J. D., and Beale, S. I. 1983. Separate physiological roles and subcellular compartments for two tetrapyrrole biosynthetic pathways in *Euglena gracilis*. *J. Biol. Chem.* 258: 6799–6807.

Whatley, J. M. 1977. The fine structure of *Prochloron*. *New Phytol.* 79: 309–13.

Whatley, J. M., and Whatley, F. R. 1981. Chloroplast evolution. *New Phytol.* 87: 233–47.

– 1984. Evolutionary aspects of the eukaryotic cell and its organelles. *Enc. Plant Physiol.* (*N.S.*) 17: 18–58.

Wheeler, W. N. 1980. Pigment content and photosynthetic rate of the fronds of *Macrocystis pyrifera*. *Mar. Biol.* 56: 97–102.

White, R. C., Jones, I. D., and Gibbs, E. 1963. Determination of chlorophylls, pheophytins, and pheophorbides in plant material. *J. Food Sci.* 28: 431–6.

White, R. C., Jones, I. D., Gibbs, E., and Butler, L. S. 1972. Fluorometric estimation of chlorophylls, chlorophyllides, pheophytins and pheophorbides in mixtures. *J. Agric. Food Chem.* 20: 773–8.

Whitfield, D. M., and Rowan, K. S. 1974. Changes in the chlorophylls and carotenoids of leaves of *Nicotiana tabacum* during senescence. *Phytochemistry* 13: 77–83.

Whitney, D. E., and Darley, W. M. 1979. A method for the determination of chlorophyll *a* in samples containing degradation products. *Limnol. Oceanogr.* 24: 183–6.

Whittingham, C. P. 1974. *The Mechanism of Photosynthesis*. London: Arnold.

Whittle, S. J. 1976. The major chloroplast pigments of *Chlorobotrys regularis* (West) Bohlen (Eustigmatophyceae) and *Ophiocytium majus* Naegeli (Xanthophyceae). *Br. Phycol. J.* 11: 111–14.

Whittle, S. J., and Casselton, P. J. 1968. Peridinin as the major xanthophyll of the Dinophyceae. *Br. Phycol. Bull.* 3: 602–3.

– 1969. The chloroplast pigments of some green and yellow-green algae. *Br. Phycol. J.* 4: 55–64.

– 1975a. The chloroplast pigments of the algal classes Eustigmatophyceae and Xanthophyceae. I. Eustigmatophyceae. *Br. Phycol. J.* 10: 179–91.

– 1975b. The chloroplast pigments of the algal classes Eustigmatophyceae and Xanthophyceae. II. Xanthophyceae. *Br. Phycol. J.* 10: 192–204.

Wiedmann, I., Wilhelm C., and Wild, A. 1983. Isolation of chlorophyll–protein complexes and quantification of electron transport components in *Synura petersenii* and *Tribonema aequale*. *Photosynth. Res.* 4: 317–29.

Wilcox, L. W. 1985. The dinoflagellate chloroplast(s). In *Proc. 2nd Int. Phycol. Cong.*, Copenhagen, August 1985, p. 176.

Wilcox, L. W., and Wedemayer, G. J. 1984. *Gymnodinium acidotum* Nygaard (Pyrrophyta), a dinoflagellate with an endosymbiotic cryptomonad. *J. Phycol.* 20: 236–42.

– 1985. Dinoflagellate with blue-green chloroplasts derived from an endosymbiotic eukaryote. *Science* 227: 192–4.

Wild, A., and Urschel, B. 1980. Chlorophyll–protein complexes of *Chlorella fusca*. *Z. Naturforsch.* 35c: 627–37.

Wilhelm, C. 1987. Purification and identification of chlorophyll c_1 from the green alga *Mantoniella squamata*. *Biochim. Biophys. Acta* 892: 23–9.

Wilhelm, C., Buchel, C., and Rousseau, B. 1988. The molecular organization of chlorophyll–protein complexes in the Xanthophycean alga *Pleurochloris meiringensis*. *Biochim. Biophys. Acta* 934: 220–6.

Wilhelm, C., and Lenarz-Weiler, I. 1987. Energy transfer and pigment composition in three chlorophyll *b*-containing light-harvesting complexes isolated from *Mantoniella squamata* (Prymnesiophyceae), *Chlorella fusca* (Chlorophyceae) and *Synapis alba*. *Photosynth. Res.* 13: 101–11.

Wilhelm, C., Lenarz-Weiler, I., and Wild, A. 1986. The light-harvesting system of *Micromonas* sp., Butcher (Prasinophyceae): The combination of three different chlorophyll species in one single chlorophyll–protein complex. *Phycologia* 25: 304–12.

Williams, E. 1967. The gamete pigments of *Hormosira banksii* (Phaeophyta). *J. Exp. Bot.* 18: 416–21.

Willstätter, R., and Page, H. G. 1914. Untersuchungen über Chlorophyll. XXIV. Uber des Pigmente de Braunalgen. *Ann. Chemie (Liebig)* 404: 237–71.

Wintermans, J. F. G. H., and De Mots, A. 1965. Spectrophotometric characteristics of chlorophylls *a* and *b* and their phaeophytins in ethanol. *Biochem. Biophys. Acta* 109: 448–53.

Withers, N. W., Alberte, R. S., Lewin, R. A., Thornber, J. P., Britton, G., and Goodwin, T. W. 1978a. Photosynthetic unit size, carotenoids and chlorophyll–protein composition of *Prochloron* sp., a prokaryotic green alga. *Proc. Natl. Acad. Sci. U.S.A.* 75: 2301–5.

Withers, N. W., Cox, E. R., Tomas, R., and Haxo, F. T. 1977. Pigments of the dinoflagellate *Peridinium balticum* and its photosynthetic endosymbiont. *J. Phycol.* 13: 254–8.

Withers, N. W., Fiksdahl, A., Tuttle, R. C., and Liaaen-Jensen, S. 1981. Carotenoids of the Chrysophyceae. *Comp. Biochem. Physiol.* 63B: 345–9.

Withers, N. W., and Haxo, F. T. 1975. Chlorophyll c_1 and c_2 and extraplastidic carotenoids in the dinoflagellate, *Peridinium foliaceum* Stein. *Plant Sci. Lett.* 5: 7–15.

Withers, N. W., Vidaver, W., and Lewin, R. A. 1978b. Pigment composition, photosynthesis and fine structure of a non-blue green prokaryotic algal symbiont (*Prochloron* sp.) in a didemnid ascidian from Hawaiian waters. *Phycologia* 17: 167–71.

Woese, C. R. 1981. Archaebacteria. *Sci. Am.* 244(6): 94–106.

– 1987. Macroevolution in the microscopic world. In *Molecules and Morphology in Evolution: Conflict or Compromise?,* ed. C. Patterson, pp. 177–202. Cambridge: Cambridge University Press.

Woese, C. R., and Fox, G. E. 1977a. Phylogeny structure of the prokaryotic domain: The primary kingdoms. *Proc. Natl. Acad. Sci. U.S.A.* 74: 5088–90.

– 1977b. The concept of cellular evolution, *J. Mol. Evol.* 10: 1–6.

Wojciechowski, I., Wojciechowska, W., Czernas, K., Galek, J., and Religa, K. 1988. Changes in phytoplankton over a ten-year period in a lake undergoing de-eutrophication due to surrounding peat bogs. *Arch. Hydrobiol. Suppl.* 78.3: 373–87.

Wollman, F.-A. 1979. Ultrastructural comparison of *Cyanidium caldarum* wild type and III-C mutant lacking phycobilisomes. *Plant Physiol.* 63: 375–81.

– 1986. Photosystem I proteins. *Enc. Plant Physiol.* (*N.S.*) 19: 487–95.

Wollman, F.-A. and Bennoun, P. 1982. A new chlorophyll–protein complex related to photosystem I in *Chlamydomonas reinhardii*. *Biochim. Biophys. Acta* 680: 352–60.

Wolters, J., and Erdmann, V. A. 1988. Cladistic analysis of ribosomal RNAs – The phylogeny of eukaryotes with respect to the endosymbiotic theory. *BioSystems* 21: 209–14.

Wood, L. W. 1985. Chloroform–methanol extraction of chlorophyll *a*. *Can. J. Fish. Aquat. Sci.* 42: 38–43.

Wright, S. W. 1987. Phytoplankton pigment data Prydz Bay region. *Australian Nat. Antarctic Res. Note,* 115 , pp. 58.

Wright, S. W., and Jeffrey, S. W. 1987. Fucoxanthin pigment markers of marine phytoplankton analysed by HPLC and HPTLC. *Mar. Ecol. Prog. Ser.* 38: 259–66.

Wright, S. W., and Shearer, J. D. 1984. Rapid extraction and high-performance liquid chromatography of clorophylls and carotenoids from marine phytoplankton. *J. Chromatogr.* 294: 281–95.

Wu, S. M., and Rebeiz, C. A. 1988. Chlorophyll biogenesis: Molecular structure of short wavelength chlorophyll *a* (E432 F662). *Phytochemistry* 27: 353–6.

Wun, C. K., Rho, J., Walker, R. W., and Litsky, W. 1979. An XAD-1 column method for the rapid extraction of phytoplankton chlorophylls. *Water Res.* 13: 645–9.

Yamada, T., and Shimaji, M. 1986. Nucleotide sequence of the 5*S* rRNA gene from the unicellular green alga *Chlorella ellipsoidea*. *Nucleic Acids Res.* 14: 9529.

Yamagishi, A., and Katoh, S. 1983. Two chlorophyll-binding subunits of the photosystem 2 reaction center complex isolated from the thermophylic cyanobacterium *Synechococcus* sp. *Arch. Biochem. Biophys.* 225: 836–46.

– 1984. A photoactive photosystem-II reaction center complex lacking a chlorophyll-binding 40 kilodalton subunit from the thermophilic cyanobacterium *Synechococcus* sp. *Biochim. Biophys. Acta* 765: 118–24.

Yamamoto, Y., Yokohama, H., and Boettger, H. 1969. Carotenoid of *Chlorella pyrenoidosa*. Pyrenoxanthin, a new carotenoid. *J. Org. Chem.* 34: 4207–8.

Yamazaki, I., Mimuro, M., Murata, T., Yamazaki, T., Yoshihara, K., and Fujita, Y. 1984. Excitation energy transfer in the light harvesting antenna system of the red alga *Porphyridium cruentum* and the blue-green alga *Anacystis nidulans:* Analysis of time-resolved fluorescence spectra. *Photochem. Photobiol.* 39: 233–40.

Yentsch, C. S., and Menzel, D. W. 1963. A method for the determination of phytoplankton chlorophyll and phaeophytin by fluorescence. *Deep-Sea Res.* 10: 221–31.

Yokohama, Y. 1981a. Distribution of the green light-absorbing pigments, siphonaxanthin and siphonein in marine green algae. *Bot. Mar.* 24: 637–40.

– 1981b. Green light-absorbing pigments in marine green algae, their ecological significance and systematic significance. *Jpn. J. Phycol.* 29: 209–22.

– 1982. The distribution of lutein and its derivatives in marine green algae. *Jpn. J. Phycol.* 30: 311–17.

– 1983. A xanthophyll characteristic of deep-water green algae lacking siphonaxanthin. *Bot. Mar.* 26: 45–8.

Yokohama, Y., Kageyama, A., Ikawa, T., and Shimura, S. 1977. A carotenoid characteristic of chlorophycean seaweeds living in deep coastal waters. *Bot. Mar.* 20: 433–6.

Yu. M.-H., and Glazer, A. N. 1982. Cyanobacterial phycobilisomes. Role of the linker polypeptides in the assembly of phycocyanin. *J. Biol. Chem.* 257: 3429–33.

Yu, M.-H., Glazer, A. N., Spencer, K. G., and West, J. A. 1981. Phycoerythrins of the red alga, *Callithamnion:* Variation in phycoerythrobilin and phycourobilin content. *Plant Physiol.* 68: 482–7.

Yu, M.-H., Glazer, A. N., and Williams, R. C. 1981. Cyanobacterial phycobilisomes. Phycocyanin assembly on the rod substructure of *Anabaena variabilis* phycobilisomes. *J. Biol. Chem.* 256: 13130–6.

Zechmeister, L. 1934. *Carotenoids.* Berlin: Springer Verlag.

– 1962. *Cis–trans Isomeric Carotenoids, Vitamins A, und Arylpolyenes.* Vienna. Springer Verlag.

Zeng, F.-J., Yang, Z.-X., and Jiang, L.-J. 1984. Isolation and characterization of R-phycocyanin from *Polysiphonia urceolata. Hydrobiologia* 116/117: 594–6.

Zeng, F., Yang, Z., Liu, H., and Jiang, L. 1986. The study of phycobiliproteins in *Porphyra yezoensis.* The physical and immunochemical properties of R-phycoerythrin. *Scientia Sinica* [*B*] 29: 824–31.

Zhang, X., Liu, M., Liu, Q., Wang, H., and Li, S. 1983. Studies of the pigments of blue-green *Gymnodinium.* 1. Preliminary separation and characteristics of phycocyanin. In *Proc. Joint China–U.S. Phycological Symp.,* ed. C.-K. Tseng; pp. 431–6. Beijing: Science Press.

Zhou, B.-C., Zheng, S.-Q., and Tseng, C.-K. 1974. Comparative studies on the absorption spectra of some green, brown and red algae. *Acta Bot. Sinica* 16: 146–55.

Zhou, P., Zhou, B., Zhang, Z., Tseng, C.-K., Ma, G., Pan, Z., Li, L., and Zhi, X. 1983. Chlorophyll–protein complexes of a marine benthic green alga *Codium fragile.* In *Proc. Joint China–U.S. Phycological Symp.* ed. C.-K. Tseng, pp. 409–24. Beijing: Science Press.

Ziegler, R., and Egle, K. 1965. Zur quantitative Analyse der Chloroplastenpigmente. 1. Kritische Uberprüfung des spectralphotometrischen Chlorophylls-Bestinunung. *Beitr. Biol. Pflanzen* 41: 11–37.

Zilinskas, B. A. 1982. Isolation and characterization of the central component of the phycobilisome core of *Nostoc* sp. *Plant Physiol.* 70: 1060–5.

Zilinskas, B. A., and Greenwald, L. S. 1986. Phycobilisome structure and function. *Photosynth. Res.* 10: 7–35.

Zilinskas, B. A., Greenwald, L. S., Bailey, C. L., and Kahn, P. C. 1980. Spectral analysis of allophycocyanin I, II, III and B from *Nostoc* sp. phycobilisomes. *Biochim. Biophys. Acta* 592: 267–76.

Zilinskas, B. A., and Howell, D. A. 1983. Role of the colorless polypeptides in phycobilisome assembly in *Nostoc* sp. *Plant Physiol.* 71: 579–87.

– 1986. The immunologically conserved phycobilisome–thylakoid linker polypeptide. *Plant Physiol.* 80: 829–33.

Zilinskas, B. A., Zimmerman, B. K., and Gantt, E. 1978. Allophycocyanin forms isolated from *Nostoc* sp. phycobilisomes. *Photochem. Photobiol.* 27: 587–95.

Zscheile, F. P., and Comar, G. L. 1941. Influence of preparative procedure on the purity of chlorophyll components as shown by absorption spectra. *Bot. Gaz.* 102: 463–81.

Zscheile, F. P., and Harris, D. G. 1943. Studies on the fluorescence of chlorophyll: The effect of concentration, temperature and solvent. *J. Phys. Chem.* 47: 623–37.

Index